# Green Energy and Technology

Climate change, environmental impact and the limited natural resources urge scientific research and novel technical solutions. The monograph series Green Energy and Technology serves as a publishing platform for scientific and technological approaches to "green"—i.e. environmentally friendly and sustainable—technologies. While a focus lies on energy and power supply, it also covers "green" solutions in industrial engineering and engineering design. Green Energy and Technology addresses researchers, advanced students, technical consultants as well as decision makers in industries and politics. Hence, the level of presentation spans from instructional to highly technical. **Indexed in Scopus**.

More information about this series at http://www.springer.com/series/8059

Cataldo De Blasio

# Fundamentals of Biofuels Engineering and Technology

Springer

Cataldo De Blasio
Laboratory of Energy Technology
Faculty of Science and Engineering
Åbo Akademi University
Vaasa, Finland

ISSN 1865-3529 ISSN 1865-3537 (electronic)
Green Energy and Technology
ISBN 978-3-030-11598-2 ISBN 978-3-030-11599-9 (eBook)
https://doi.org/10.1007/978-3-030-11599-9

Library of Congress Control Number: 2019934516



This Springer imprint is published by the registered company Springer Nature Switzerland AG
The registered company address is: Gewerbestrasse 11, 6330 Cham, Switzerland

*A Daniel Lucas Ettore e Stefan Louis Emanuel.*

*Siete la mia vita,*

*Papà.*

# Preface

Biomass and waste have become of much interest in recent years for their usage in power conversion processes. Biomass is considered as a renewable source of energy, and therefore, it is assumed usually as environmentally friendly to utilize biomass in power plants or for biofuels production, even if this has some constraints in reality. The topic is enormously large, and it would require an entire encyclopedia to treat all the related subfields. In this case, the manuscript is driven mainly by the questions:

- Why biomass would be a suitable choice for energy conversion?
- How biomass is generated and assessed? (This is important to understand that nothing is coming for free, and time is needed to have suitable yields).
- How biomass is utilized for converting power and producing biofuels?

This first edition is divided into two main parts (Parts I and II) and an additional part (Part III) which includes some additional material mainly meant as support for the reader.

The manuscript is addressed at a large audience, and it is suitable for students at bachelor and master levels of their university studies. Naturally, researchers who are working in these fields might find this book of interest. In addition, the aim is at a broad range of readers (engineers, technicians, professionals) who are looking for some specific topics within the area of biomass and energy. This book would be suitable as textbook for university courses and lectures given within the field treated here.

The general outline of the manuscript is to give information following a "pedagogical style," in the same way as lessons are given by this author. At the same time, the author was trying to give more simplified explanations for some of the treated arguments, and instead, for some other topics, he is explaining them into very details. Steps and derivations of equations given normally in the literature, especially in the case of mass balances, are given here in full. To give a better understanding of the passages involved, equations are repeated and numbered more than one time, when needed, to give a better understanding and flow to the comprehension.

The author of this book would like to welcome the reader to the first edition of this manuscript. The readers of this book are also encouraged to give any comments and suggestions which could improve the current version of the manuscript.

Vaasa, Finland Cataldo De Blasio

# Acknowledgements

Eng. Marco Simonetti, Eng. Gaetano Lucca, Dr. Mauro Prestipino, and Prof. Antonio Galvagno are greatly acknowledged for their collaboration and support. I am proud to work with you.

Many thanks go to my colleagues at the Laboratory of Energy Technology in Vaasa; Professor Margareta Björklund-Sänkiaho and Dr. Jessica Tuuf, for the great working environment we have.

The colleagues at the Department of Chemical Engineering in Turku are also greatly acknowledged. I'd like to thank Prof. Henrik Saxén, Prof. Tapio Westerlund, and Prof. Frank Petterson for their suggestions, support, and comments to the manuscript.

The companies UPM, Metsä, and Vaasa Water are acknowledged for giving permissions to use some of their figures.

Many thanks to my wife Leena Matilda and my children, you are my strength and joy in this life. Thanks to my sister, Maria Teresa, brother, Luigi, and my parents Donato and Teresa who are living in the beautiful village of Brienza in Italy. They are always of great support and sustainment. My mother, Teresa Giallorenzo, is greatly acknowledged also for providing information on some agrofuels yield and rate.

# About This Book

The scope of this book is to deliver suitable information related to the "generation" of power and its conversion from biomass sources. This would include also the production of biofuels. The author of this book wanted to keep a more "teaching style" during the adopted descriptions, and therefore, this would make the manuscript suitable to students and persons interested on the subject of biomass, how it is generated, and how it is converted to get power and fuels. The book is divided mainly into two parts with an additional part related to the explanation of support information.

If the intention is to utilize biomass or bioproducts for energy conversion, it would be beneficial to understand the underlying principles behind the biomass generation itself.

For this reason, during the first part, the main principles of biomass generation will be explained with a particular focus on how plants harvest the energy from the sun. The photosynthetic process is described in some more details to give an example of the main principles common to many kinds of microorganisms including micro-algae. These arguments are also beneficial if evaluations should be done on feasibility studies and process integration which concern biomass.

The main analogies between this kind of biomass generation ($CO_2$ fixation processes) and processes utilized by microorganisms for reproducing themselves will be pointed out clearly. After explaining the principal concepts of biomass generation, some sections of the first part of the book will be dedicated to the principles of biomass analysis and assessment for the purpose of evaluating its quality before the utilization in power plants. Particular focus will be on proximate and ultimate analysis methods.

If the book is utilized in a course at university, the goal of the first part of the book is to put students in the position of:

- having a sufficient knowledge of the principles behind biomass generation;
- possess sufficient knowledge to understand the principles behind the main types of chemical analysis done on biomass and biofuels;

- have knowledge on what is the link between the physics of biomass generation and some other engineering problems. Recognize that there are many analogies between the working principles of nature and technical solutions.

The driving questions of this first part of the manuscript are: How biomass is generated and what are the principles behind it? What kind of feedstock we should use for energy conversion and to which purposes? How we assess the suitability of biomass and biofuels made for energy conversion purposes? Bio-organisms are demonstrated to have many analogies with technical devices converting energy, and therefore, the thermodynamic efficiencies of these processes can be evaluated. The most important part here is shown to be the formation of an electrochemical potential together with a difference in concentration related to the $H^+$ ions across the membranes. The $H^+$ ions are coming from water oxidation (in the case of photosynthesis); clear is the resemblance with the galvanic cells and fuel cells principles. Aspects related to light harvesting performed by living organisms are also covered along with the evaluation of the degree of response in relation to the converting device. The treatment of the analyses done on biomass is covered at first by the Kjeldahl method. Then, an accurate description of the proximate analysis (calorimetric bomb) with calculation of the higher heating value is given. The analysis related to the composition of biomass is done by describing elemental analysis, thermogravimetric analysis, and chromatography. There are statistical implications which are descriptive of the methods and the outcomes.

The second part of the manuscript focuses on the production methods for certain biofuels. The part is mostly focused on giving the basic knowledge required to understand the details of the biorefinery methods and on technological aspects. Theoretical aspects are covered as well, especially for the case of biochemical reactions engineering methods.

The sections start with a preliminary introduction of biorefineries, and then the principles of biochemical reactor engineering are given in detail. The reader will differentiate between the different types of reactor configurations and the related modeling of the kinetics involved.

The main biological pathways of processes like fermentation and anoxic digestion are discussed along with the main parameters utilized for bioreactors design. The sections continue with the discussion on biogas, bio-ethanol, biodiesel production, and Fischer–Tropsch methods.

This part of the book will give the students the possibility to:

- have an understanding of what are the principles behind biochemical reactors design;
- recognize some of the most important methods used for biofuels production;
- differentiate the main types of reactors used for biofuels production. In addition, the reader will be capable of analyzing the main related parameters involved;
- distinguish and summarize the different routes performed for biogas, bio-ethanol, biodiesel production, and Fischer–Tropsch synthesis;
- possess critical thinking about the different technologies.

The main focus of the second part is on the questions:

After knowing the main principles of biomass generation, what are the best solutions applied in bioreaction engineering? What are the different routes involved for biofuels production? How we could utilize the biofuel generated for the purpose of energy conversion?

# Contents

# Acronyms

| | |
|---|---|
| A0 | Chlorophyll |
| A1[Q] | Phylloquinone |
| AAE | Average absolute error |
| ADH | Alcohol dehydrogenase |
| AFEX | Ammonia fiber expansion |
| AFW | Animal fat waste |
| aq | Aqueous (if used as subscript in equations) |
| ASF | Anderson–Schulz–Flory (Distribution) |
| ASTM | American Society for Testing and Materials |
| ASU | Air separation unit |
| ATP | Adenosine triphosphate |
| ATR | Autothermal reforming |
| BAL | Balanced |
| BOCLE | Ball on cylinder lubricity evaluator |
| BPG | 1,3-biphosphoglycerate |
| BSTR | Batch stirred tank reactor |
| CALB | Candida antarctica lipase B |
| CBP | Consolidated bioprocessing |
| Chl | Chlorophyll |
| CMM | Computational molecular modeling |
| CN | Cetane number |
| CPA | Centrally planned Asia and China |
| CSSC | Chlorophyll-sensitized solar cell |
| Cyt | Cytochrome |
| Cyt bLp and Cyt bHp | Cytochrome b low- and high-potential forms |
| Cyt f | Cytochrome f |
| DDGS | Dried distillers grain solubles |
| DG | Diacylglycerides |
| DSC | Differential scanning calorimetry |

| | |
|---|---|
| DTA | Differential thermal analysis |
| DTG | Derivative thermogravimetry |
| EA | Elemental analyzer |
| EMF | Electromotive force |
| EMP | Embden–Meyerhof Pathway |
| ENO | Enolase |
| EoS | Equation of state |
| FAME | Fatty acid methyl ester |
| FBA | Fructose bisphosphate aldolase |
| Fd | Ferredoxin |
| FFA | Free fatty acids |
| FID | Flame ionization detector |
| FNR | Flavo-protein ferredoxin–NADP reductase |
| FSU | Former Soviet Union |
| FT | Fischer–Tropsch |
| Fx | Iron–sulfur center |
| GC | Gas chromatography |
| GCV | Gross calorific value |
| GDP | Gross domestic product |
| HC | Hydrocarbons |
| HFRR | High-frequency reciprocating ring |
| HHV | Higher heating value |
| HPLC | High-pressure liquid chromatography |
| HRT | Hydraulic residence time |
| HXK | Hexokinase |
| IAPWS | International Association for the Properties of Water and Steam |
| IEA | International Energy Agency |
| IGCC | Integrated gasification combined cycle |
| IIASA | International Institute for Applied Systems Analysis |
| IR | Infrared |
| LAM | Latin America |
| LHC | Light-harvesting complex |
| LHV | Lower heating value |
| MEA | Middle East and North Africa |
| MG | Monoacylglycerides |
| MS | Mass spectroscopy |
| NADP | Nicotinamide adenine dinucleotide phosphate |
| NADP* | Oxidized nicotinamide adenine dinucleotide phosphate |
| NAM | North America |
| NCV | Net calorific value |
| NEXBTL | Next-generation biomass to liquid (By Neste) |

| | |
|---|---|
| NIST | National Institute of Standards and Technology |
| OECD | Organisation for Economic Co-operation and Development |
| OY | Osakeyhtiö (Limited Company) |
| P680 | Primary electron donor of PSII composed of chlorophyll (Chl) |
| P700 | Primary electron Chl donor of PSI |
| PAO | Pacific OECD |
| PAR | Photosynthetic active radiation |
| PAS | Other Pacific Asia |
| PC | Plastocyanine |
| PDC | Pyruvate decarboxylase |
| PFK | Phosphofructokinase |
| PFR | Plug flow reactor |
| PGA | 3-phosphoglycerate |
| PGAL | Phosphoglyceraldehyde |
| PGI | Phosphoglucose isomerase |
| PGK | 3-phosphoglycerate kinase |
| PGM | Phosphoglycerate mutase |
| Pheo | Pheophytin |
| PK | Pyruvate kinase |
| POX | Partial oxidation |
| PSI | Photosystem I |
| PSII | Photosystem II |
| $Q_a$ | Plastoquinone A |
| $Q_b$ | Plastoquinone B |
| QHE | Quantum heat engine |
| QYP | Quantum yield of photosynthesis |
| R&D | Research and Development |
| RC | Reaction center |
| RI | Refractive index |
| RuBP | Ribulose 1,5 bi-phosphate |
| RUE | Radiation use efficiency |
| s | Solid (if used as subscript in equations) |
| SAS | South Asia |
| SBL | Strong black liquor |
| SCWG | Supercritical water gasification |
| SCWO | Supercritical water oxidation |
| SHF | Separate hydrolysis and fermentation |
| SRK | Redlich–Kwong |
| SSF | Simultaneous saccharification and fermentation |
| SSFF | Simultaneous saccharification, filtration, and fermentation |
| STR | Stirred tank reactor |
| TAG | Triacylglycerols |
| TAN | Total acid number |

| | |
|---|---|
| TCD | Thermal conductivity detector |
| TDH | Triose phosphate dehydrogenase |
| TDI | Triose phosphate isomerase |
| TG | Triacylglycerides |
| TGA | Thermogravimetric analysis |
| VFA | Volatile fatty acid |
| WBL | Weak black liquor |
| WCO | Waste cooking oils |
| WEC | World Energy Council |
| WEU | Western Europe |
| WFGD | Wet flue gas desulfurization |
| WGS | Water-gas shift |
| WGSR | Water-gas shift reaction |
| Yo | Symmetrically related tyrosine to YZ |
| YZ | Tyrosine |

# List of Figures

# List of Tables

# Part I
# Notions of Biomass Formation and Development with Some Analytical Methods

# Chapter 1
# Introduction

*Give me a lever and I will move the Earth.*
Sentence generally attributed to Archimedes of Syracuse, Sicily. 287–212 BC. The author of the: *On the Equilibrium of Planes* and father of the first laws of mechanics.

## 1.1 Biomass as a "Almost" Renewable Energy Source

It is well known that alternative source of energy is required and necessary because of the increasing pollution and global warming. The world population is reaching 8 billion, and the world energy demand is constantly increasing, as can be seen from Figs. 1.1 and 1.4.

Up to now, we cannot rely on a sustainable system of energy conversion. This means that our resources are going to be depleted within decades according to some forecasts. As a matter of fact, the end of "cheap" oil has been already predicted in the last decades and inevitably this will happen sooner or later (Campbell and Laherrere 1998).

The increased amount of $CO_2$ in the atmosphere is to be taken seriously into account since the current situation is that we have 350–450 ppm of $CO_2$ in the atmosphere and the acceptable levels are at less than 600 ppm. The very dangerous situation is that the $CO_2$ increases the temperature of our atmosphere and an increase of temperature reflects on higher solubility of calcium carbonate rocks (De Blasio et al. 2009). Now the question is, how much $CO_2$ is actually stored within sedimentary rocks or oceans? The amount is surely scary, considering that a big portion of the earth rocks is actually constituted by this kind of formation. Also to be noticed that enormous amounts of $CO_2$ is actually stored within the oceans, and if the temperature of the oceans will increase, this will cause the release of huge quantities of $CO_2$, and in return, cause the dissolution of the sedimentary rocks which are sited at the bottom of the oceans. This gives terrible results.

Let us look at these facts a little closer. In the literature (Houghton and Woodwell 1989), it is possible to find data on the main carbon flows at global level and some



C. De Blasio, *Fundamentals of Biofuels Engineering and Technology*,
Green Energy and Technology, https://doi.org/10.1007/978-3-030-11599-9_1

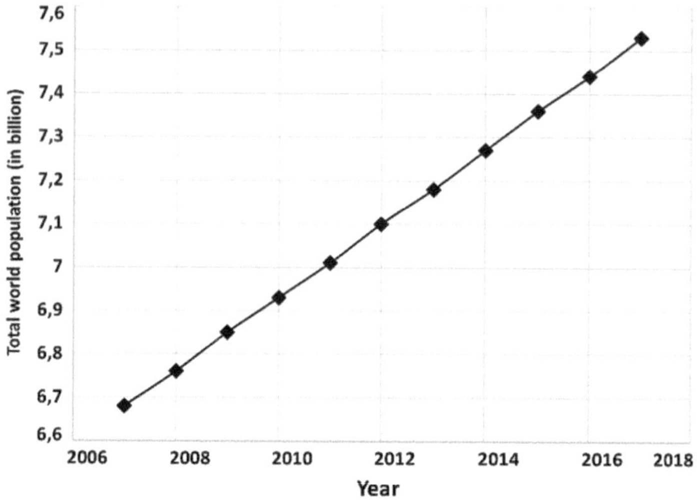


**Fig. 1.1** World total population by year. Data from Statista (2018)

more data will be also given later on in this manuscript. Carbon is stored in the world vegetation, living animals and organisms, the soil, the atmosphere, sedimentary rocks, oceans and of course fossil fuels. Now if we consider that the amount of $CO_2$ stored in biomass would be 500 Gton (we do not need to be precise here), the amount of $CO_2$ stored within the oceans would be five times more! And we are just looking at the oceans. On the other hand, if the $CO_2$ released by burning fossil fuels is around 5 Gton per year, then the amount of $CO_2$ transferred from and to the oceans will be 20 times more. Plants help in capturing around 50 Gton of carbon per year. Fortunately at the present times, $CO_2$ is captured by oceans and because of human destruction of the environment and fossil fuels, the net increase of carbon in this cycle is around 3 Gton per year. On the basis of these reflections, the carbon dioxide equilibrium is dependent mostly on solid–liquid reactions systems as demonstrated here (De Blasio et al. 2013).

$$MgCO_{3(s)} \leftrightarrow Mg^{2+} + CO_3^{2-} \tag{1.1}$$

$$CaCO_{3(s)} \leftrightarrow Ca^{2+} + CO_3^{2-} \tag{1.2}$$

$$H_3O^+ + CO_3^{2-} \leftrightarrow HCO_3^- + H_2O \tag{1.3}$$

$$H_3O^+ + HCO_3^- \leftrightarrow CO_{2(aq)} + 2H_2O \tag{1.4}$$

$$CO_{2(aq)} \leftrightarrow CO_{2(g)} \tag{1.5}$$

In addition to the equilibrium reactions given previously, the dissociation of water should also be taken into consideration.

$$2H_2O \leftrightarrow H_3O^+ + OH^- \tag{1.6}$$

Equilibrium constants and related thermodynamics for the above mentioned reactions can be found in the literature, Sillén and Martell (1965) or Millero (1979). Here, it is sufficient to point out that there is indeed huge potential for catastrophic events if these equilibria are distorted. It should be also acknowledged that indeed the ocean water is working as a buffer in this sense and it would be not that easy to affect these equilibrium states, however the increment of temperature could play a major role.

Some compounds which are contributing to the degree of alkalinity in oceans are shown in Fig. 1.2. Here, the carbon dioxide is in equilibrium with the ocean water which is absorbing it. The carbon dioxide which is solubilized in water is also in equilibrium with the carbonic acid, the bicarbonate, and the carbonate ions. In addition, there are a number of compounds which are contributing further to the alkalinity of water.

The world is also subject to normal and very regular oscillations; for instance, every year we have winter and summer and the change of the seasons heavily affects the biomass production. As a consequence, the $CO_2$ captured and stored by plants and microorganisms is also subject to cycles. This reflects on the atmospheric concentration of carbon dioxide as well; as a matter of fact, it is reported in the literature that the amplitude of the $CO_2$ oscillation within the year between the north and south hemisphere is 15 times the accumulation rate (Box 1988; D'Arrigo et al. 1987; Fung et al. 1987). It is also reported that this amplitude is increasing with time (Bacastow et al. 1985). Photosynthesis and emissions play a major role within this equilibrium,

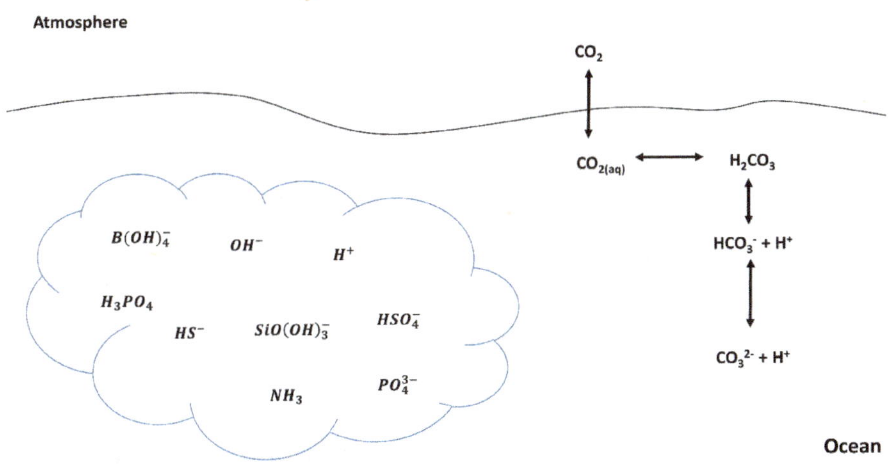


**Fig. 1.2** Alkalinity of oceans and the $CO_2$ equilibrium

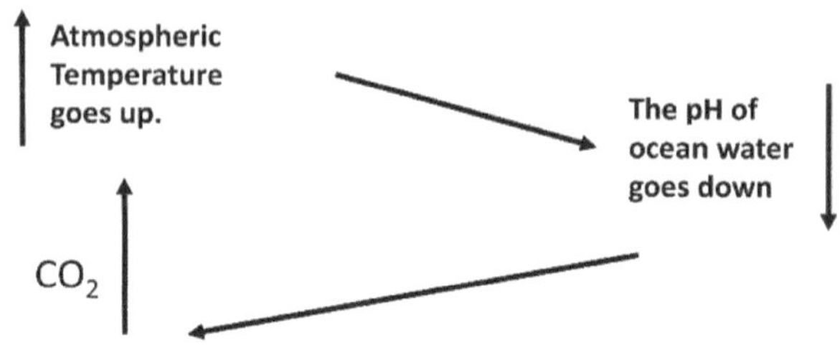


**Fig. 1.3** A simple schematics on the relation between $CO_2$ concentration, atmospheric temperature, and oceans pH

and the alkalinity of water is also something that should be really monitored. A simple representation of this cycle is given in Fig. 1.3.

To be noticed that a decrease of pH of the oceans within a 0.1 range could give an enormous increase of the concentration of $CO_2$ in the atmosphere.

Hydrocarbons (including biofuels) are light compared to other kind of fuels. For instance, it is well known that batteries are quite heavy when we buy them at the shop and to give a further example, the battery of my electric scooter weights around 12 kg and I mostly can do 10 km when it is fully charged. To be specified that the battery I am using is not among the best on the market, but still there is no doubt that they are heavy. Compared to the metals of batteries, the molecular weight of carbon and hydrogen is very low. Furthermore, the amount of energy in hydrocarbons is very high compared to their weight and also it has to be noticed that it is very easy to store liquid fuels and they are mostly not explosive. Hydrogen is very powerful, but it is not that safe to use it. In fact, if hydrogen comes into contact with oxygen at high temperatures, it can give some explosions.

For this reason, when hydrogen is used for experiments in laboratory it is a good idea to have extra precautions concerning safety. For instance, a Plexiglas protection is needed/suggested to be put as protection prior the reactor area.

Aircraft are well known to consume enormous quantity of fuel when operating; therefore, also for this process, the utilization of renewable kinds of feedstock has been proposed. One example is the Neste MY Renewable Jet fuel and, in addition, there is an increasing will from the European Union to increase the production of biofuels for the aircraft sector. Concerning the aviation and the related fuels, in this case, there are not much of alternative choices, since the fuels should be in any case liquid at the flying conditions and they should still contain a sufficient amount of energy per unit mass.

Biomass is considered in the literature as a renewable type of energy resource for the reason that plants, for instance, utilize solar energy and $CO_2$ from the atmosphere to develop themselves. In this way, the total carbon emissions are considered to be zero and biomass is said to not contribute to the global warming effect. Nevertheless, processes like photosynthesis (which will be considered in the following sections) are not limited to plants, but are extensively used by algae (Thomas et al. 1984).

It is important to keep this in mind, especially when algae are considered to fuel the third-generation biorefineries of the future and that biodiesel can be produced by using them. The concepts of microorganisms' growth and the kinetics involved will be considered in the second part of this manuscript along with some examples of algae utilization.

Algae which utilize photosynthesis can produce diverse kinds of oils and ethanol, and it has also referred that they represent the energy of the future by representatives of some companies working in this field (Mascarelli 2009). The very good thing with algae is that they could grow in spaces where they are not competing with other biomass sources, for instance they could be cultivated off shore. However, at this moment, the production of algae for these purposes is indeed expensive, and it is difficult for this technology to compete with fossil fuels considering the actual prices (Mascarelli 2009).

It is not always feasible to utilize wood-based biomass for energy conversion. As a matter of fact, it is indeed true that water and the energy of the sun are available in large quantities (this is true mostly for the second statement); however, plants do need to have inorganic materials in order to function and grow. For this reason, the depletion of land and soil should be taken into account when planning of producing crops with the purpose of energy conversion since there is a diminishing availability of resources for these scopes (Breure et al. 2018).

Of course, if the public do not have knowledge of how the plant actually works to grow themselves up, it is very probable that they will consider trees as a very nice source of energy and propose their utilization for large-scale energy production (Fig. 1.4). Of course politicians are no exception. Knowing the fundamental concepts of biomass generation and the related simple mass balances, will give the foundations for a critical thinking on the matter of feasibility of biomass and "power production."

Biomass can be produced constantly, even if some types of biomass are seasonal, and they can offer short-term sequestration of carbon in non-harvested aboveground

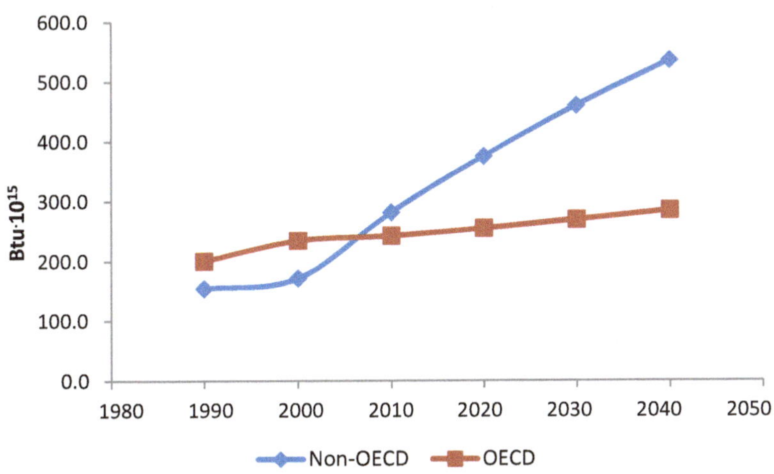


**Fig. 1.4** World energy consumption, 1990–2040 (Btu·$10^{15}$). Data from US Energy Information Administration (2013)

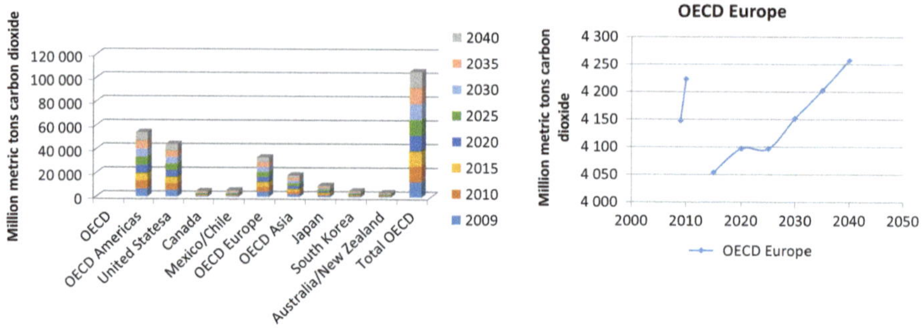


**Fig. 1.5** World carbon dioxide emissions by region, left: all OECD countries; right: OECD Europe. Data from US Energy Information Administration (2013)

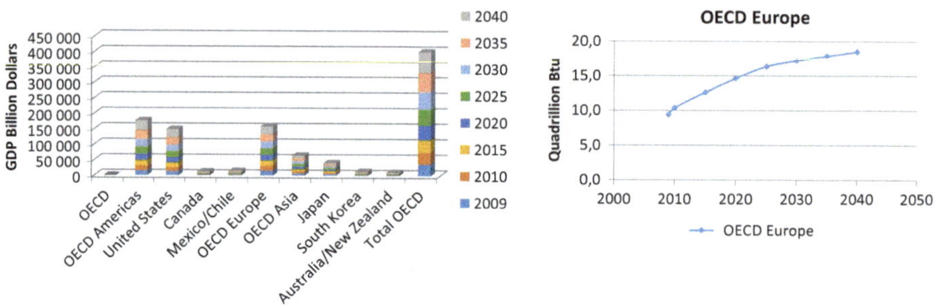


**Fig. 1.6** World gross domestic product (GDP) by region. Data from US Energy Information Administration (2013)

growth and medium-term sequestration in belowground root biomass. The world amount of $CO_2$ emissions is reported in Fig. 1.5.

Figure 1.6 shows instead an updated version from 2013, of the gross domestic product, GDP, for OECD countries divided by continents. Also in this case, it is possible to notice that the GDP projection to 2040 is steadily increasing; this means that the production will also increase and for this reason the energy demand.

Following the need for alternative forms of energy, the usage of renewables is also increasing; this is shown in Fig. 1.7 where the overall consumption of renewables is reported by regions and for only the European OECD countries.

However, there are more recent studies which show that the extensive and large use of biomass for industrial purposes of producing energy is not feasible and it would not give a null net carbon emission balance. One study is reported here as an example of this theory (Schulze et al. 2012); the study done at the Max Planck Institute together with other important institutions refers that with an extensive usage of biomass, we will have younger forest and less forest pools, depleted soil nutrients together with loss of the ecosystem function of the forests themselves. The depletion of the soil fertility will make the extensive use of biomass unsustainable.

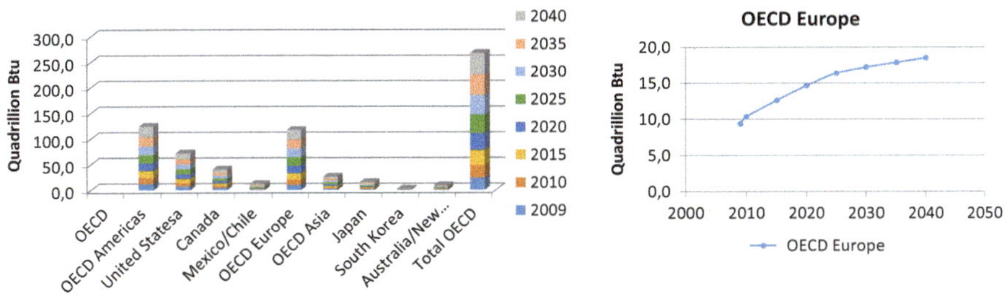


**Fig. 1.7** World consumption of renewable energy by region. Data from US Energy Information Administration (2013)

**Table 1.1** Some studies concerning biomass supply and its usage in energy conversion

| Study | References | Study | References |
|---|---|---|---|
| WEC | World Energy Council (1994) | LESS/BI | Williams (1995) |
| IIASA-WEC | Grubler et al. (1996) | LESS/IMAGE | Leemans et al. (1996) |
| FFES | Lazarus et al. (1993) | BATTJES | Battjes (1994) |
| EDMONDS | Edmonds et al. (2003) | GLUE | Yamamoto et al. (1999) |
| SWISHER | Swisher (1993) | FISHER | Fischer and Schratten-holzer (2001) |
| USEPA | Lashof and Tirpak (1990) | DESSUS | Dessus et al. (1992) |
| SØRENSEN | Soerensen et al. (1999) | SHELL | Shell (1995) |
| HALL | Hall (1993) | SRES/IMAGE | Nakicenovic et al. (2000) |
| RIGES | Johansson (1993) | | |

Other than utilizing only biomass, the solution would be to utilize all the possible alternatives and renewable energy sources and at the minimum possible rate. Meaning that the overall consumption of energy must be reduced.

The usage and the contribution of biomass supply for energy conversion has been studied by several authors, and different scenarios have been provided. Table 1.1 shows some of the most important references on this matter. In the literature, there is information about the approach and the time frame taken by these authors and organizations.

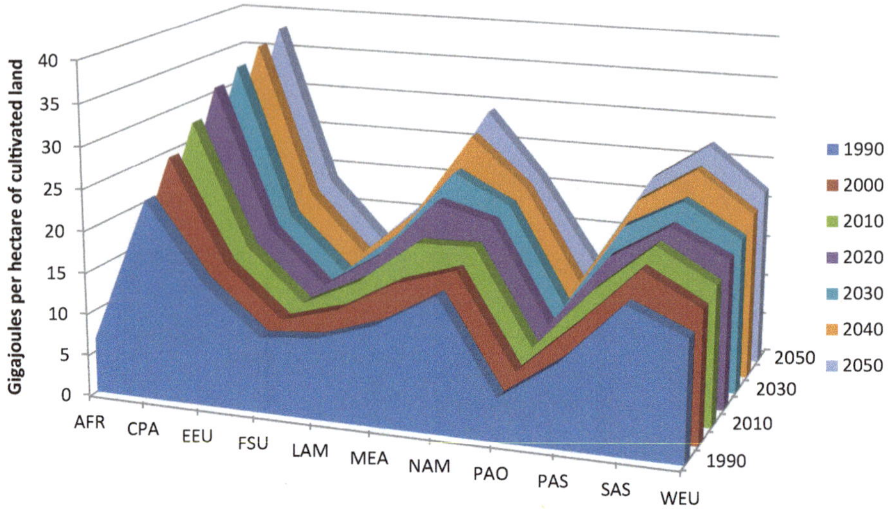


**Fig. 1.8** Potential bio-energy of crop residues by world region, gigajoules per hectare of cultivated land, NAM, North America; LAM, Latin America; WEU, Western Europe; FSU, former Soviet Union; MEA, Middle East and North Africa; CPA, Centrally planned Asia and China; SAS, South Asia; PAS, other Pacific Asia; PAO, Pacific OECD

An estimated bio-energy potential for the crop residues is given in Fig. 1.8; data were retrieved from Fischer and Schrattenholzer (2001).

From Fig. 1.8, it is possible to notice that there will be an increase of the bio-energy potential, and this is because there will be agricultural progress, at least according to the forecasts provided.

In a report given by the IIASA (Nakicenovic and Riahi 2001), the economic evolution of the world is given following different scenarios according to:

- Scenario A1: Rapid economic growth;
- Scenario A2: A differentiated world with conservation of local entities;
- Scenario B1: Convergent world with a rapid change;
- Scenario B2: The focus here is on local sustainability solutions.

The report gives a detailed analysis and forecast about future energy demand and economic growth.

In this text book, the focus will be not only dedicated to energy crops, but also on other types of biomass utilized for power generation; however, since there are many options and the topic is really vast, it was preferred to privilege only some types of biomass. For this reason in the first part of the manuscript, the focus will be on light harvesting and how plants generate work in order to bind $CO_2$ during photosynthesis. Only the main routes for this process will be covered, and after this, the main procedures used for the assessment of biomass before utilization will be treated with particular concentration on proximate and ultimate analysis.

## References


Bacastow, R. B., Keeling, C. D., & Whorf, T. P. (1985). Seasonal amplitude increase in atmospheric $CO_2$ concentration at Mauna Loa, Hawaii, 1959–1982. *Journal of Geophysical Research: Atmospheres, 90*(D6), 10529–10540. https://doi.org/10.1029/JD090iD06p10529.

Battjes, J. J. (1994). *Global options for biofuels from plantations according to IMAGE simulations* (No. IVEM-SR-77). Rijksuniversiteit Groningen (Netherlands). Interfacultaire Vakgroep Energie en Milieukunde. Retrieved from http://inis.iaea.org/Search/search.aspx?orig_q=RN:26031212.

Box, E. O. (1988). Estimating the seasonal carbon source-sink geography of a natural, steady-state terrestrial biosphere. *Journal of Applied Meteorology, 27*(10), 1109–1124. https://doi.org/10.1175/1520-0450(1988)027%3c1109:ETSCSS%3e2.0.CO;2.

Breure, A. M., Lijzen, J. P. A., & Maring, L. (2018). Soil and land management in a circular economy. *Science of the Total Environment, 624*, 1125–1130. https://doi.org/10.1016/j.scitotenv.2017.12.137.

Campbell, C. J., & Laherrere, J. (1998). The end of cheap oil. *Scientific American, 278*(3). Retrieved from https://www.jstor.org/stable/26057679.

D'Arrigo, R., Jacoby, G. C., & Fung, I. Y. (1987). Boreal forests and atmosphere–biosphere exchange of carbon dioxide. *Nature, 329*(6137), 321–323. https://doi.org/10.1038/329321a0.

De Blasio, C., Ahlbeck, J., & Westerlund, T. (2009). Modeling the hydrodynamics and mass-transfer phenomena for sedimentary rocks used for flue gas desulfurization. The effect of temperature. In R. M. de Brito Alves, C. A. O. do Nascimento, & E. C. Biscaia (Eds.), *Computer aided chemical engineering* (Vol. 27, pp. 411–416). Elsevier. https://doi.org/10.1016/S1570-7946(09)70289-5.

De Blasio, C., Carletti, C., Westerlund, T., & Järvinen, M. (2013). On modeling the dissolution of sedimentary rocks in acidic environments. An overview of selected mathematical methods with presentation of a case study. *Journal of Mathematical Chemistry, 51*(8), 2120–2143. https://doi.org/10.1007/s10910-013-0202-3.

Dessus, B., Devin, B., & Pharabod, F. (1992). World potential of renewable energies actually accessible in the nineties and environmental impacts analysis. *Houille Blanche, 47*(1), 21–70.

Edmonds, J. A., Wise, M. A., Sands, R., Brown, R., & Kheshgi, H. (2003). Agriculture, land use, and commercial biomass energy: A preliminary integrated analysis of the potential role of biomass energy for reducing future greenhouse related emissions. In *Proceedings of the 6th Greenhouse Gas Control Technologies Conference* (pp. 0-08-044045–2). Oxford UK: Elsevier Inc.

Fischer, G., & Schrattenholzer, L. (2001). Global bioenergy potentials through 2050. *Biomass and Bioenergy, 20*(3), 151–159. https://doi.org/10.1016/S0961-9534(00)00074-X.

Fung, I. Y., Tucker, C. J., & Prentice, K. C. (1987). Application of advanced very high resolution radiometer vegetation index to study atmosphere-biosphere exchange of $CO_2$. *Journal of Geophysical Research: Atmospheres, 92*(D3), 2999–3015. https://doi.org/10.1029/JD092iD03p02999.

Grubler, A., Jefferson, M., & Nakicenovic, N. (1996). *Global energy perspectives: A summary of the joint study by IIASA and world energy council* (Monograph). Retrieved July 9, 2018, from http://pure.iiasa.ac.at/id/eprint/4860/.

Hall, D. O. (1993). Biomass for energy: Supply prospects. In *Renewable energy: Sources for fuels and electricity* (pp. 593–651). Washington D.C.: Island Press.

Houghton, R. A., & Woodwell, G. M. (1989). Global climatic change. *Scientific American, 260*(4), 36–47.

Johansson, T. B. (1993). A renewables-intensive global energy scenario. In *Renewable energy: Sources for fuels and electricity* (pp. 1071–1143). Washington D.C.: Island Press.

Lashof, D. A., & Tirpak, D. A. (1990). *Policy options for stabilizing global climate*. U.S.: Environmental Protection Agency.

Lazarus, M., Greber, L., Hall, J., Bartels, C., Bernow, S., Hansen, E., … Von Hippel, D. (1993). *Towards a fossil free energy future. The next energy transition*. Stockholm Environment Institute Boston Center.

Leemans, R., van Amstel, A., Battjes, C., Kreileman, E., & Toet, S. (1996). The land cover and carbon cycle consequences of large-scale utilizations of biomass as an energy source. *Global Environmental Change, 6*(4), 335–357. https://doi.org/10.1016/S0959-3780(96)00028-3.

Mascarelli, A. L. (2009). Gold rush for algae. *Nature, 461*(7263), 460–461. https://doi.org/10.1038/461460a.

Millero, F. J. (1979). The thermodynamics of the carbonate system in seawater. *Geochimica et Cosmochimica Acta, 43*(10), 1651–1661. https://doi.org/10.1016/0016-7037(79)90184-4.

Nakicenovic, N., & Riahi, K. (2001). *An assessment of technological change across selected energy scenarios* (Monograph). Retrieved July 9, 2018, from http://pure.iiasa.ac.at/id/eprint/6521/.

Nakicenovic, N., Alcamo, J., Grubler, A., Riahi, K., Roehrl, R. A., Rogner, H.-H., & Victor, N. (2000). *Special report on emissions scenarios (SRES), a special report of working group III of the intergovernmental panel on climate change*. Cambridge: Cambridge University Press. Retrieved from http://pure.iiasa.ac.at/id/eprint/6101/.

Schulze, E.-D., Körner, C., Law, B. E., Haberl, H., & Luyssaert, S. (2012). Large-scale bioenergy from additional harvest of forest biomass is neither sustainable nor greenhouse gas neutral. *GCB Bioenergy, 4*, 611–616. https://doi.org/10.1111/j.1757-1707.2012.01169.x.

Shell. (1995). *Evolution of the world's energy system 1860–2060*. London: Shell Center.

Sillén, L. G., & Martell, A. E. (1965). Stability constants of metal-ion complexes (Sillen, Lars Gunnar; Martell, Arthur E.). *Journal of Chemical Education, 42*(9), 521. https://doi.org/10.1021/ed042p521.1.

Soerensen, B., Meibom, P., & Kuemmel, B. (1999). *Long-term scenarios for global energy demand and supply. Four global greenhouse mitigation scenarios*. Final Report (No. IMFUFA-TEKST--359). Roskilde Universitetscenter (Denmark): Inst. for Studiet af Matematik og Fysik samt deres Funktioner i Undervisning. Retrieved from http://inis.iaea.org/Search/search.aspx?orig_q=RN:30029769.

Statista. (2018). *World—Total population 2007–2017 | statistic*. Retrieved December 31, 2018, from https://www.statista.com/statistics/805044/total-population-worldwide/.

Swisher, J. (1993). Renewable energy potentials, Chap. 3. *Energy, 18*(5), 437–459. https://doi.org/10.1016/0360-5442(93)90022-6.

Thomas, W. H., Seibert, D. L. R., Alden, M., Neori, A., & Eldridge, P. (1984). Yields, photosynthetic efficiencies and proximate composition of dense marine microalgal cultures. I. Introduction and *Phaeodactylum tricornutum* experiments. *Biomass, 5*(3), 181–209. https://doi.org/10.1016/0144-4565(84)90022-2.

U.S. Energy Information Administration. (2013). *International energy outlook 2013*. Retrieved from http://www.eia.gov/forecasts/ieo/.

Williams, R. H. (1995). *Variants of a low $CO_2$-emitting energy supply system (LESS) for the world.* Prepared for the IPCC Second Assessment Report Working Group IIa, Energy Supply Mitigation Options.

World Energy Council. (1994). *New renewable energy resources*. Kogan Page Ltd.

Yamamoto, H., Yamaji, K., & Fujino, J. (1999). Evaluation of bioenergy resources with a global land use and energy model formulated with SD technique. *Applied Energy, 63*(2), 101–113. https://doi.org/10.1016/S0306-2619(99)00020-3.

# Chapter 2
# Light Harvesting and Biomass Generation

*So when we consider the Photosystem II and their reaction centers, they are producing oxygen and oxygen in combination with light is very often very toxic. What happens there is that you get an oxidative degradation of one protein unit and nature has developed a way to exchange these damaged proteins every 20 minutes. So every 20 minutes a repair has to be made and I do not think that we can mimic this process in a technical manner.*

Michel Hartmut, Nobel Prize in Chemistry 1988, talking about renewable energy in 2008.

## 2.1 Preliminary Data Related to Light Harvesting and Biomass Generation

The question of how the biomass is produced is central to understand the concept of feasibility of this feedstock in energy conversion processes. All biological processes utilize mechanisms of oxidation of substrate molecules and reduction of final acceptors, and photosynthesis is one good example which allows for understanding many useful concepts. In the photosynthetic process, there is oxidation of a substrate molecule, water, followed by the generation of an electrochemical potential and a related "current" of electrons. The basic explanation of this process is reported here for pedagogical purposes to demonstrate the analogy between biological mechanisms and some technical solutions.

According to Ksenzhek and Volkov (1998), 105 net Gton of carbon is fixed by photosynthesis on land and into the oceans every year. While Kell (2012) refers that soil could accommodate 50 ton $ha^{-1}$ of carbon in the same period. On the other hand, 260 Gton of oxygen are released into the atmosphere every year (Ksenzhek and Volkov 1998). We can consider that the amount of oxygen given by plants could be increasing with time considering that today, plants store 25% of all the carbon emissions produced by the utilization of fossil fuels (Owen 2017). In the world, there are more than 390,000 known species of plants (Chapman 2009), and it is reported in the literature (Ksenzhek and Volkov 1998) that the atmosphere contains $1.2 \times 10^6$ Gton oxygen.



C. De Blasio, *Fundamentals of Biofuels Engineering and Technology*,
Green Energy and Technology, https://doi.org/10.1007/978-3-030-11599-9_2

**Table 2.1** Carbon release per GJ for some biomass types and fossil fuels, data from Ksenzhek and Volkov (1998)

| Fuel | kg C/GJ |
|---|---|
| Wood | 26.1–29.9 |
| Peat | 30.0 |
| Coal | 23.9–25.8 |
| Crude oil | 19.0–21.4 |
| Natural gas | 13.6–15.4 |

Furthermore in all OECD countries, the amount of $CO_2$ released by fuel combustion was 11,720 Mton in 2015 (IEA 2017), and it can be noticed that 1 GJ of energy from oil results in the net release of 73 kg $CO_2$ while natural gas gives 52 kg of $CO_2$ for 1 GJ (Ksenzhek and Volkov 1998). In all the world, the amount of $CO_2$ released by fuel combustion was 32,294.2 Mton in 2015 (IEA 2017).

Concerning the carbon release from biomass and fossil fuels, Table 2.1 gives some data related to the most common types of feedstock.

Some estimations for the potential biomass production give that the power which could be obtained by photosynthetic processes could be between 1500 and 2250 TW, while the total power coming from the sun is estimated to be 178,000 TW (Pisciotta et al. 2010). It is also estimated that 220 billion ton of dry biomass can be produced per year, and this would be equal to $4 \times 10^{18}$ MJ of energy. In addition, the standing biomass is estimated to represent $36 \times 10^{18}$ MJ (Ksenzhek and Volkov 1998).

From the previously presented data, it can be noticed that there is huge potential concerning the utilization (and the production) of biomass for the purposes of power generation. There are, however, positive and negative aspects with biomass production for the purposes of energy conversion. Among the positive aspects, we have as follows:

- The environmental impact varies in relation to the source and the applied technology; however, there is much less impact than the fossil fuels in particular for the emissions of greenhouse gases.
- It is possible to evaluate the total reduction of $CO_2$ and $CH_4$ emissions taking into consideration the analysis of the life cycle for the renewable sources.
- The renewable alternatives have a low environmental impact and offer a possibility for economic and social development; new work places will be created, and there will be a positive impact on the territory.

Among the limits of biomass production and with reference to energy crops:

- They are available at intervals of time with no continuity.
- It is necessary to utilize a large surface in order to have a significant amount of potential energy.
- The climate of course plays a fundamental role in biomass growth.

Some examples of biomass which could be utilized for power conversion:

- wood and wood waste;
- municipal organic solid waste (one example of activities which could be done in this case is the incineration and methane utilization from landfills);
- herbaceous biomass and agricultural residues;
- aquatic biomass;
- industrial solid organic wastes;
- municipal bio-solids;
- products from farm animal manures;
- algae.

Among the lignocellulosic kinds of biomass, we have:

- agricultural residues, residues from forests, paper industry, and wood industry;
- energetic cultivations.

Of course there is also a series of microorganisms to be considered as biomass, and these microbes can indeed be utilized for energy conversion. For instance, microalgae and several microorganisms can drive fermentation and production of biogas. However, these concepts will be covered within the second part of this manuscript.

## 2.2 Light Harvesting

In a very general sense, biofuels are derived from the energy coming as radiation from the sun. The incoming energy is "harvested" by plants and microorganisms, and by using the existing $CO_2$ of the atmosphere, new biomass can be then produced. In order to develop biomass (grow), a plant has to perform a series of activities which are listed as follows:

- The photosynthetic process needs water and light to work. Now in case we consider a tree, the water should be brought up where the leaves are. For this reason, water should be taken from the soil.
- The same process in order to work needs minerals and therefore, extraction of mineral components from soil is also necessary;
- The nitrogen should be bond into the process as well, for instance activities related to its storage and cycling (Li and Coleman 2018);
- Finally, the carbon dioxide of course should be taken from air. This can be also directly related to a novel trend of research which includes the artificial photosynthesis (Liu et al. 2016).

In addition to all these operations, there are all a series of other activities to be performed and these activities have the goal of synthesizing the organic matter:

- The plant should obtain the photosynthesis products from a larger surface (the leaves).
- Transport of metabolites and metals through the system (Kobayashi et al. 2019).

- Intracellular movements (for instance, in plant cells and among which chloroplasts or nucleus) (Nagai 1993).
- Maintenance of an electrochemical potential difference on cellular membranes (De Vrieze et al. 2018).
- Building of plant structures on a cellular and macroscopic level.
- Plant tropism (Gilroy and Masson 2008).

In order to do all this work, plants need energy. The source of all this energy is the sun: Solar radiation is collected and used by plants to perform the photolysis of water (this is also referred as photochemical water treatment) (Furatian and Mohseni 2018), produce a transmembrane electrochemical potential, and finally bond $CO_2$ molecules with the help of ATP molecules. This process is called photosynthesis.

The incoming radiation is of course measured in terms of energy and power, and for a determinate incoming radiation, there will be some production of biomass. Now, not all the plants for instance have the same capability of producing biomass from incoming radiation from the sun and therefore, they are classified and divided by their radiation use efficiency, RUE. One preliminary example is given here for the type of cultivation Nicknejad Cultivar, and the value of the RUE in this case is 0.84 g (of biomass produced) on MJ of radiation coming on 1 $m^2$ of land (Miranzadeh et al. 2011). Several examples of RUE for diverse kinds of biomass will be given later in this manuscript.

Coming back to the introduction of the photosynthesis, this process produces sugar and oxygen from $CO_2$ and water with the help of solar radiation following the general reaction:

$$6CO_2 + 6H_2O \rightarrow C_6H_{12}O_6 + 6O_2 \tag{2.1}$$

There are two kinds of photosynthesis: oxygenic and anoxygenic (Blankenship et al. 2007); in oxygenic photosynthesis, there is production of oxygen (Nelson 2011), while the anoxygenic photosynthesis is carried on by oxygen-intolerant prokaryotic organisms or phototrophic bacteria (Qi et al. 2018) which utilize sun radiation and bacteriochlorophyll.

Light harvesting is done only at determinate wavelength or frequencies, and this is because only a part of the photosynthetic spectrum has the necessary energy to carry on the required steps for water oxidation (McCree 1981). This can be described more simply by the following example: Let us assume that we want to break a stone by letting the stone to fall down from a determinate altitude; we could see that the altitude has to be necessary in order for the stone to gain the required kinetic energy from its potential energy. We could see, for instance, that the energy required to break the stone could not be sufficient if the object does not gain the required energy. In the same way, the photons should have a sufficient energy to allow for the process of oxidation.

It is good to remember that oxidation means that we are taking electrons from a substrate, for instance if we oxidize iron means that electron are taken from it. On the other hand, reduction indicates that we are actually giving electrons to a substrate, for this reason an object receiving electrons is said to be reduced. The radiations utilized for the photosynthetic processes are included in the so-called photosynthetic active radiation, PAR (Sun et al. 2017); this means that radiations with an energy less than PAR are discarded and not used. An electromagnetic wave, such as light, is produced by the movement of electrons in the source of the radiation itself. Electrons in atoms are situated in energy levels (Jones and Fleming 2014), and when they move from one level to another, they release the difference in energy between the two levels; this energy is released as an electromagnetic wave. There is much theory about this, and students interested more on the argument can read further references of physics and even better, quantum mechanics. Here only some basic information about the atoms will be given, and more information will be given during the course of the manuscript when the oxidation state of atoms and molecule will be described. For a more detailed description of where electrons are located around the nucleus of atoms and how they are involved in chemical bonds, I invite the readers to refer to textbooks like Brown et al. (2014) or Jones and Fleming (2014).

The velocity of an electromagnetic wave, $c$, is directly proportional to its frequency, $\nu$, and its wavelength, $\lambda$, and it is given by:

$$c = \nu \cdot \lambda \tag{2.2}$$

and the energy associated with it is:

$$E = h \cdot \nu \tag{2.3}$$

where $h$ is the Planck constant ($6.626 \times 10^{-34}$ J s) and $\nu$ is the frequency of the wave. So as seen from the above equations, the higher the frequency, the lower is the wavelength, and the more is the value of the frequency the more is the energy carried by the wave. The constant $c$ is the speed of the wave; now since light is considered as an electromagnetic wave, $c$ is the speed of light by definition. The wave velocity is a function of the medium where the wave propagates; in the vacuum, the speed of light is $c = 299.8 \times 10^3$ km/s, and in water for instance, the speed of light is different and equal to $c = 224.9 \times 10^3$ km/s (Resnick et al. 2001). The speed of the electromagnetic wave in the medium is a function of the index of refraction of the same medium in the form:

$$n = \frac{c}{v} \tag{2.4}$$

where $n$ is the index of refraction, $c$ is the speed of light in vacuum, and $v$ is the velocity of the wave in the medium considered. Table 2.2 gives some values for the refraction index for selected materials, with references.

Figure 2.1 shows the visible spectrum among a range of wavelengths.

It is possible to notice from Fig. 2.1 that the visible spectrum represents only a very small region of all the possible values for the frequency and the wavelengths; the infrared region is at lower frequency (larger wavelength), and the ultraviolet is

**Table 2.2** Values for the refraction index for selected materials, with references

| Material | *n* | References | Material | *n* | References |
|---|---|---|---|---|---|
| Vacuum | 1 | | Fused silica glass | 1.45282 | Malitson (1965) |
| Air | 1.0003 | Ciddor (1996) | Sodium chloride | 1.52894 | Bass et al. (2009) |
| Water | 1.42433 | Daimon and Masumura (2007) | Cellulose | 1.4608 | Kasarova et al. (2007) |
| Ethyl alcohol | 1.36652 | Rheims et al. (1997) | Gold | 0.28 | Palik (1985) |
| Calcite | 1.63703 | Ghosh (1999) | Glycerol | 1.4631 | Rheims et al. (1997) |

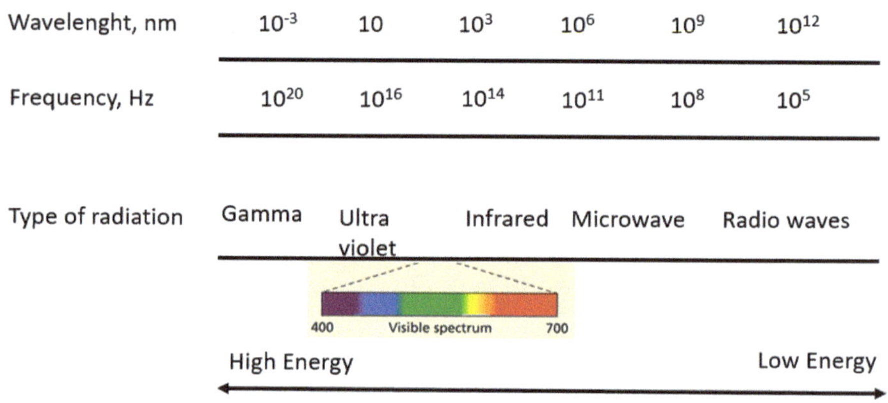


**Fig. 2.1** Visible spectrum among the total wavelength spectrum. Adapted from Taiz et al. (2014)

at higher frequency and shorter wavelengths. Energy is coming from a source as electromagnetic waves, a part of this energy is reaching a determinate object, and a simple energy balance on the object would give that a part of the incoming energy is absorbed; a part is transmitted through the object, and a part is reflected. So the total incoming radiation would then be:

$$\mathrm{TR} = \alpha \cdot (\mathrm{TR}) + \tau \cdot (\mathrm{TR}) + \rho(\mathrm{TR}) \tag{2.5}$$

where TR, $\alpha$, $\tau$, $\rho$ are, respectively, the total incoming radiation, the absorbance, the transmittance, and the reflectance coefficient.

For this reason, when the radiation arrives on the earth, it has to go through the atmosphere in order to reach its surface. Figure 2.2 shows the energy arriving on the earth surface as a function of the wavelength.

It is possible to notice from Fig. 2.2 that the energy arriving on the surface is lower at larger wavelengths and it is not a smooth curve; this is due to the fact that at larger

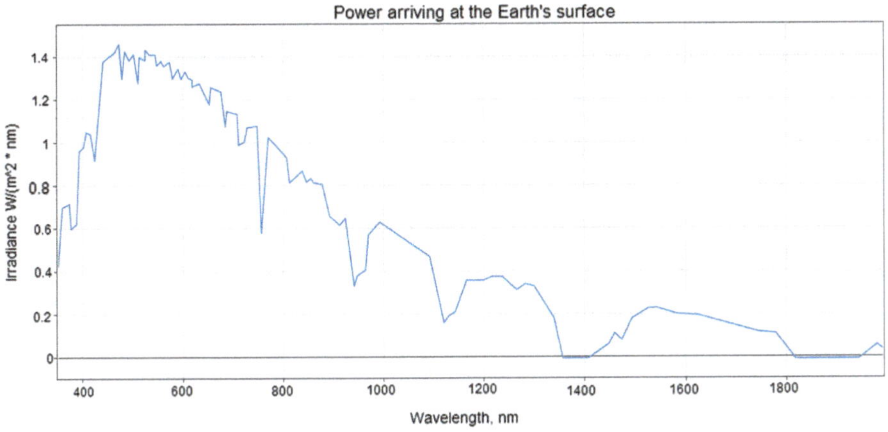


**Fig. 2.2** Energy at the earth's surface. Adapted from Taiz et al. (2014)

wavelengths, we have lower energy of the wave (because we have less frequency) and also because the atmosphere does not absorb in the same way at all wavelengths. This is mostly due to the absorbance of the steam in the atmosphere.

Like the atmosphere, also plants absorb incoming sun radiation. The incoming radiation absorbed by chlorophyll is reported in the literature as a function of the wavelength. It is possible to see that between 550 and 600 nm, chlorophyll does not absorb and instead it reflects light at that wavelength. This is just the wavelength of the green color, and this is why chlorophyll appears to be green at our eyes. This is still valid for chlorophyll-sensitized solar cells (CSSCs) for which chlorophyll shows a similar behavior; see Fig. 2.3.

Figure 2.4 describes the possible states of chlorophyll. Radiation is arriving at an energy level, and it is giving energy at the surface of the plant. The incoming energy at higher level (more energy per photon) is then absorbed, lost as heat or reflected as, for instance, luminescence.

The most common method used by plants to grow is the photosynthesis carried out by means of chlorophyll. Chlorophylls are only some of the photosynthetic units; they constitute the light-harvesting (LH) antenna system. The energy in the antenna complex (or light-harvesting complex, LHC) is transferred by resonance transfer, and in the literature, it is also mentioned that energy transfer is done by "jumping" of energy among discrete energy levels (Blankenship 2013; Mirkovic et al. 2017; Scholes and Fleming 2005). We could do here the hypothesis that 95–99% of the photons absorbed by the antennae are transferred to the reaction center. Within the LHC, the photosystem II is composed by diverse structures; for instance, it is referred that it is constituted by six types of light-harvesting complexes (Xu et al. 2017).

In the first step of the photosynthetic process, light is absorbed by the pigments associated with the antenna proteins (the LHCs) which are organized as hetero-oligomers.

The way in which the pigments are organized and linked to the LHC determines the way in which the energy is transferred. This energy leads to the formation of chemical reactions, and this phenomenon is called photochemistry. As referred by Xu and collaborators, the most studied LHC is the one related to the PSII while less information is available for the PSI structures.

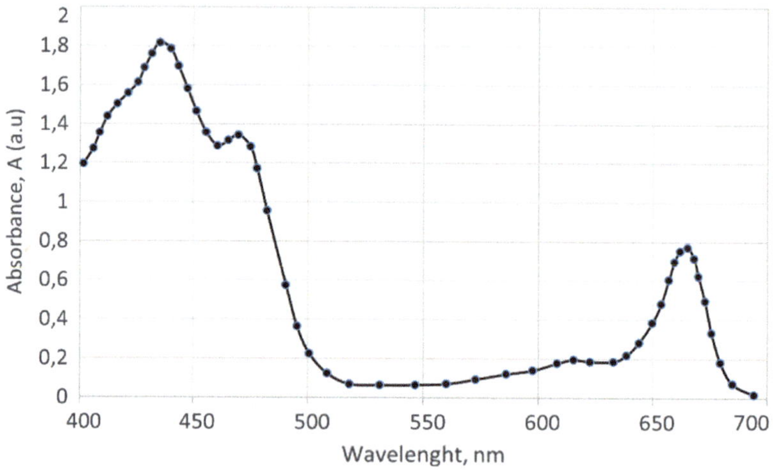


**Fig. 2.3** Absorption of chlorophyll-sensitized solar cells (CSSCs). Chlorophyll, main regions where the chlorophyll absorbs the most and the less energy in. Data from Hassan et al. (2016)

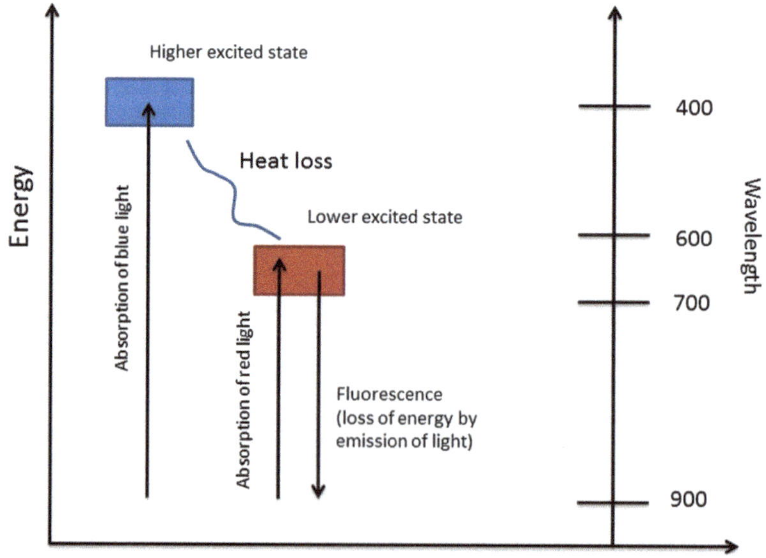


**Fig. 2.4** Excited states of chlorophyll, a qualitative representation

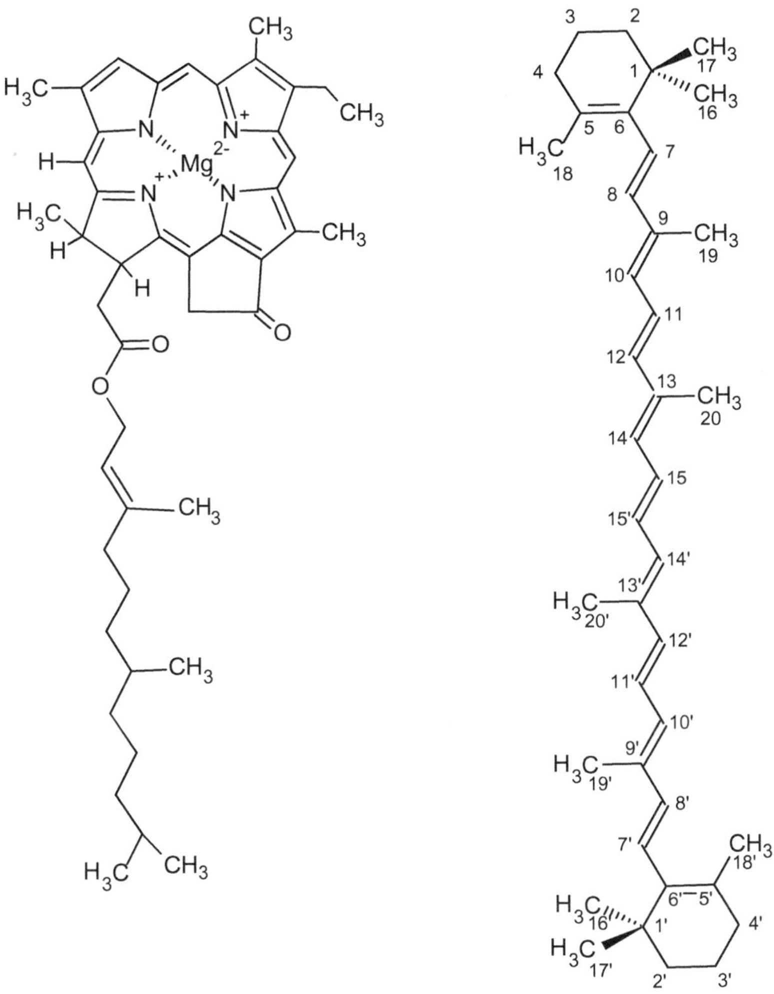

**Fig. 2.5** Chlorophyll and beta-carotene molecules

The thylakoid membrane contains chlorophyll and the electron transport system that carries out the initial light energy. Figure 2.5 describes the most common components of the light-harvesting complex, basically the most common are chlorophylls and carotenoids.

A simplification of the main reaction of photosynthesis is given as follows:

$$CO_2 + H_2O \rightarrow (CH_2O) + O_2 \tag{2.6}$$

In case we have wavelengths at 0.68 μ, the energy absorbed is 1760 kJ/mol ($O_2$) and the energy necessary to carry on the reaction given previously is 467 kJ/mol.

This is simply evaluated as the heat of combustion of glucose (which is 2801 kJ/mol) divided by six.

In the reaction center, the most important step of all the process takes place: the conversion of solar energy to chemical and electrical energy. Within this step, the solar energy (the one given by the photons) is converted in a stable charge flow of electrons and this is done by ultrafast electron transfer. However, as also reported in other parts of this manuscript, this "near unity" quantum efficiency of the process is still unknown (Romero et al. 2014). The reaction center works also as a quantum heat engine (QHE) since it converts solar energy into useful work, and this is done by the charge separation and the formation of the necessary fuels (ATP molecules) with the related chemical energy.

Nevertheless, it has been studied that quantum coherence could enhance the photosynthetic efficiency of the charge separation (the oxidation of water) in these QHE, and it has also been showed that, even if quantum coherence could play a major role in this process, photosynthesis does not require coherent light to work (Dorfman et al. 2013). This enhancement suggests to reproduce natural solutions to improve photovoltaic cells or related technical QHE. To be remembered here that with coherent light, it is meant a beam of light where the photons have all the same frequency and wavelength (a laser) and it is remarkable that plants have been adapted even to use not coherent light, like the one coming from the sun. What a great engine is a leaf of a plant!

On the other hand, it is left to us to define the quantum yield of photosynthesis, QYF, which is (Sommer et al. 2015):

$$\mathrm{QYF} = \frac{n_{\mathrm{mol}}}{n_{\mathrm{q}}}; \tag{2.7}$$

where $n_{\mathrm{mol}}$ is the number of molecules converted and $n_{\mathrm{q}}$ is the number of quanta absorbed. The amount of molecules converted is a function of the energy carried by the electromagnetic waves (light). Figure 2.6 describes, qualitatively, the quantum yield of photosynthesis as a function of the wavelength (energy) of the sun radiation.

Plants and microorganisms which utilize photosynthesis to reproduce or growth their biomass can be considered as energy-converting machines. Since these machines could be considered as black boxes, energy balances and thermodynamic rules apply also here. The next paragraphs describe more in detail the so-called reaction centers of photosynthesis; these are very complex molecular structures and as it will be mentioned, many of the steps involved during the oxidation of water are still not well understood. It will be shown in the following chapters that the creation of transmembrane electrochemical potential is a key step for the generation of the necessary work to function the ATP synthase enzyme. This functioning is equivalent to the same principles applied in galvanic cells or batteries (van Rotterdam et al. 2002), and this is explained here since it has pedagogical value, at least according to this author opinion.

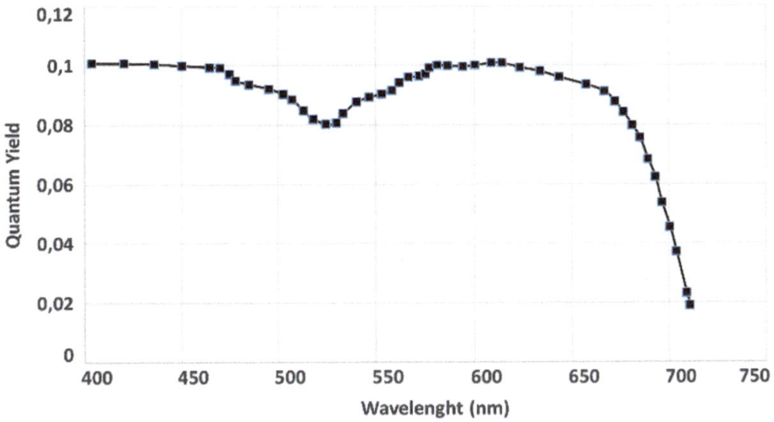


**Fig. 2.6** Quantum yield of photosynthesis, in the visible spectrum. Adapted from Zeiger and Taiz (1991)

Attempts to simplify the photosynthetic process and to give a thermodynamic treatment of the photosynthetic energy-converting devices (the plants) have been carried on in the literature (Albarrán-Zavala and Angulo-Brown 2007; Brittin and Gamow 1961). These studies often utilize data given in the literature for the standard state Gibbs free energy of overall biochemical reactions (Madigan et al. 2000; Dean 1992; Conn and Stumpf 1972). Superior plants and cyanobacteria produce glucose molecules from $CO_2$ and water; however, alternative methods exist to form glucose from different sources of hydrogen; purple and green bacteria are some examples.

## 2.3 Energy Transfer and Reaction Centers

As previously mentioned, the focus is given here to superior plants and their way to generate the building blocks for the development of the biomass. This is also because, it has been derived that they are among the most efficient energy-converting living organisms. The chlorophyll and its reaction centers are responsible for the necessary oxidation of water. All the chlorophyll is contained within the thylakoid membrane in particular cell structures called chloroplast. The reactions of carbon reduction, catalyzed by enzymes which are soluble in water, take place in the stroma. This is the membrane sited in the chloroplast outside the thylakoids. The stroma lamellae are membranes which extend then to the stacks of the thylakoids (Yahia et al. 2019). The light-driven reactions are taking place mostly within the thylakoids; however, also within the photosystem I there are light-driven reactions and here the energy produced within the thylakoids in the form of ATP and NADPH molecules is utilized to reduce and utilize the $CO_2$ which is coming from the pores of the chloroplasts.

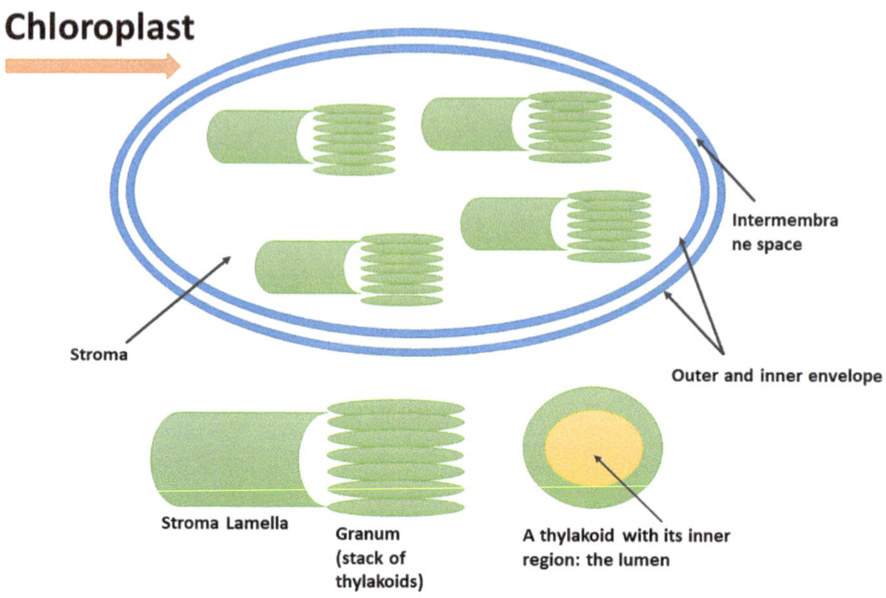


**Fig. 2.7** Chloroplast structure, a qualitative representation

The exact structure of the light-harvesting complexes proteins has not been determined yet; however, this is not the only topic that still needs research; as it will be shown within the manuscript, there are many principles yet to be discovered concerning light harvesting. The purpose of this section is to introduce the reader to the general principles of power (and biomass) generation in plants and how these processes can have analogy with the working principles of galvanic cells and batteries. As a matter of fact, there is an electron flow within the chloroplast structure and the creation of the transmembrane electrochemical potential. The figures that follow are made by the author of this book, and they might be not the most beautiful ones; however, these figures were created with the purpose of being simple and to give a conceptual description of the subject at hand.

Figure 2.7 describes the structure of the chloroplasts and the inner thylakoids, grana lamellae and stroma lamellae.

The reaction centers, the antenna pigment–protein complexes, and most of the electron transport enzymes are all integral membrane proteins (Kumari 2017). The internal part of the thylakoid is called lumen. More in detail, the photosystem II, PSII, with the related reaction center and chlorophylls, is situated in the grana lamellae, formed by the thylakoids. The photosystem I (PSI) instead, with the related reaction center and the electron transport system, is situated in the stroma lamellae. The cytochrome b6f is situated in between the stroma and grana (Nicholls and Ferguson 2013). In PSII, the oxidation of two molecules of water produces four protons, four electrons, and one $O_2$. Figure 2.8 gives a simple representation of how the energy

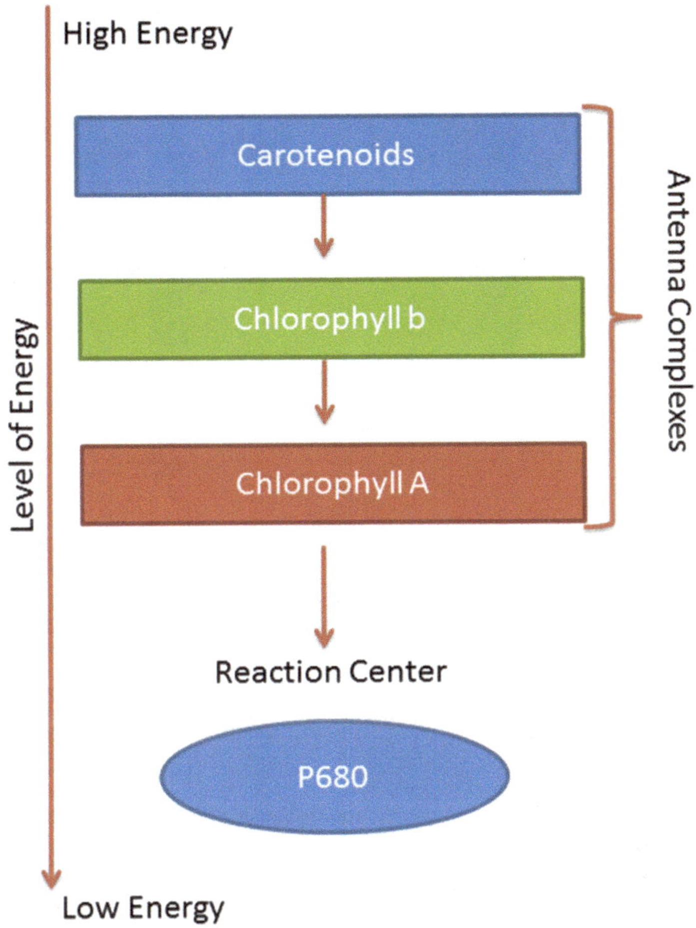


**Fig. 2.8** Transfer of energy in light-harvesting antennas

is transferred among the antenna complexes and arrives to the reaction center of the PSII.

The reaction centers of the two photosystems act as a catalyzer; the water is transported firstly by the tyrosine molecule to a manganese–oxygen-evolving complex, and here the reactions of oxidation are taking place. The tyrosine is then coupled to the Mn cluster (Noguchi et al. 1997).

The Step 3 illustrated in Fig. 2.9 releases oxygen and from this, two atoms of hydrogen are available together with electrons to create the necessary transmembrane electrical potential. The exact working mechanism for the manganese–oxygen-evolving complex and the release of the oxygen is not yet understood exactly. However, in this manuscript, the goal is to just give general information related to these phenomena and for more detailed information, the reader could refer to more spe-

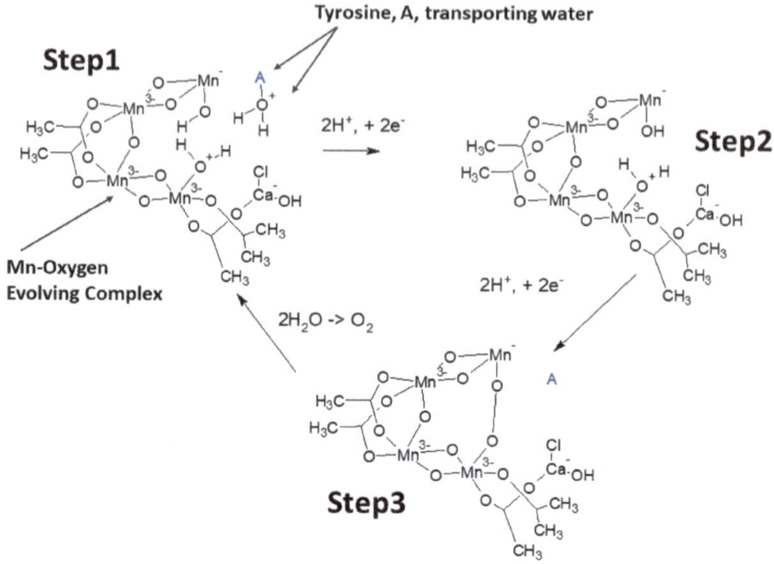


**Fig. 2.9** Oxidation of water into the reaction center and transported by tyrosine molecule

cialized literature. The goal here is to understand that the water is oxidized (removal of electrons) and this is because the plant needs a "current" (flow of electrons) and a transmembrane electrochemical potential to function some of the very important enzymes like for instance the ATP synthase.

In the Photosystem II, the oxidation of water and the release of electrons together with $H^+$ ions are taking place by means of a reaction center P680 which is naturally a strong oxidant (we have to oxidize water and this is the most difficult task). How photosynthetic reaction centers control the oxidation power in the canters P680, P700, and P870 is explained more in detail in the literature (Ishikita et al. 2006, p. 870). After this, the reaction center has some more electrons and therefore its electrochemical potential is increased. By means of an electron transport chain (imagine this as a wire), the electrons are transported to a second reaction center which is situated in the PSI this time, the reaction center P700 where 680 and 700 refer to the wavelength of the incident electromagnetic wave at which the reaction center absorbs the most of the energy. At the reaction center P700, the electrochemical potential should increase further to carry on subsequent actions and therefore the P700 reaction center should be a strong reductant. This is illustrated generally in Fig. 2.10 where the so-called Z-scheme of photosynthesis is demonstrated. The Z-scheme represents a fundamental process if light harvesting is utilized for biomass production and generation of power within the system, as a matter of fact today's trend is to develop direct Z-scheme photocatalysts (Xu et al. 2018). Artificial photosynthesis is naturally a trend as well (El-Khouly et al. 2017).

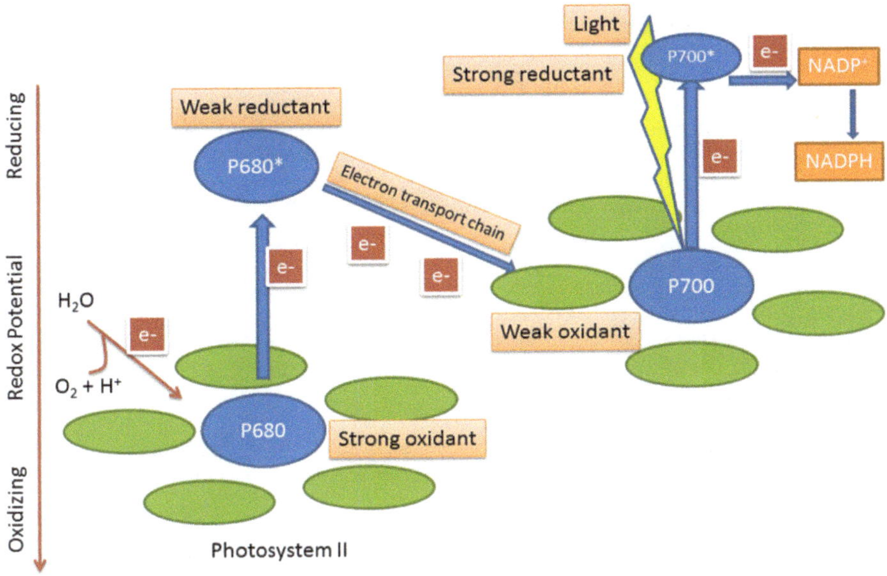


**Fig. 2.10** Z-scheme of photosynthesis

# References


Albarrán-Zavala, E., & Angulo-Brown, F. (2007). A simple thermodynamic analysis of photosynthesis. *Entropy, 9*(4), 152–168. https://doi.org/10.3390/e9040152.

Bass, M., DeCusatis, C., Enoch, J., Lakshminarayanan, V., Li, G., MacDonald, C., … Van Stryland, E. (2009). *Handbook of optics, volume IV: Optical properties of materials, nonlinear optics, quantum optics* (3rd ed.). McGraw-Hill Education.

Blankenship, R. E. (2013). *Molecular mechanisms of photosynthesis*. Wiley.

Blankenship, R. E., Sadekar, S., & Raymond, J. (2007). The evolutionary transition from anoxygenic to oxygenic photosynthesis, Chap. 3. In P. G. Falkowski & A. H. Knoll (Eds.), *Evolution of primary producers in the sea* (pp. 21–35). Burlington: Academic Press. https://doi.org/10.1016/B978-012370518-1/50004-7.

Brittin, W., & Gamow, G. (1961). Negative entropy and photosynthesis. *Proceedings of the National Academy of Sciences of the United States of America, 47*(5), 724–727.

Brown, T. E., LeMay, H. E., Bursten, B., Murphy, C., Woodward, P., & Stoltzfus, M. E. (2014). *Chemistry: The central science* (13th ed.). Pearson.

Chapman, A. D. (2009). *Numbers of living species in Australia and the world* (2nd ed.). Canberra, AU: Australian Government, Department of the Environment, Water, Heritage and the Arts.

Ciddor, P. E. (1996). Refractive index of air: New equations for the visible and near infrared. *Applied Optics, 35*(9), 1566–1573. https://doi.org/10.1364/AO.35.001566.

Conn, E. E., & Stumpf, P. K. (1972). *Outlines of biochemistry* (3rd ed.). New York, USA: Wiley.

Daimon, M., & Masumura, A. (2007). Measurement of the refractive index of distilled water from the near-infrared region to the ultraviolet region. *Applied Optics, 46*(18), 3811–3820. https://doi.org/10.1364/AO.46.003811.

Dean, J. A. (1992). *Lange's handbook of chemistry*. New York, USA: McGraw-Hill.

De Vrieze, J., Arends, J. B. A., Verbeeck, K., Gildemyn, S., & Rabaey, K. (2018). Interfacing anaerobic digestion with (bio)electrochemical systems: Potentials and challenges. *Water Research, 146,* 244–255. https://doi.org/10.1016/j.watres.2018.08.045.

Dorfman, K. E., Voronine, D. V., Mukamel, S., & Scully, M. O. (2013). Photosynthetic reaction center as a quantum heat engine. *Proceedings of the National Academy of Sciences of the United States of America, 110*(8), 2746–2751. https://doi.org/10.1073/pnas.1212666110.

El-Khouly, M. E., El-Mohsnawy, E., & Fukuzumi, S. (2017). Solar energy conversion: From natural to artificial photosynthesis. *Journal of Photochemistry and Photobiology C: Photochemistry Reviews, 31*(June), 36–83.

Furatian, L., & Mohseni, M. (2018). Temperature dependence of 185 nm photochemical water treatment—The photolysis of water. *Journal of Photochemistry and Photobiology A: Chemistry, 356,* 364–369. https://doi.org/10.1016/j.jphotochem.2017.12.030.

Ghosh, G. (1999). Dispersion-equation coefficients for the refractive index and birefringence of calcite and quartz crystals. *Optics Communications, 163*(1), 95–102. https://doi.org/10.1016/S0030-4018(99)00091-7.

Gilroy, S., & Masson, P. H. (2008). *Plant tropism*. Blackwell Publishing.

Hassan, H. C., Abidin, Z. H. Z., Chowdhury, F. I., & Arof, A. K. (2016). A high efficiency chlorophyll sensitized solar cell with quasi solid PVA based electrolyte. *International Journal of Photoenergy, 2,* 1–9. https://doi.org/10.1155/2016/3685210.

IEA. (2017). *Emissions from fuel combustion highlights*. International Energy Agency IEA. Retrieved from https://www.iea.org/publications/freepublications/publication/CO2EmissionsfromFuelCombustionHighlights2017.pdf.

Ishikita, H., Saenger, W., Biesiadka, J., Loll, B., & Knapp, E.-W. (2006). How photosynthetic reaction centers control oxidation power in chlorophyll pairs P680, P700, and P870. *Proceedings of the National Academy of Sciences, 103*(26), 9855–9860. https://doi.org/10.1073/pnas.0601446103.

Jones, M., & Fleming, S. A. (2014). *Organic chemistry* (5th ed.). W. W. Norton & Company.

Kasarova, S. N., Sultanova, N. G., Ivanov, C. D., & Nikolov, I. D. (2007). Analysis of the dispersion of optical plastic materials. *Optical Materials, 29,* 1481–1490. https://doi.org/10.1016/j.optmat.2006.07.010.

Kell, D. B. (2012). Large-scale sequestration of atmospheric carbon via plant roots in natural and agricultural ecosystems: Why and how. *Philosophical Transactions of the Royal Society B: Biological Sciences, 367*(1595), 1589–1597. https://doi.org/10.1098/rstb.2011.0244.

Kobayashi, T., Nozoye, T., & Nishizawa, N. K. (2019). Iron transport and its regulation in plants. *Free Radical Biology and Medicine*, *133*, 11–20. https://doi.org/10.1016/j.freeradbiomed.2018.10.439.

Ksenzhek, O. S., & Volkov, A. G. (1998). *Plant energetics*. San Diego, California: Academic Press.

Kumari, A. (2017). Electron transport chain, Chap. 3. In *Sweet biochemistry remembering structures, cycles, and pathways by mnemonics* (1st ed.). Academic Press.

Li, G., & Coleman, G. D. (2018). Nitrogen storage and cycling in trees. In *Advances in botanical research*. Academic Press. https://doi.org/10.1016/bs.abr.2018.11.004.

Liu, S.-Q., Zhou, S.-S., Chen, Z.-G., Liu, C.-B., Chen, F., & Wu, Z.-Y. (2016). An artificial photosynthesis system based on $CeO_2$ as light harvester and N-doped graphene Cu(II) complex as artificial metalloenzyme for $CO_2$ reduction to methanol fuel. *Catalysis Communications, 73,* 7–11. https://doi.org/10.1016/j.catcom.2015.10.004.

Madigan, M. T., Martinko, J. M., & Brock, P. J. (2000). *Biology of microorganisms* (8th ed.). Prentice Hall.

Malitson, I. H. (1965). Interspecimen comparison of the refractive index of fused silica*,†. *JOSA, 55*(10), 1205–1209. https://doi.org/10.1364/JOSA.55.001205.

McCree, K. J. (1981). Photosynthetically active radiation. In O. L. Lange, P. S. Nobel, C. B. Osmond, & H. Ziegler (Eds.), *Physiological plant ecology I: Responses to the physical environment* (pp. 41–55). Berlin, Heidelberg: Springer Berlin Heidelberg. https://doi.org/10.1007/978-3-642-68090-8_3.

Miranzadeh, H., Emam, Y., Sayyed, H., & Zare, S. (2011). Productivity and radiation use efficiency of four dryland wheat cultivars under different levels of nitrogen and chlormequat chloride. *Journal of Agricultural Science and Technology, 13,* 339–351.

Mirkovic, T., Ostroumov, E. E., Anna, J. M., van Grondelle, R., Govindjee, & Scholes, G. D. (2017). Light absorption and energy transfer in the antenna complexes of photosynthetic organisms. *Chemical Reviews, 117*(2), 249–293. https://doi.org/10.1021/acs.chemrev.6b00002.

Nagai, R. (1993). Regulation of intracellular movements in plant cells by environmental stimuli. In K. W. Jeon & J. Jarvik (Eds.), *International review of cytology* (Vol. 145, pp. 251–310). Academic Press. https://doi.org/10.1016/S0074-7696(08)60429-5.

Nelson, N. (2011). Photosystems and global effects of oxygenic photosynthesis. *Biochimica et Biophysica Acta (BBA)—Bioenergetics, 1807*(8), 856–863. https://doi.org/10.1016/j.bbabio.2010.10.011.

Nicholls, D. G., & Ferguson, S. J. (2013). Photosynthetic generators of proton motive force, Chap. 6. In *Bioenergetics* (4th ed., pp. 159–196). Academic Press.

Noguchi, T., Inoue, Y., & Tang, X.-S. (1997). Structural coupling between the oxygen-evolving Mn cluster and a tyrosine residue in photosystem II as revealed by fourier transform infrared spectroscopy. *Biochemistry, 36*(48), 14705–14711. https://doi.org/10.1021/bi971760y.

Owen, A. (2017). *Plants release up to 30 per cent more $CO_2$ than previously thought, study says*. ABC News. Retrieved from http://www.abc.net.au/news/2017-11-18/plant-respiration-co2-findings-anu-canberra/9163858.

Palik, E. D. (1985). *Handbook of optical constants of solids*. Boston: Academic Press.

Pisciotta, J. M., Zou, Y., & Baskakov, I. V. (2010). Light-dependent electrogenic activity of cyanobacteria. *PLoS ONE, 5*(5), e10821. https://doi.org/10.1371/journal.pone.0010821.

Qi, X., Ren, Y., Liang, P., & Wang, X. (2018). New insights in photosynthetic microbial fuel cell using anoxygenic phototrophic bacteria. *Bioresource Technology, 258,* 310–317. https://doi.org/10.1016/j.biortech.2018.03.058.

Resnick, R., Halliday, D., & Krane, K. S. (2001). *Physics* (5th ed.). Wiley.

Rheims, J., Köser, J., & Wriedt, T. (1997). Refractive-index measurements in the near-IR using an Abbe refractometer. *Measurement Science & Technology, 8*(6), 601. https://doi.org/10.1088/0957-0233/8/6/003.

Romero, E., Augulis, R., Novoderezhkin, V. I., Ferretti, M., Thieme, J., Zigmantas, D., et al. (2014). Quantum coherence in photosynthesis for efficient solar-energy conversion. *Nature Physics, 10*(9), 676–682. https://doi.org/10.1038/nphys3017.

Scholes, G. D., & Fleming, G. R. (2005). Energy transfer and photosynthetic light harvesting. In *Adventures in chemical physics* (pp. 57–129). Wiley Ltd. https://doi.org/10.1002/0471759309.ch2.

Sommer, M. E., Elgeti, M., Hildebrand, P. W., Szczepek, M., Hofmann, K. P., & Scheerer, P. (2015). Structure-based biophysical analysis of the interaction of rhodopsin with G protein and arrestin, Chap. 26. In A. K. Shukla (Ed.), *Methods in enzymology* (Vol. 556, pp. 563–608). Academic Press. https://doi.org/10.1016/bs.mie.2014.12.014.

Sun, Z., Liang, H., Liu, J., & Shi, G. (2017). Estimation of photosynthetically active radiation using solar radiation in the UV–visible spectral band. *Solar Energy, 153,* 611–622. https://doi.org/10.1016/j.solener.2017.06.007.

Taiz, L., Zeiger, E., Møller, I. M., & Murphy, A. (2014). *Plant physiology and development*. Sinauer Associates, Oxford University Press.

van Rotterdam, B. J., Crielaard, W., van Stokkum, I. H. M., Hellingwerf, K. J., & Westerhoff, H. V. (2002). Simplicity in complexity: The photosynthetic reaction center performs as a simple 0.2 V battery. *FEBS Letters, 510*(1–2), 105–107. https://doi.org/10.1016/S0014-5793(01)03210-0.

Xu, P., Roy, L. M., & Croce, R. (1858). Functional organization of photosystem II antenna complexes: CP29 under the spotlight. *Biochimica et Biophysica Acta (BBA)—Bioenergetics, 10,* 815–822. https://doi.org/10.1016/j.bbabio.2017.07.003.

Xu, Q., Zhang, L., Yu, J., Wageh, S., Al-Ghamdi, A. A., & Jaroniec, M. (2018). Direct Z-scheme photocatalysts: Principles, synthesis, and applications. *Materials Today*. https://doi.org/10.1016/j.mattod.2018.04.008.

Yahia, E. M., Carrillo-López, A., Barrera, G. M., Suzán-Azpiri, H., & Bolaños, M. Q. (2019). Photosynthesis, Chap. 3. In E. M. Yahia (Ed.), *Postharvest physiology and biochemistry of fruits and vegetables* (pp. 47–72). Woodhead Publishing. https://doi.org/10.1016/B978-0-12-813278-4.00003-8.

Zeiger, E., & Taiz, L. (1991). *Plant physiology*. Redwood City, CA: Benjamin-Cummings.

# Chapter 3
# Redox Potential and Galvanic Cells

*You know..., they were all things which were born spontaneously. One did what he could do and according to his personal attitudes and the possessed abilities. One was doing one task and another one was doing some other tasks and there was very much friendship among us. And then, there was enormous passion in what we were doing.*

A suitable description of a successful research group. Freely translated from the words of Emilio Segrè, Nobel Prize in Physics, 1959.

## 3.1 Functioning Principles

The schematics given in Fig. 2.10 can be also described as a galvanic cell (van Rotterdam et al. 2002). Since, and as we will see later within this text, many biochemical reactions follow the purpose of oxidizing a substrate and producing a quantity of electrons which are acting as a current. This will give the power to carry on several biochemical functions. Later on, tables will be given including biochemical half-cell reactions. To better understand this, in this section, an explanation of the principles for the electrochemical power generation and the Nernst law is given. This will provide the reader with the basics and the understanding of how the electrochemical potential can be actually estimated for biological redox couples. This is also a topic that mostly relates to engineers, since they are focused on the quantification of processes and phenomena.

In this kind of electronic generators of power (the galvanic cell), a reaction is normally taking place on the anode of the cell. Here, electrons are produced and sent to flow in a wire. The electrons flow as a function of a difference in electrochemical potential between the anode and the cathode give the power to do work according to the relation:

$$1\text{W} = 1\text{V} \cdot 1\text{A} \tag{3.1}$$



C. De Blasio, *Fundamentals of Biofuels Engineering and Technology*,
Green Energy and Technology, https://doi.org/10.1007/978-3-030-11599-9_3

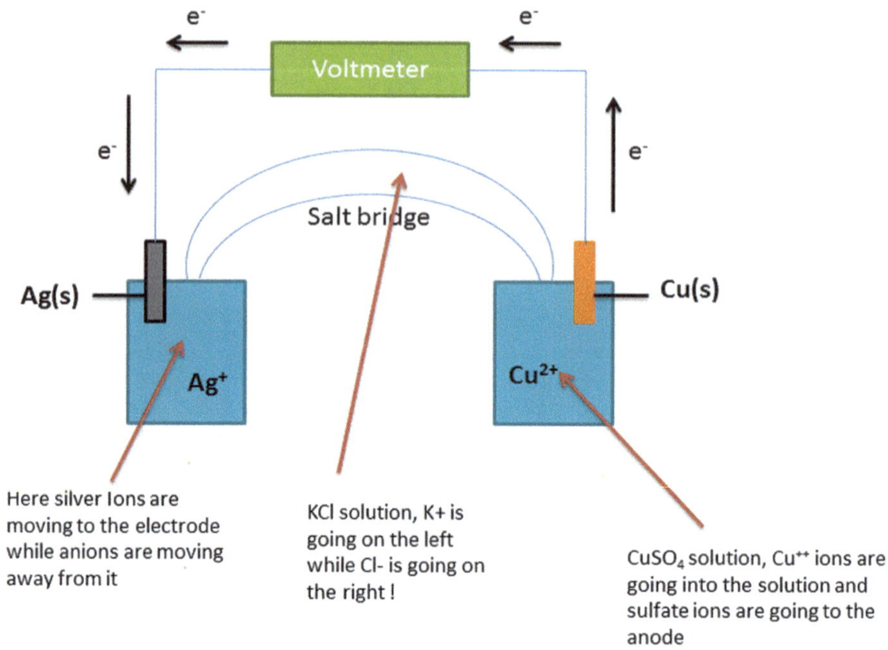


**Fig. 3.1** Functioning of a galvanic cell where a salt bridge is employed

To give one example, in Fig. 3.1, the flow of electrons is given from the right to the left. At the right side and in the figure presented, a $CuSO_4$ solution is positioned at the anode where a Cu metallic plate is situated. Here, from the $CuSO_4$ solution, the $SO_4^{2-}$ ions transfer electrons to the anode side. As a consequence, electrons start to flow in the direction of the cathode. A salt bridge is positioned between the anode and the cathode. In Fig. 3.1, a salt bridge composed by KCl solution is given; the $CL^-$ ions produced will compensate for the ones lost in the anode by the $CuSO_4$ solution. In a similar way, at the cathode, there will be an accumulation of silver anions and therefore this is compensated by the $K^+$ ions from the salt bridge. The galvanic cell is not working anymore when equilibrium is reached.

To be noticed that it is important to understand the flow of ions and electrons within the solutions and the salt bridge. The flows follow the driving force given by the difference in concentrations: This means that anions are going from higher concentrations to lower concentrations, and the same is with cations and even electrons are no exception in this sense since they move from a region at higher electrochemical potential to a lower one. For a more accurate treatise of the galvanic cells thermodynamics, I invite the reader to consider books of physical chemistry like (Levine 2009).

Nevertheless, it is pointed out here that, when a thermodynamic analysis should be carried out on these particular systems, a standard should be defined. This will include parameters like pressure and temperature, and this also concerns the kind of

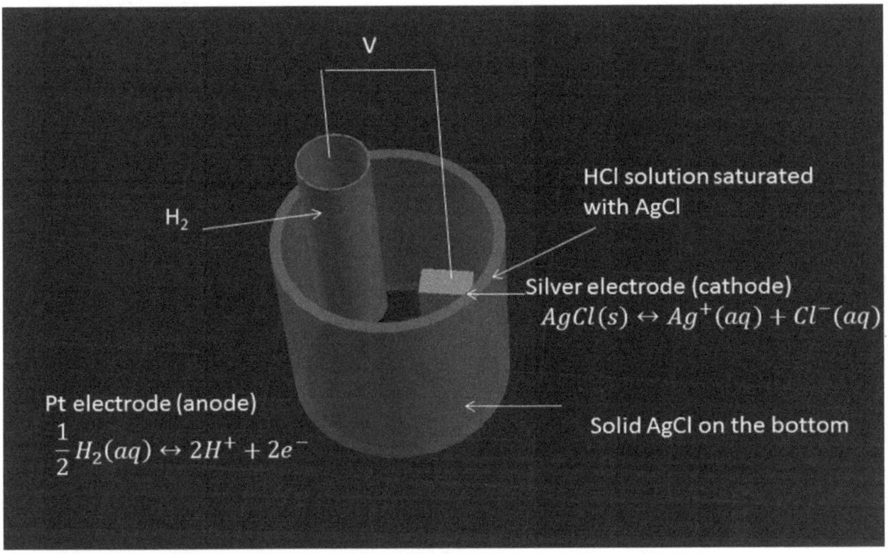


**Fig. 3.2** Galvanic cell with no liquid junction. A representation of the hydrogen anode; hydrogen is pumped on to the electrode by using a particular casing

cell itself. In case we consider the standard electrochemical potential, this is taken at 25 °C, 1 Atm and when the effective concentration for the ions involved is 1 M.

A standard galvanic cell is drafted in Fig. 3.2 where hydrogen is pumped onto the electrode by using a particular casing. From the reactions:

$$H_2(g) \leftrightarrow H_2(aq) \tag{3.2}$$

$$H_2(aq) \leftrightarrow 2H^+ + 2e^- \tag{3.3}$$

electrons are released from the hydrogen molecule, collected by a wire, and transferred to a silver electrode where the following reaction is taking place:

$$AgCl(s) \leftrightarrows Ag^+ + Cl^- \tag{3.4}$$

In this case, the silver cations react with the electrons and precipitate as Ag(s) according to the equation:

$$Ag^+ + e^- \leftrightarrow Ag(s) \tag{3.5}$$

The concept of redox potential needs to be explained further at this point. The redox potential consists of an electrical measurement. This measurement shows the tendency of a object to transfer electrons to a reference electrode or from a reference electrode.

When Ag is giving electrons, the reaction taking place at the standard electrode is:

$$Ag \rightleftarrows Ag^{+} + e^{-} \tag{3.6}$$

Even in this case the previous equation is written as a reversible reaction. When the reaction is more directed on the right hand side, the electronegativity of the reference electrode becomes higher than the point of measure; for this reason, the voltmeter will measure less electronegativity for the sample measured.

Let us consider another example here; the following is the Zn–Cu cell.

$$Zn(s) + Cu^{2+}(aq) \rightarrow Zn^{2+}(aq) + Cu(s), \tag{3.7}$$

where at the cathode the electrons are accepted as

$$Cu^{++} + 2e^{-} \rightarrow Cu; \tag{3.8}$$

and the anode is given by:

$$Zn \rightarrow Zn^{++} + 2e^{-}. \tag{3.9}$$

The standard overall electrochemical potential $E^0$ is calculated by the formula:

$$E^0 = E^0(\text{cathode}) + E^0(\text{anode}) \tag{3.10}$$

Cu and Zn are very much used in galvanic cells, and even it is proposed to boost them by inserting anode–cathode couples within the same electrolyte (Lacina et al. 2018).

Since the standard state is defined as the state at which the system is at 25 °C, the liquid reactant is at 1 M concentration, and the gaseous reactant is at 1 Atm, the standard reaction of $H^+$ reduction is taken at that condition.

$$2H^{+}(aq) + 2e^{-} \rightarrow H_2(g) \quad E^0 = 0 \tag{3.11}$$

For this reason, we chose to evaluate the standard potential for the anode and the cathode with respect to the hydrogen electrode. The following is the representation of the overall cell reaction as a sum of the two half-cell reactions taking place, respectively, at the anode, Eq. (3.12), and cathode, Eq. (3.13).

$$Zn(s)|Zn^{2+}(aq)||H^{+}(aq)|H_2(g)|Pt \quad E^0 = 0.76\ V \tag{3.12}$$

$$Pt|H_2(g)|H^{+}(aq)||Cu^{2+}(aq)|Cu(s) \quad E^0 = 0.34\ V \tag{3.13}$$

**Table 3.1** Evaluation of the standard cell potential

| Electrode | Reaction |
|---|---|
| Anode | $Zn(s) \rightarrow Zn^{2+}(aq) + 2e^{-} \quad E^{0} = +0.76$ V |
| Cathode | $Cu^{2+}(aq) + 2e^{-} \rightarrow Cu(s) \quad E^{0} = +0.34$ V |
| Overall | $Cu^{2+}(aq) + Zn(s) \rightarrow Cu(s) + Zn^{2+}(aq) \quad E^{0} =$ 1.10 V |

Equations (3.12) and (3.13) are a typical way of writing the electrodes, in this case as a reference to the standard electrode of hydrogen. The left side represents the anode side (where electrons are released) while on the right side there is the cathode which is receiving electrons. The standard Gibbs free energy is given by (Levine 2009):

$$\Delta G^{0} = -n(F)\left(\Delta E^{0}\right) \tag{3.14}$$

where $F$ is the Faraday constant which is equal to 96,485.33 C $mol^{-1}$ (NIST Physical Measurement Laboratory 2018), and $n$ is the number of electron moles moving. Equation (3.14) will give dimensions of Coulomb times Volt which will be equivalent to joules, and this is the correct dimension for the energy. The author of this book suggests the readers to do often this kind of reasoning on the dimensions to check for the consistency of the formulas at hand.

The overall cell reaction for the Zn(s)–Cu(s) system is given by:

$$Zn(s)\left|Zn^{2+}(aq)\right|\left|Cu^{2+}(aq)\right|Cu(s) \tag{3.15}$$

In Table 3.1, a clear example of how to calculate the overall standard electrochemical potential of a cell is given.

To understand better the calculations, it is convenient to remember that, when a half-cell reaction is encountered, the sign of the standard electrochemical potential determines if the reaction is spontaneous or not. In fact, when the given standard potential is positive, then the reaction is spontaneous. It is not the same case when the reaction gives negative values. However, there are some conventions to take into account when data are retrieved.

## 3.2 Conventions on the Electrodes

With respect to Table 3.1, it is important to notice that when a reaction is in reality taking place in the way it is written, then the potential that is read on the reading instrument is always positive. In the case presented, the Zn(s) is indeed an anode and the Cu(s) is indeed receiving electrons. However, by convention the electrode potentials are a measure of the driving force for the two half reactions when the half

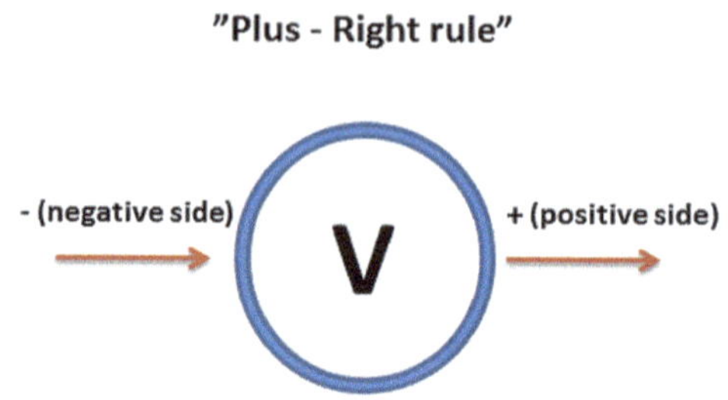


**Fig. 3.3** Plus–Right Rule

reactions are written as reductions; this would mean that the electrons are always written on the left side of the equation:

$$2AgCl(s) + 2e^- \leftrightarrow 2Ag(s) + 2Cl^- \tag{3.16}$$

$$2H^+ + 2e^- \leftrightarrow H_2(g) \tag{3.17}$$

This means also that the anode side is written on the left and the cathode side is written on the right. Figure 3.3 gives a simple description of the convention sides for the galvanic cell.

Considering the reference electrode as in Fig. 3.3, when the voltage is positive, the electrons are flowing on the right side, so it means that the right side is the cathode and the left side is the anode. If written according to the convention (with the electrons on the left side), the anode will have a negative electrochemical potential and this would also be in accordance with the fact that (always by convention) electrons are indicated as having a negative charge. The more electronegative is a site, the more will be the electrons at that site and therefore the more electrons will move on the basis of the difference in concentration.

## 3.3 Calculation of the Redox Potential at Non-standard Conditions, the Nernst Law

If the reaction is written as:

$$pP + qQ + \cdots + ne^- \leftrightarrow rR + sS + \cdots \tag{3.18}$$

where the capital letters refer to general components and the non-capital letters refer to the stoichiometric coefficients. The overall electrochemical potential will be expressed by the Nernst Law as follows (Levine 2009):

$$E = E^0 - \frac{RT}{nF} \cdot \ln \frac{\left[(a_R)^r \cdot (a_S)^s \cdots\right]}{\left[(a_P)^p \cdot (a_Q)^q \cdots\right]} \tag{3.19}$$

where $E$ is the overall electrochemical potential, $E^0$ is the standard electrochemical potential evaluated as previously. The terms $a_x^y$ refer to the activities of the reactants and products. The standard potential is defined as the potential of a half-cell reaction when all the reactant and products have activity = 1. In the standard galvanic cell, the reference hydrogen electrode is on the left side (so if the electrode to be measured has positive potential, then it will be a cathode accordingly to what was mentioned previously).

We remember that the total Gibbs free energy will be evaluated as:

$$\Delta G = -nFE \tag{3.20}$$

where this time $E$ is the overall electrochemical potential or also mentioned as electromotive force (EMF) under non-standard conditions. As an example, consider the following reaction, related to the Ag (silver) electrode (Holze 2009):

$$\mathrm{Ag}^+ + \mathrm{e}^- \leftrightarrow \mathrm{Ag(s)} \text{ with } E^0 = +0.799\ \mathrm{V}; \tag{3.21}$$

this would be a cathode and we want to evaluate the Nernst equation for one molar concentration of Ag(s). Remember that:

$$R = 8.316 \frac{\mathrm{J}}{\mathrm{mol\,K}}; \tag{3.22}$$

and

$$F = 96{,}487 \frac{\mathrm{C}}{\mathrm{mol}}; \tag{3.23}$$

at 25 °C the quantity $\frac{RT}{F} = 0.0592$ for this reason:

$$E = 0.799 - \frac{0.0592}{1} \log \frac{1}{a_{\mathrm{Ag}^+}}; \tag{3.24}$$

and this is because $n = 1$ (we have one electron mole transferred). Remember also that activities are taken as 1 for solids, for dilute solutions the activity coefficient is equal to 1 (do not confuse the activity coefficient with the activity values themselves), so the activity can be approximated with the molar concentration. The activity of saturated solutions is also considered to be 1 (Levine 2009). The equation above is obtained by changing the base of the logarithm from natural to decimal.

For the sake of clarity, a simple derivation of the change of logarithm base is reported step by step; we start from the definition of the general logarithm on base $a$:

$$x = \log_a C; \tag{3.25}$$

which implies:

$$a^x = C; \tag{3.26}$$

for this reason:

$$\log_b a^x = \log_b C; \tag{3.27}$$

and from one of the properties of the logarithms:

$$x \log_b a = \log_b C; \tag{3.28}$$

therefore we have:

$$\log_a C = \frac{\log_b C}{\log_b a}. \tag{3.29}$$

Consider the following cell reaction:

$$\mathrm{AgCl(s)} + \mathrm{e}^- \leftrightarrow \mathrm{Ag(s)} + \mathrm{Cl}^-; \tag{3.30}$$

and suppose that the concentration of $Cl^-$ ions is known. We want to calculate the potential of the cell by the Nernst equation rewritten here:

$$E = E^0 - \frac{RT}{nF} \cdot \ln \frac{\left[(a_\mathrm{R})^r \cdot (a_\mathrm{S})^s \cdots\right]}{\left[(a_\mathrm{P})^p \cdot (a_\mathrm{Q})^q \cdots\right]} \tag{3.31}$$

The activity of a specie $x$ is: $a_x = \gamma_x c_x$, and the activity coefficient $\gamma_x$ is calculated from the Debye–Hückel equation (Hückel and Debye 1923) which is fully derived by (Bockris and Reddy 1998) and a simplified version is given in the literature as follows (Levine 2009):

$$-\ln \gamma_A = \frac{Z_A^2 A \sqrt{\mu}}{1 + B\alpha_A \sqrt{\mu}} \tag{3.32}$$

where $A = 1.1744\sqrt{\left(\frac{\mathrm{kg}}{\mathrm{mol}}\right)}$ and $B = 3.285 \times 10^{-9}\sqrt{\left(\frac{\mathrm{kg}}{\mathrm{mol}}\right)} \cdot \mathrm{m}^{-1}$. Substituting these values in the previous equation and changing the base of the logarithm to the decimal one, this will be written as:

$$-\log \gamma_A = \frac{0.510 Z_A^2 \sqrt{\mu/m^0}}{1 + 0.328(\alpha_A/\mathrm{\AA})\sqrt{\mu/m^0}} \tag{3.33}$$

and it could be further simplified as (not specifying for the dimensions involved):

$$-\log\gamma_A = \frac{0.510Z_A^2\sqrt{\mu}}{1 + 3.28\alpha_A\sqrt{\mu}} \tag{3.34}$$

where $Z_A$ is the charge of the $A$ specie, $\alpha_A$ is the effective diameter of the hydrated specie (this is given in nanometers in Eq. (3.36), $\mu$ is the ionic strength of the solution.

The ionic strength is calculated from:

$$\mu = \frac{1}{2}\left(c_1 Z_1^2 + \cdots + c_n Z_n^2\right) \tag{3.35}$$

where the subscript numbers (1,... $n$) are referring to the species involved. For example, if we have an HCl solution:

$$\mathrm{HCl} \leftrightarrow \mathrm{H}^+ + \mathrm{Cl}^- \tag{3.36}$$

$Z$ would be 1 for both $H^+$ and $Cl^-$ and the ionic strength will be $2 \times \frac{1}{2}\left(c \cdot Z^2\right)$, this is because c (the molar concentration) is equal for both ions and $Z$ is also 1 for both.

The effective diameter of the hydrated specie in nanometers can be found in the literature, for instance, the data given in Table 3.2 are taken from Kielland (1937).

Table 3.3 gives standard electrochemical potential for some half-cell reactions; at this point, the reader is invited to notice that these values are related to the reactants which are present in biological systems. In fact, we should remember here that the purpose of this treatise is to give the fundamental notions of how the biomass is generated.

To give some additional description of how the redox potential relates with the Z-scheme of photosynthesis given previously, Fig. 3.4 represents the Z-scheme in relation to the redox potential.

Notice that in Fig. 3.4 the higher is the potential the more negative is its value. This is because the system will be at a more "electronegative" state. It is good to remember that it is commonly referred to electrons to have a negative charge and therefore, with this in mind it will be easier to understand the Z-scheme. The redox potential is a measure of the affinity of a substance to attract electrons.

When an electron proceeds "downhill", there is the formation of free energy, $\Delta G$, which is calculated according to Eq. (3.20) (Acrivos 1988), which is repeated here:

$$\Delta G = -n \cdot F \cdot (\Delta E) \tag{3.37}$$

We remember that $n$ is the moles of electrons moving; $F$, is the Faraday constant, 23.062 kcal/V per mole, and it is the energy released when a mole of electron is moved by a potential of 1 V.

**Table 3.2** Diameters of non-hydrated, value on top, and hydrated, second value, ions

| | | | | | | | | | | | | | | | | | |
|---|---|---|---|---|---|---|---|---|---|---|---|---|---|---|---|---|---|
| $H^+$<br>–<br>900 | | | | | | | | | | | | | | | | | He |
| $Li^+$<br>80<br>600 | $Be^{+2}$<br>60<br>80 | | | | | | | | | | | B | C | N | O | $F^-$<br>150<br>350 | Ne |
| $Na^+$<br>100<br>450 | $Mg^{2+}$<br>90<br>800 | | | | | | | | | | | $Al^{3+}$<br>80<br>900 | Si<br>80<br>900 | P | | $Cl^-$<br>190<br>300 | Ar |
| $K^+$<br>160<br>300 | $Ca^{2+}$<br>140<br>600 | $Sc^{3+}$<br>140<br>900 | Ti | V | $Cr^{3+}$ | $Mn^{2+}$ | $Fe^{2+}$ | $Co^{2+}$ | $Ni^{2+}$ | $Cu^{2+}$ | $Zn^{2+}$ | $Ga^{3+}$ | $Ge^{4+}$ | $As^{5+}$ | Se | $Br^-$ | Kr |
| $Rb^+$<br>180<br>250 | $Sr^{2+}$<br>170<br>500 | $Y^{3+}$<br>160<br>900 | $Zr^{4+}$<br>160<br>1100 | Nb | Mo | Tc | Ru | Rh | Pd | $Ag^+$<br>150<br>250 | $Cd^{2+}$<br>140<br>500 | $In^{3+}$<br>140<br>900 | $Sn^{2+}$<br>–<br>600 | $Sb^{3+}$ | Te | $I^-$<br>220<br>300 | Xe |
| $Cs^+$<br>210<br>250 | $Ba^{2+}$<br>210<br>500 | Lu | Hf | Ta | W | Re | Os | Ir | Pt | Au | $Hg^{2+}$<br>150<br>500 | $Tl^{+1}$<br>140<br>250 | $Pb^{2+}$<br>–<br>590 | $Bi^{3+}$ | Po | $At^-$ | Rn |
| $NH_4^+$<br>–<br>250 | | $La^{3+}$<br>200<br>900 | $Ce^{3+}$<br>–<br>900 | $Pr^{3+}$<br>–<br>900 | $Nd^{3+}$<br>–<br>900 | – | $Sm^{3+}$<br>–<br>900 | | | | | | $Sn^{4+}$<br>–<br>1100 | $NO_{3-}$<br>–<br>300 | $OH^-$<br>–<br>350 | | |
| | | | $Ce^{4+}$<br>–<br>1100 | | | | $Fe^{3+}$<br>–<br>900 | | | | | | $CO_3^{2-}$<br>–<br>450 | $NO_2^-$<br>–<br>300 | $SO_4^{2-}$<br>–<br>400 | | |

Adapted from Kielland (1937). Values are expressed in picometers

**Table 3.3** Standard potentials of selected half reactions of biological importance, 25 °C and pH 7 (Ksenzhek and Volkov 1998; Ross 1991)

| Half reaction | $E^0$ (V) |
|---|---|
| $\frac{1}{2}O_2 + 2H^+ + 2e^- \rightarrow H_2O$ | 0.816 |
| Ferredoxin $(Fe^{3+}) + e^- \rightarrow$ ferredoxin $(Fe^{2+})$ | −0.432 |
| $2H^+ + 2e^- \rightarrow H_2$ | −0.414 |
| $NO_3^- + 2H^+ + 2e^- \rightarrow NO_2^- + H_2O$ | 0.421 |
| $\alpha -$ ketoglutarate $+ CO_2 + 2H^+ + 2e^- \rightarrow$ isocitrate | −0.38 |
| Cytochrome $f(Fe^{3+}) + e^- \rightarrow$ cytochrome $f(Fe^{2+})$ | 0.365 |
| Acetoacetate $+ 2H^+ + 2e^- \rightarrow \beta -$ hydroxybutyrate | −0.38 |
| $Fe(CN)_6^{3-} + e^- \rightarrow Fe(CN)_6^{4-}$ | 0.36 |
| $NAD^+ + H^+ + 2e^- \rightarrow$ NADH | −0.32 |
| $NADP^+ + H^+ + 2e^- \rightarrow$ NADPH | −0.324 |
| $2H^+ + 2e^- \rightarrow H_2$ at $pH = 0$ | 0.00000 |
| cytochrome $a(Fe^{3+}) + e^- \rightarrow$ cytochrome $a(Fe^{2+})$ | 0.29 |
| cytochrome $c(Fe^{3+}) + e^- \rightarrow$ cytochrome $c(Fe^{2+})$ | 0.254 |
| cytochrome $b(Fe^{3+}) + e^- \rightarrow$ cytochrome $b(Fe^{2+})$ | 0.077 |
| Ubiquinone $+ 2H^+ + 2e^- \rightarrow$ Ubiquinol $+ H_2$ | 0.045 |
| $Fumarate^{2-} + 2H^+ + 2e^- \rightarrow succinate^{2-}$ | 0.031 |
| $O_2 + 2H^+ + 2e^- \rightarrow H_2O_2$ | 0.295 |
| Crotonyl − CoA $+ 2H^+ + 2e^- \rightarrow$ butyryl − CoA | −0.015 |
| $Oxaloacetate^{2-} + 2H^+ + 2e^- \rightarrow malate^{2-}$ | −0.166 |
| $Pyruvate^- + 2H^+ + 2e^- \rightarrow lactate^-$ | −0.185 |
| Acetaldehyde $+ 2H^+ + 2e^- \rightarrow$ ethanol | −0.197 |
| FAD $+ 2H^+ + 2e^- \rightarrow FADH_2$ | −0.219 |
| Glutathione $+ 2H^+ + 2e^- \rightarrow$ 2 reduced glutathione | −0.23 |
| $S + 2H^+ + 2e^- \rightarrow H_2S$ | −0.243 |
| Lipoic acid $+ 2H^+ + 2e^- \rightarrow$ dihydrolipoic acid | −0.29 |

To give one example here, if for one molecule of glucose we need 24 electrons from oxygen in water and the associated electrochemical potential at non-standard conditions will be −1.24 V, this will give:

$$\Delta G = -24\left(\frac{23.062\ \text{kcal}}{\text{V}}\right)(-1.24)\text{V} \tag{3.38}$$

To be noticed that the value of the Gibbs free energy is positive here because we have calculated it for the process of gathering the necessary electrons from $O_2$;

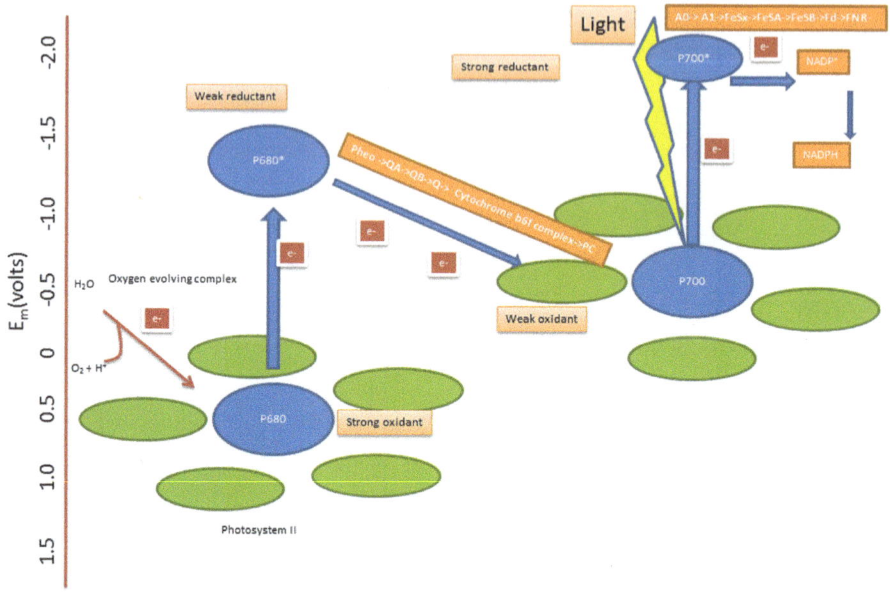


**Fig. 3.4** Z-scheme in terms of the redox potential

therefore, we need energy to do that. In the case presented here and referring to literature data (Sinha 2013):

$$E_{\text{oxygen}} = +0.82 \text{ V},$$

and

$$E_{\text{carbon in carbohydrates}} = -0.42 \text{ V}$$

Therefore:

$$\Delta E = -0.42 - 0.82 = -1.24.$$

Some examples of redox potentials in biological systems are shown in the following equations which are reported to demonstrate different kinds of electron acceptors in biological systems:

$$CO_2 + 4H^+ + 4e^- \rightarrow CH_2O + H_2O \tag{3.39}$$

$$NO_3^- + 6H^+ + 5e^- \rightarrow \frac{1}{2}N_2 + 3H_2O \tag{3.40}$$

$$O_2 + 4H^+ + 4e^- \rightarrow 2H_2O \tag{3.41}$$

Equation (3.39) shows how the $CO_2$ is actually reduced within the process of photosynthesis; Eq. (3.40) shows how nitrogen oxides (nitrate) also accept electrons and the last equation refers to the respiration of living organisms including humans. This is done also to point out that when an electron flow is generated, there has to be also a ultimate electron acceptor. In anaerobic respiration, for instance, the ultimate electron acceptor can be organic and also inorganic and more in detail we can have nitrate, sulfate $\left(SO_4^{2-}\right)$ and $Fe^{3+}$ as ultimate acceptors. Because the $E^0$ values are calculated for $n = 1$ and given in the literature in this way, the reactions' Eqs. (3.39)–(3.41) should be divided by 4, 5, 4, respectively.

## 3.4 A Remark on the Artificial Splitting of Water to Produce Hydrogen, Electrolysis

From the previous sections, it is possible to get an understanding of why biological systems have many things in common with galvanic cells and the principles of electrochemical potential, electron flow, and the related power obtained. Here, we take into consideration very briefly the process of water reduction and as a consequence the production of hydrogen by artificial electrolysis. Let us consider the following figure.

In the figure, there are two electrodes, one (Electrode 1) is connected to the cathode of a battery and the second (Electrode 2) is connected to the anode of a battery. When the battery will be on, there will be the generation of charges on the electrodes and for this reason, the electrode on the right of Fig. 3.5 will have positive charges on it. Within the cell, there is a solution of NaOH which will give $OH^-$ ions; now, because of the generated electric field, the $OH^-$ ions near the positively charged electrode will be attracted by the electrode and releasing electrons to it. Because of this, the section 1 will be the anode (where electrons are taken). The electrons will flow from the anode to the cathode (section 1) and the section 2 in Fig. 3.5 will be charged negatively. Water is a polar compound, which means that the molecule has a partial positive charge on the hydrogen atoms and a partial negative charge on the oxygen atoms. Because of the electric field generated by the battery, the positive side of the water molecules will be attracted by the negative electrode when they are near the electrode itself. At the negative electrode, the water molecule is reduced to give the formation of hydrogen according to the reaction:

$$4H_2O + 4e^- \rightarrow 2H_2 + 4OH^-, \quad \Delta E^0 = -0.83\ \text{V} \tag{3.42}$$

where this would represent an half-cell reaction with a standard electrochemical potential equivalent to −0.83 V. This means that this reaction (written from the left to the right as above) is not spontaneous, but in this case the power required to have the reaction written in this way is given by the battery.

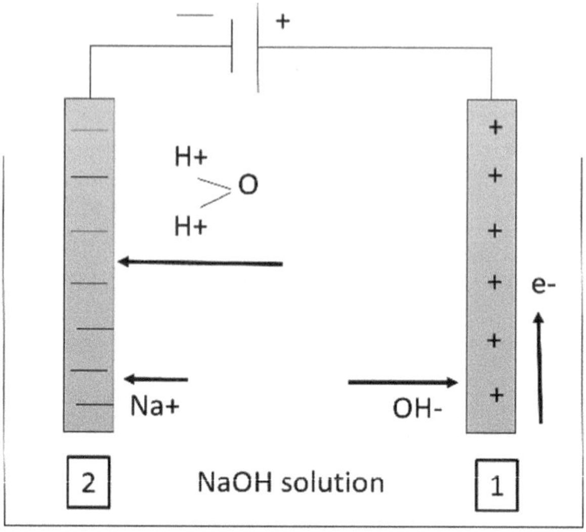


**Fig. 3.5** Schematic representation of an electrolytic cell

On the other hand, at the anode we have:

$$4OH^- \rightarrow O_2 + 2H_2O + 4e^-, \ \Delta E^0 = -0.4 \text{ V} \quad (3.43)$$

To be noticed that a reasonable question would be if the sodium ions would also react with the negative electrode and give the formation of molecular sodium. However, as we can notice, the following reaction is difficult to achieve since it is highly not spontaneous

$$Na^+ + e^- \rightarrow Na, \ \Delta E^0 = -2.7 \text{ V} \quad (3.44)$$

Therefore, the reaction (3.42) is the one mostly taking place and which gives the formation of $OH^-$ contributing to the equilibrium between the $Na^+$ and $OH^-$. Equations (3.42) and (3.43) can be summed to give:

$$2H_2O \rightarrow O_2 + 2H_2, \quad \Delta E^0 = -1.23 \text{ V} \quad (3.45)$$

which is the minimum required voltage necessary to carry on the steps above by the battery.

## References

Acrivos, J. (1988). Physical chemistry, third edition (Levine, Ira N.). *Journal of Chemical Education, 65*(12), A335. https://doi.org/10.1021/ed065pA335.3.

Bockris, J. O. M., & Reddy, A. K. (1998). *Modern electrochemistry* (2nd ed.). Berlin: Springer.

Holze, R. (2009). *Experimental electrochemistry: A laboratory textbook* (1st ed.). Wiley.

Hückel, E., & Debye, P. (1923). The theory of electrolytes: I. lowering of freezing point and related phenomena. *Physikalische Zeitschrift, 24,* 185–206.

Kielland, J. (1937). Individual activity coefficients of ions in aqueous solutions. *Journal of the American Chemical Society, 59*(9), 1675–1678. https://doi.org/10.1021/ja01288a032.

Ksenzhek, O. S., & Volkov, A. G. (1998). *Plant energetics*. San Diego, California: Academic Press.

Lacina, K., Sopoušek, J., Skládal, P., & Vanýsek, P. (2018). Boosting of the output voltage of a galvanic cell. *Electrochimica Acta, 282,* 331–335. https://doi.org/10.1016/j.electacta.2018.06.080.

Levine, I. N. (2009). *Physical chemistry* (6th ed.). McGraw-Hill Education.

National Institute of Standards and Technology. (2018). *NIST chemistry webbook*. Retrieved October 3, 2018, from https://webbook.nist.gov/chemistry/.

Ross, J. R. (1991). Practical handbook of biochemistry and molecular biology. *Biochemical Education, 19*(2), 95–96. In G. D Fasman (Ed.) (pp. 601). Boca Raton, Florida, USA: CRC Press, 1989. $00 ISBN 0-8493-3705-4. https://doi.org/10.1016/0307-4412(91)90020-9.

Sinha, R. K. (2013). *Modern plant physiology* (2nd ed.). Alpha Science International Ltd.

van Rotterdam, B. J., Crielaard, W., van Stokkum, I. H. M., Hellingwerf, K. J., & Westerhoff, H. V. (2002). Simplicity in complexity: the photosynthetic reaction center performs as a simple 0.2 V battery. *FEBS Letters, 510*(1–2), 105–107. https://doi.org/10.1016/S0014-5793(01)03210-0.

# Chapter 4
# Overview of the Main Mechanisms of Photosynthesis

*Lastly, I believe that a man who sees the light could admire the sun itself and not only its image reflected in the water or on any other surface.*

Freely translated from Plato, The Republic Book VII, 360 b.c.

## 4.1 Transport Phenomena in Reaction Centers

After a little parenthesis on how the biological mechanisms described previously can be compared to the functioning of Galvanic cells, an overview of the most important systems in photosynthesis (aerobic) is given here. More information can be found also in literature (Nelson and Cox 2012; Hall and Rao 1999; Hoptkins 1999).

We recall that the principal parts needed to carry on the process of photosynthesis are called photosynthetic systems. The main photosynthetic systems are:

- The photosystem II, PSII, which oxidizes water to $O_2$ in structures called thylakoids (that is why we need here a strong oxidant) and in this process releases protons into a region called lumen.
- The cytochrome $b_6f$ receives electrons from the previously named PSII and delivers them to the photosystem I, PSI. It also transports additional protons into the lumen from the stroma.
- The photosystem I, PSI, reduces $NADP^+$ to NADPH in the stroma (that is why we need here a strong reductant) by the action of ferredoxin (Fd) and the flavoprotein ferredoxin–NADP reductase (FNR) (Fultz and Durst 1982).
- The ATP synthase is responsible for producing ATP which is a highly energetic molecule. The ATP synthase is driven by the diffusion of protons which diffuse back through it from the lumen (where they had higher concentration) into the stroma. To be noticed here that, a lot of work is done to obtain a difference in concentration concerning the $H^+$ ions and this is with reason.

The cytochrome pumping of $H^+$ ions is shown in Fig. 4.1.

From the figure above, it can be observed that the reduced quinone, Q complex takes $2H^+$ from the stroma and completely reduces the quinone itself.



C. De Blasio, *Fundamentals of Biofuels Engineering and Technology*,
Green Energy and Technology, https://doi.org/10.1007/978-3-030-11599-9_4

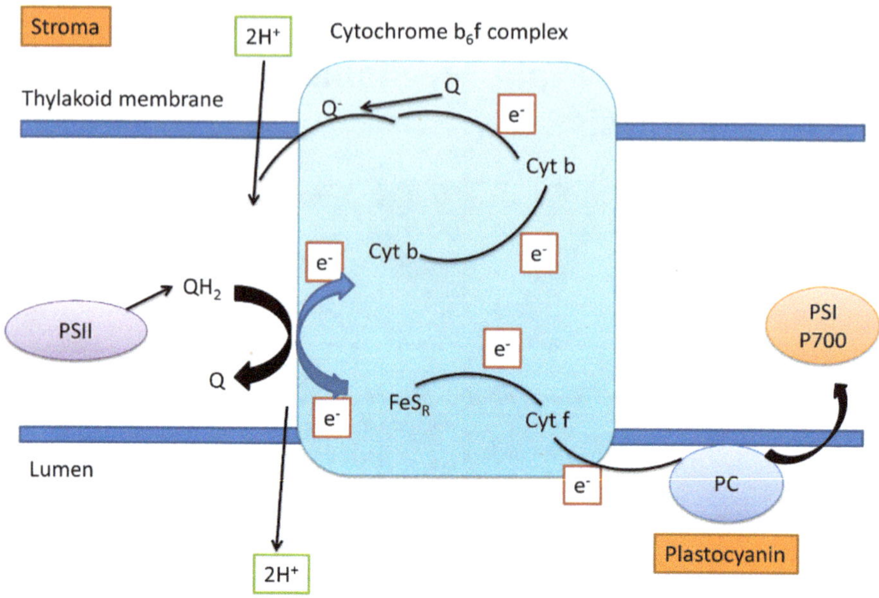


**Fig. 4.1** Cytochrome $b_6f$ complex pumping $H^+$ ions. Adapted from Taiz et al. (2014)

From the photosystem II, it can be stated that:

- the pheophytin takes the electrons;
- gives them to the quinones, and completely reduce them ($QH_2$);
- the quinones are oxidized and two electrons are available;
- one goes to photosystem I;
- one goes to a cycle which takes $2H^+$ from the stroma to the lumen.

Figure 4.2 shows the overall process of photosynthesis with the main systems involved. To be noticed here that, in order for the ATP synthase to work, a substantial amount of hydrogen ions is necessary on the lumen side. The difference in electrochemical potential and the chemical diffusion of hydrogen ions through the ATP synthase channel drives the ATP synthase enzyme to produce ATP molecules.

In the first stage, light-dependent reactions, or also named as light reactions, capture the energy of light and use it to make the energy storage molecules (ATP). It will be repeated here that the ATP molecules represent a font of energy storage in the sense that when these molecules react to form ADP, an important quantity of energy is released.

During the second stage of photosynthesis, the process accounts for light-independent reactions. These reactions use these products to capture and reduce carbon dioxide by means of NADPH.

As a summary, the main reaction centers (RC) are:

- photosystem II;
- cytochrome $b_6f$;

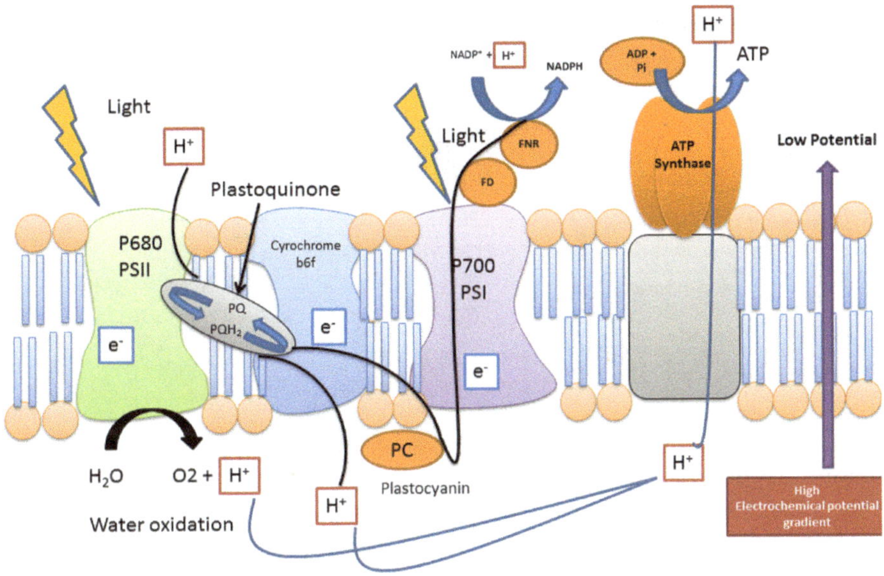


**Fig. 4.2** Summary of the main mechanisms of photosynthesis. The FNR refers to the ferredoxin—NADP+ reductase. Adapter from Holtzegel (2016)

- photosystem I;
- ferredoxin–NADP reductase (FNR);
- the ATP synthase.

## 4.2 Light and Dark Reactions of Photosynthesis

Light reactions can be described by the following overall reaction (Arnon, 1971):

$$2H_2O + 2NADP^+ + 3ADP + 3Pi + \text{light} \rightarrow 2NADPH + 2H^+ + 3ATP + O_2 \tag{4.1}$$

While dark reactions can be written as follows:

$$ATP + H_2O \rightarrow ADP + Pi + nH^+ \quad \Delta G = -34 \text{ kJ per mole at } pH\ 7 \tag{4.2}$$

Table 4.1 gives the main steps of photosynthesis. As can be noticed, we have mainly six steps while the overall reaction is demonstrated at the end of the table.

It is referred in literature (Klass 1998) that for each of the two light reactions, one or two photons are required to transfer each electron from the oxidation of the water; according to this calculation, a total of eight photons is, for this reason, required to

**Table 4.1** Reaction steps in photosynthesis

| Step | Characteristic reaction |
|---|---|
| 1st | $2H_2O(l) \rightarrow 4H^+ + 4e + O_2(g)$ |
| 2nd | $4Fd^{+3} + 4e \rightarrow 4Fd^{+2}$ |
| 3rd | $3ADP + 3P \rightarrow 3ATP$ |
| 4th | $4Fd^{+2} \rightarrow 4Fd^{+3} + 4e$ |
| 5th | $2NADP + 4H^+ + 4e \rightarrow 2NADPH_2$ |
| 6th | $CO_2(g) + 3ATP + 2NADPH_2 \rightarrow (CH_2O) + 3ADP + 2NADP + 3P + H_2O(l)$ |
| Overall | $CO_2(g) + H_2O(l) \rightarrow (CH_2O) + O_2(g)$ |

*Fd* ferredoxin

fix one molecule of $CO_2$. This is not really important now to argue if it is one photon, two or many photons required to reduce water and take one electron as this seems to be difficult to really predict and of course we could do many speculations about it. It is important now to understand that there are many photos arriving on the plant with a sufficient energy and within the photosynthetic active radiation required and these photos are responsible for the oxidation of water.

We assume that $CO_2$ is in the gaseous phase and the initial product is glucose; consider now the reverse reaction between the water (six molecules) and glucose to give six molecules of $CO_2$. In this reaction, 24 atoms of hydrogen are produced. Now, considering that each of these atoms could be represented as $H^+$, for each of these atoms, one electron is then available and therefore, a total of 24 electrons are produced. Going backward, for each water molecule, four electrons are then derived.

Just to have an idea of the energies involved in this process: the standard Gibbs free energy change at standard conditions (25 °C) is +0.48 MJ (+114 kcal) per mole of $CO_2$ assimilated (Klass 1998) and the corresponding enthalpy change is +0.47 MJ (+112 kcal) (Othmer 2005, p. 12).

The main terms of Fig. 4.3 are explained as follows:

- YZ = tyrosine;
- Fx = Iron–Sulfur center;
- P680 = primary electron donor of PSII composed of chlorophyll (Chl);
- Pheo = pheophytin;
- Qa Qb = plastoquinone A and B (they receive electrons from pheophytin) Qb receives two protons from the stroma side;
- cytochrome f (Cyt f). cytochrome b low- and high-potential forms (Cyt bLp and Cyt bHp);
- PC = plastocyanin;
- P700 = primary electron Chl donor of PSI;
- A0 = chlorophyll; A1[Q] = phylloquinone;
- Fd = ferredoxin; FNR = ferredoxin NADP reductase;
- NADP* = oxidized nicotinamide adenine dinucleotide phosphate;
- Yo = symmetrically related tyrosine to YZ.

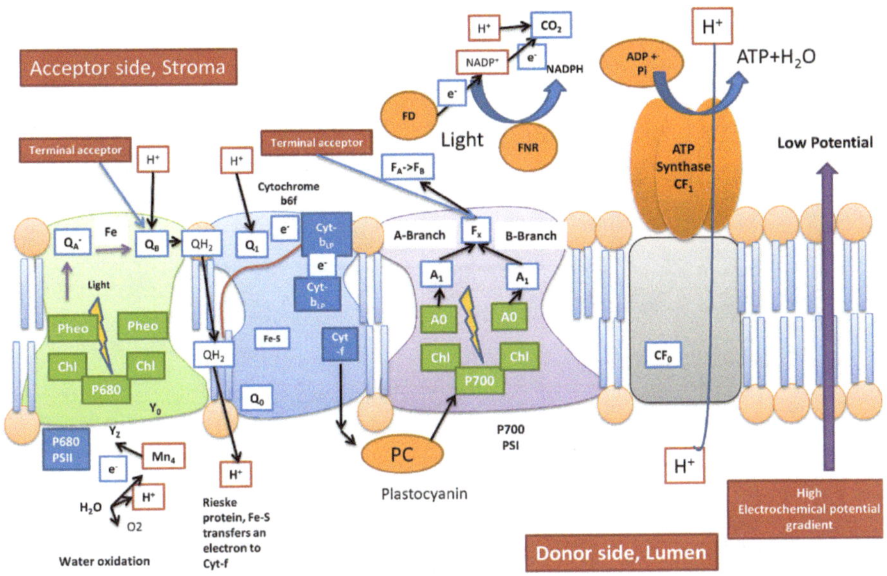


**Fig. 4.3** Oxygenic photosynthesis, the light side. Adapted from US Department of Energy (2018)

The oxidation of water is described by Eq. (4.3) which is Eq. (3.41) with opposite direction:

$$2H_2O \rightarrow O_2 + 4e^- + 4H^+ \quad (4.3)$$

Water is oxidized by four states of excitation of manganese ions (Kolling et al. 2012):

$$S_0 \rightarrow S_1 \rightarrow S_2 \rightarrow S_3 \rightarrow S_4$$

These states are probably four due to the fact that, as mentioned previously, four electrons are taken for each water molecule and at each of these steps, one electron will be released. As a summary, four protons are transported across the membrane for every two electrons delivered to P700.

Concerning the common structure of Type II and Type I reaction center, proteins of the Type II and Type I reaction centers show a pseudo-twofold axis symmetry (Tsiotis et al. 1997). The reasons for this particular structure are still not fully understood. Additionally, it is still not clear how Type II centers are able to differentiate their active and inactive branches. Figure 4.4 shows a schematic view of how the light reactions are connected with the dark reactions to form glucose molecules.

Electrons from the light process in the thylakoid membrane are responsible for the reduction of nicotinamide adenine dinucleotide (NAD*) or its phosphorylated form (NADP*) (Archer and Barber 2004). We need the electron flow and the production

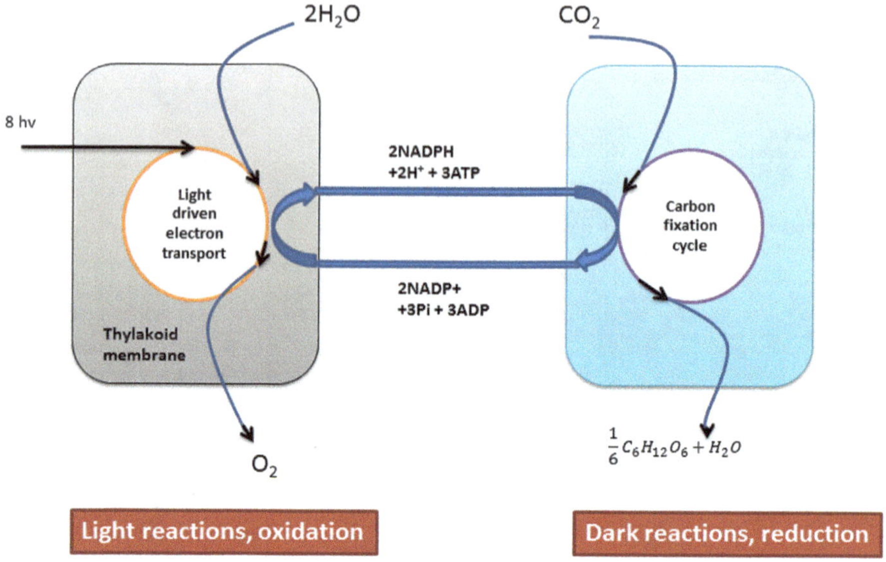


**Fig. 4.4** Summary of the oxygenic photosynthesis, light and dark reactions. Adapted from Ksenzhek and Volkov (1998)

of electrons to provide the reduced forms NADH or NADPH which in turn will give the necessary electrons (reducing power) necessary for the $CO_2$ fixation. The ATP is the molecule which will help in this process by giving additional energy generated by photosynthetic phosphorylation.

The reaction of water with ATP is shown as follows. This reaction is then used in the process of binding the $CO_2$ and formation of carbohydrates.

$$ATP + H_2O \rightarrow ADP + P_{inorganic} + nH^+ \tag{4.4}$$

This reaction is reported to have a standard Gibbs free energy of −34.5 kJ/mol at pH 7.0 (Ksenzhek and Volkov 1998).

## 4.3 Binding of Carbon Dioxide and Production of Sugar Molecules

Figure 4.5 shows the Calvin–Benson cycle. This cycle is a common way used by living organisms to produce sugars from $CO_2$ and therefore, it is very important to study it and to analyze it in the most accurate way. In literature, many models are given on this cycle and also quantitative comparisons are made between different models (Arnold and Nikoloski 2011). From the figure, it should be noted that from ribulose-

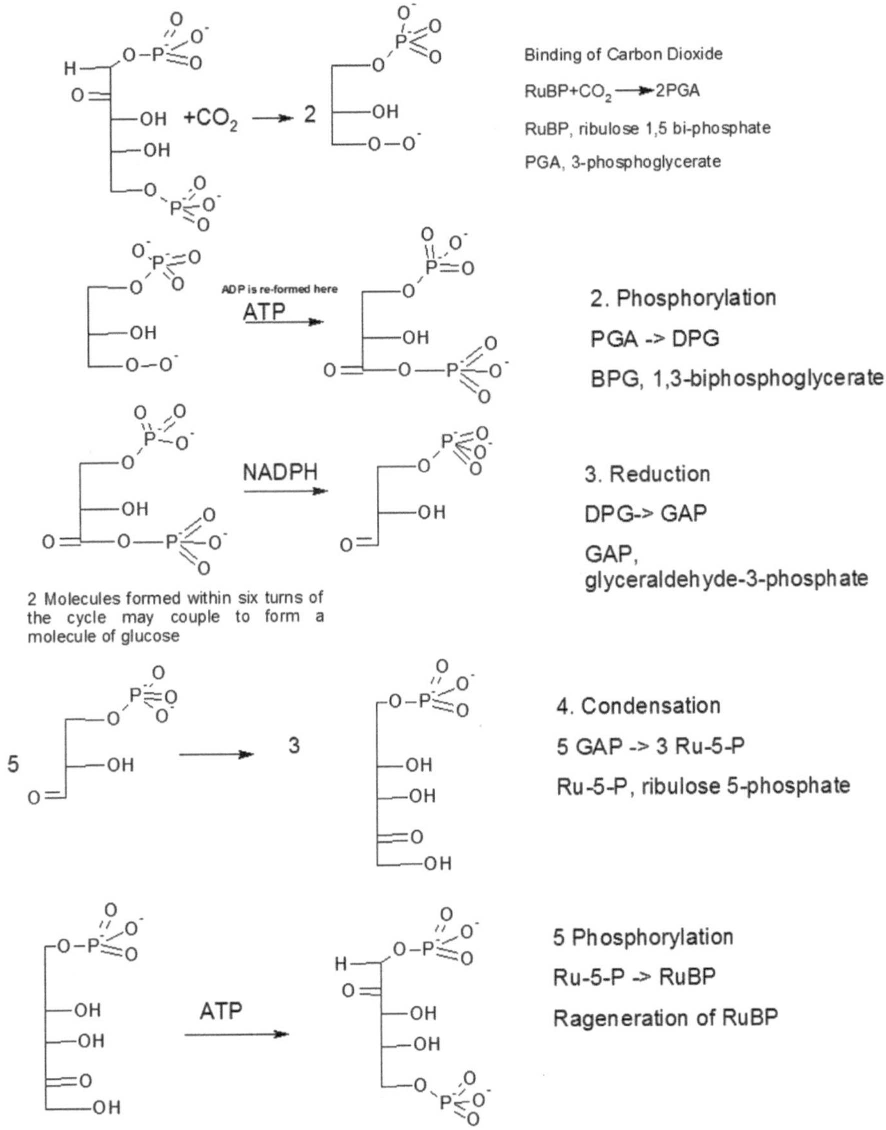


**Fig. 4.5** Binding the carbon dioxide, the Calvin–Benson cycle. Adapted from Ksenzhek and Volkov (1998)

1,5-biphosphate, it is possible to obtain glyceraldehide 3-phosphate molecules which condense to form ribulose 5-phosphate. For each five turns of the cycle, we have six ribulose 5-phosphates. This molecule, in turns, forms again the ribulose-1,5-biphosphate. It is shown here that ATP and NADP-H molecules are effectively used during these steps.

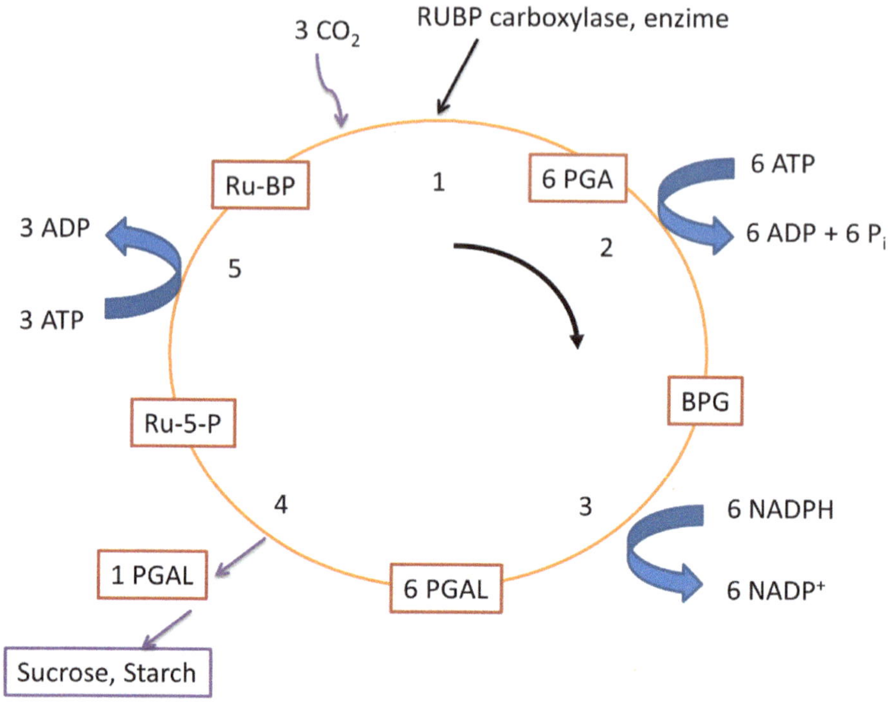


**Fig. 4.6** Calvin–Benson cycle, a simplified scheme. The numbers 1–5 refer to the steps of the cycle. PGAL is the phosphoglyceraldehyde. Adapted from Starr et al. (2016)

A more simplified diagram is shown in Fig. 4.6. This version is usually more frequent in literature.

As a summary, it should be remembered here that photosynthesis is the only biochemical system able to do water oxidation rewritten here as:

$$2H_2O \rightarrow O_2 + 4H^+ + 4e^- \tag{4.5}$$

The chemical–physical mechanisms of photosynthesis are not yet exactly known even if a large quantity of information is provided in literature. Regarding the molecule of ATP, this molecule is an extremely fast and unstable compound; it stays in this form for a very short time and it reacts to give inorganic phosphorus, $H^+$ ions, and energy (Ingermann et al. 1997). The equation is rewritten here as:

$$ATP + H_2O \leftrightarrow ADP + P_i + nH^+ \tag{4.6}$$

With a $\Delta G = -34.5$ kJ/mol at pH 7.

This process and the production of ATP are done in the thylakoid membrane while NADPH is used during the reduction of $CO_2$. This is more clearly written as:

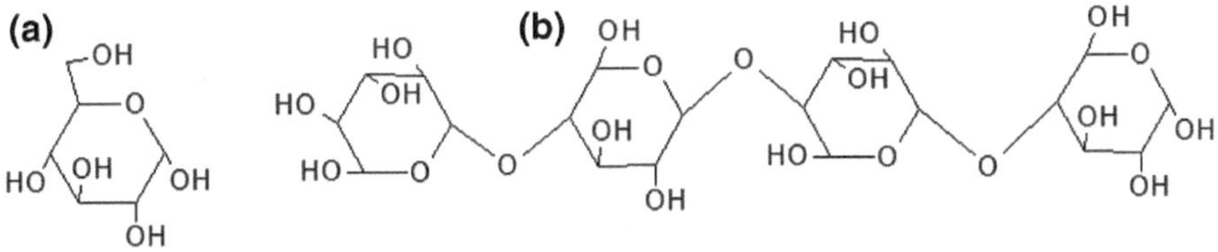


**Fig. 4.7** Structure of D-glucose and the derived linear combination to give cellulose

$$NADP^{+} + H^{+} + 2e \leftrightarrow NADPH \tag{4.7}$$

The molecules which are storing carbohydrates are D-glucose and the linear combination of D-glucose which forms the cellulose. Their structure is shown here in Fig. 4.7.

These molecules are split with the normal digestion and completely oxidized in $CO_2$ and $H_2O$ by means of the process denominated as glycolysis. The energy released is used to produce 36 molecules of ATP, the energy-carrying molecule in all living organisms.

$$\begin{aligned} 6CO_2 + 12NADPH + 18ATP \rightarrow C_6H_{12}O_6 + 12NADP + 18ADP \\ + 18P_{\text{inorganic}} + 6H_2O \end{aligned} \tag{4.8}$$

## References


Archer, M. D., & Barber, J. (2004). *Molecular To global photosynthesis*. London: Imperial College Press.

Arnold, A., & Nikoloski, Z. (2011). A quantitative comparison of Calvin-Benson cycle models. *Trends in Plant Science, 16*(12), 676–683. https://doi.org/10.1016/j.tplants.2011.09.004.

Arnon, D. I. (1971). The Light reactions of photosynthesis. *Proceedings of the National Academy of Sciences, 68*(11), 2883–2892.

Fultz, M. L., & Durst, R. A. (1982). Mediator compounds for the electrochemical study of biological redox systems: A compilation. *Analytica Chimica Acta, 140*(1), 1–18. https://doi.org/10.1016/S0003-2670(01)95447-9.

Hall, D. O., & Rao, K. (1999). *Photosynthesis*. Cambridge: Cambridge University Press.

Holtzegel, U. (2016). The Lhc family of *Arabidopsis thaliana. Endocytobiosis and Cell Research, 27*(2), 71–89.

Hoptkins, W. G. (1999). *Introduction to plant physiology* (2nd ed.). New York, USA: Wiley Inc.

Ingermann, R., Bencic, D., & Herman, J. (1997). Stability of nucleoside triphosphate levels in the red cells of the snake. *Journal of Experimental Biology, 200*(7), 1125–1131.

Klass, D. L. (1998). *Biomass for renewable energy, fuels, and chemicals*. Academic Press.

Kolling, D. R. J., Cox, N., Ananyev, G. M., Pace, R. J., & Dismukes, G. C. (2012). What are the oxidation states of manganese required to catalyze photosynthetic water oxidation? *Biophysical Journal, 103*(2), 313–322. https://doi.org/10.1016/j.bpj.2012.05.031.

Ksenzhek, O. S., & Volkov, A. G. (1998). *Plant energetics*. San Diego, California: Academic Press.

Nelson, D. L., & Cox, M. M. (2012). *Lehninger principles of biochemistry* (6th ed.). W.H: Freeman.

Othmer, K. (2005). *Kirk-othmer encyclopedia of chemical technology* (5th ed., Vol. 12). Wiley-Interscience.

Starr, C., Evers, C. A., & Starr, L. (2016). *Biology: Concepts and applications* (10th ed.). Cengage Learning and National Geograohic Learning.

Taiz, L., Zeiger, E., Møller, I. M., & Murphy, A. (2014). *Plant physiology and development*. Sinauer Associates, Oxford University Press.

Tsiotis, G., Hager-Braun, C., Wolpensinger, B., Engel, A., & Hauska, G. (1997). Structural analysis of the photosynthetic reaction center from the green sulfur bacterium Chlorobium tepidum. *Biochimica et Biophysica Acta (BBA)—Bioenergetics, 1322*(2), 163–172. https://doi.org/10.1016/S0005-2728(97)00073-X.

U.S. Department of Energy. (2018). *Photosynthesis production of hydrogen from water*. U.S. Department of Energy Office of Biological and Environmental Research. Retrieved from https://public.ornl.gov/site/gallery/detail.cfm?id=152&topic=&citation=&general=&restsection.

# Chapter 5
# Work from Light

*Il sole é la lampadina del mondo.*
*The sun is the light bulb on the world.*
Daniel Lucas Ettore De Blasio, 10 October 2018.

## 5.1 Incoming Radiation

It is clear now that the energy coming from the sun plays a fundamental role in biomass generation and in how plants and organisms also adapted themselves to the nature of this energy source. In this part of the manuscript, some notions will be given on how the incoming radiation from the sun affects the thermodynamics involved.

The sun is our star, and it is well known that life depends on this energy source; therefore, it is good to know a little about it and some data are given as follows (Bonanno et al. 2002; Williams 2013):

- The age of the sun is around 4.6 billion years.
- Its inner temperature is around $1.57 \times 10^7$ K.
- The inner pressure is enormous and estimated around $2.447 \times 10^{11}$ bar.
- The inner density is something like $1.622 \times 10^5$ kg/m$^3$.
- The temperature at surface is commonly assumed to be 5700 K.

It is also stated in the literature that at least 4000 years are needed for the radiation energy to reach the surface of the sun from the center (Odenwald 2018). This is because, in case we consider the main heat transfer process, heat will be put forward by radiation mainly. And radiation, as it will be explained later in this manuscript, is strongly dependent on the direction, which goes from the considered source point to the receiving unit. For instance, starting from the center, heat will be transferred to the nearest surfaces (molecules) and forwarded to the nearest facing surfaces. These surfaces will reflect, absorb, and transmit energy to the next ones and so on until the surface is reached. Conduction cannot play a major role considering the structure of the sun, its density, and also the elevated temperatures. The way radiation transferred

 
C. De Blasio, *Fundamentals of Biofuels Engineering and Technology*,
Green Energy and Technology, https://doi.org/10.1007/978-3-030-11599-9_5

in the sun can be modeled by considering the so-called drunkard's walk problem or "random walk" problem which are mathematical problems (Tijms 2003).

The incoming radiation from the sun is constituted by its photons, and the photons' density is directly proportional to the temperature of the source. Now let us consider a general number of particles with a particular energy $\Delta E$, and the overall energy will be determined as:

$$n(E)\Delta E = g(E) \cdot f(E) \cdot \Delta E; \tag{5.1}$$

where $n$ is the number of photos in this case, $\Delta E$ is the energy range considered, while $g(E)$ is denominated as the number of energy states per unit volume in the energy range considered, and it is given by Richtmyer et al. (1969):

$$g(E) = \frac{8\pi}{(h \cdot c)^3} E^2; \tag{5.2}$$

while $f(E)$ is the Bose–Einstein distribution:

$$f(E) = \frac{1}{A\mathrm{e}^{\frac{E}{KT}} - 1} \tag{5.3}$$

Therefore, the original expression, Eq. (5.1) will become:

$$n(E)\mathrm{d}E = \frac{8\pi}{(h \cdot c)^3} E^2 \cdot \frac{1}{A\mathrm{e}^{\frac{E}{KT}} - 1} \cdot \mathrm{d}E \tag{5.4}$$

Substituting

$$\frac{E}{KT} = x;$$

this will give

$$E = K \cdot T \cdot x;$$

and

$$\mathrm{d}E = K \cdot T\,\mathrm{d}x;$$

Equation (5.11) will be then:

$$n(E)\mathrm{d}E = \frac{8\pi}{(h \cdot c)^3} \cdot K^3 T^3 \int_0^\infty \frac{x^2}{A\mathrm{e}^x - 1}\mathrm{d}x \tag{5.5}$$

The integral is solved for $A = 1$ in the case when photons are considered and it is equal to the Riemann zeta function which multiplies the Euler beta function:

$$\int_0^\infty \frac{x^2}{Ae^x - 1}\mathrm{d}x = \zeta(p) \cdot \Gamma(p); \tag{5.6}$$

where $p$ in our case is equal to 3.

Values of the Riemann zeta function and for the Euler beta function are tabulated, and in this case:

$$\zeta(p) \cdot \Gamma(p) = 2.404 \tag{5.7}$$

This will mean that the number of photos per unit volume, which is their density, is equal to:

$$n^* = 2.404\frac{8\pi}{c^3}\left(\frac{kT}{h}\right)^3; \tag{5.8}$$

and in terms of number of moles of photons:

$$D = 2.404\frac{8\pi}{N_A c^3}\left(\frac{kT}{h}\right)^3 \tag{5.9}$$

where

$D$ = moles of photons ($E$)/$m^3$.
$h = 6.626 \times 10^{-34}$ J s is the Planck constant.
$k = 1.38 \times 10^{-23}$ J/K is the Boltzmann constant.
$c = 2.998 \times 10^8$ m/s is the speed of light.
$N_A = 6.02 \times 10^{23}$ is the Avogadro number (the photons in one photon mole in this case).

## 5.2 A Simple Derivation of the Stefan–Boltzmann Law

It is also known that the radiative energy (Joules) is directly proportional to the temperature of the source, the solid angle between the source and the receiving body, and a constant value which is the usual proportionality constant. This is expressed by the Stefan–Boltzmann law (Incropera et al. 2006):

$$E_s = \omega_s a T_s^4; \tag{5.10}$$

where the proportionality constant $a$ is expressed by:

**Table 5.1** Common quantities in electromagnetic transport energy

| Name of the quantity | Symbol | Dimensions |
|---|---|---|
| Energy density | $u$ | $ML^{-1}T^{-2}$ |
| Velocity of light | $c$ | $LT^{-1}$ |
| Electron mass | $m$ | M |
| Charge of electron | $e$ | $M^{1/2}L^{3/2}T^{-1}$ |
| Absolute temperature | $\theta$ | $\theta$ |
| Gas constant | $k$ | $ML^{2}T^{-2}\theta^{-1}$ |

With M = mass; L = length; T = time; θ = temperature

$$a = \frac{2\pi^4 k^4}{15c^2 h^3} \tag{5.11}$$

The following is a derivation of the Stefan–Boltzmann law done by dimensional analysis. This is done here to demonstrate a simpler derivation for the formula and also to give some consideration on the reasoning on the dimensions of the problem at hand. As a matter of fact, just by reasoning on the dimensions of a problem, this could give an important number of information and also influence our decisions before we start actually to do our complicated balances. Notions of dimensional analysis are reported in the literature (Bridgman 1963), and here, the analyzed problem is discussed with additional comments.

Consider the quantities shown in Table 5.1.

We would like to find the function of the energy density from the other variables. At this point, we have four dimensions and six relations given in the previous table, and this means that we can obtain two non-dimensional numbers. We choose arbitrarily:

$$\text{First non-dimensional number} = uk^{\alpha}\theta^{\beta}e^{\gamma}m^{\delta} \tag{5.12}$$

$$\text{Second non-dimensional number} = k\theta^{\alpha 1}u^{\beta 1}m^{\gamma 1}c^{\delta 1} \tag{5.13}$$

Since the two numbers have no dimensions, it follows that

$$\begin{array}{ll} 1+\alpha+0+\frac{1}{2}\gamma+\delta=0 & \text{condition on M} \\ -1+2\alpha+0+\frac{3}{2}\gamma+0=0 & \text{condition on L} \\ -2-2\alpha+0-\gamma+0=0 & \text{condition on T} \\ \beta-\alpha=0 & \text{condition on } \theta \end{array}$$

which gives:

$$\alpha=-4,\ \beta=-4,\ \gamma=6,\ \delta=0$$

Solving also for Eq. (5.13), we have that the two non-dimensional numbers are expressed as:

$$\text{First non-dimensional number} = ue^6k^{-4}\theta^{-4} \tag{5.14}$$

$$\text{Second non-dimensional number} = k\theta \cdot m^{-1}c^{-2} \tag{5.15}$$

Notice that if we would have put together the terms ($u$, $c$, $m$, $e$, $\theta$) in the system of equation, we would have obtained $1 = 0$ which is impossible. This is why by inspecting we notice that $\theta$ and $m$ cannot stay alone, because they have only one dimension so for instance $\theta$ should be with $k$.

From Eqs. (5.14) and (5.15), it can be stated that

$$u = k^4e^{-6}\theta^4 f\left(k \cdot \theta \cdot m^{-1}c^{-2}\right) \tag{5.16}$$

And this is because, since the two numbers are non-dimensional, they are simply numbers and therefore any number can be expressed as a function of another number in the same way the number 3 can be derived as 6/2.

Because the argument of the function is very small (notice the power of the velocity of light at the denominator), and it is not varying much with the temperature, then the solution becomes:

$$u = \text{Const}\, k^4e^{-6}\theta^4 \tag{5.17}$$

And this leads to

$$u = a\theta^4 \tag{5.18}$$

which is a version of the Stefan–Boltzmann law.

In reference to our thermodynamic system, the plant, it was shown previously that plants utilize energy from the sun to bind $CO_2$ and form organic matter. In case we consider the plant as an energy converter, a simple energy balance would be:

$$W = E_s - E_c - Q \tag{5.19}$$

where $E_c$ is the energy absorbed by the converter and has a similar expression of $E_s$ (energy of the source) while $Q$ is the heat lost by the converter at the temperature of the converter.

## 5.3 The Solid Angle

The energy coming from an object and transported by radiation is highly dependent on what is the inclination of the surface source or the inclination of the receiving surface. This is very well understood if we consider the sun as a point and the earth surface. As a matter of fact, it is very well known that at the equator it is much warmer than at the north or south pole. Concerning the evaluation and the understanding of the solid angle concept, the author will try to give also in this case a simplified derivation of the solid angle expression.

Let us give a look at Fig. 5.1. The radiation is coming from a distance (let us consider a large distance; in this way, the source can be considered as a point) on a surface. Now let us assume that the surface on which the incident radiation is going is not perpendicular to the direction of the incoming photons. The surface d$A$ where the photons are going should be multiplied by cos($\Phi$) to obtain the perpendicular surface.

Consider again Fig. 5.1 and the following proportion:

$$\frac{\mathrm{d}W}{\mathrm{d}\omega} = \frac{s^2}{\sigma^2} \tag{5.20}$$

And the additional proportion:

$$\frac{s}{\sigma} = \frac{r}{1} \tag{5.21}$$

From Eqs. (5.20) and (5.21), we have:

$$\frac{\mathrm{d}W}{\mathrm{d}\omega} = \left(\frac{r}{1}\right)^2 = r^2; \tag{5.22}$$

and

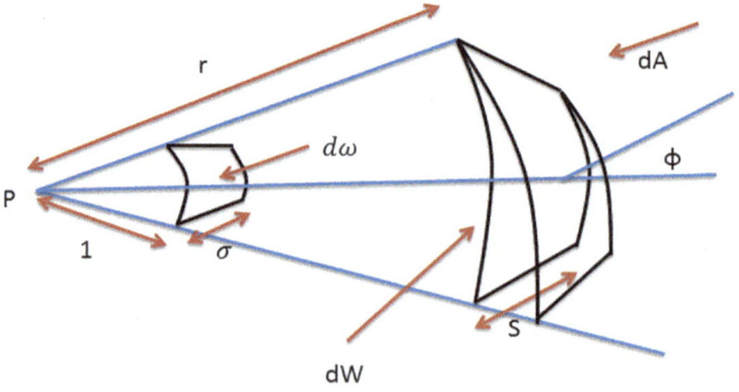


**Fig. 5.1** Solid angle evaluation

$$d\omega = \left(\frac{\cos\Phi}{r^2}\right)dA; \tag{5.23}$$

which gives

$$\omega = \int_{A}^{A_1}\left(\frac{\cos\Phi}{r^2}\right)dA; \tag{5.24}$$

where $A$ refers to the differential area shown in the figure. Now, let us suppose that the incoming radiation is going to plants and consider the plants as an infrared thermometer working in the photovoltaic mode. In this case, the responsivity of the detecting body is (DeWitt and Nutter 1988):

$$R(\lambda) = \frac{dS}{d\Phi(\lambda)} \tag{5.25}$$

where $dS$ is the signal output (Ampere) and $d\Phi(\lambda)$ is the quantity of radiation arriving on the detector (Watt). Equation (5.25) can be also written as:

$$R(\lambda) = \frac{i}{d\Phi(\lambda)} \tag{5.26}$$

where $i$ is the current. The relation is valid for linearly responding photodiode. The energy of the photons is (Liddle 2015):

$$E = h \cdot \nu \text{ (J)} \tag{5.27}$$

where $h = 6.626176 \times 10^{-34}$ J s is the Plank constant and $\nu$ is the wavelength frequency $\nu = \frac{c_0}{\lambda}$, $c_0$ is the speed of the wavelength in the considered medium, and $\lambda$ is the wavelength at hand

$$E = \frac{h \cdot c_0}{\lambda} \tag{5.28}$$

The radiant flux is given by

$$\Phi(\lambda) = \frac{n \cdot E}{t}; \tag{5.29}$$

where $n$ is the number of photons and $t$ is the time; therefore, the radiant flux is expressed by:

$$\Phi(\lambda) = \frac{n \cdot h \cdot c_0}{t \cdot \lambda} \tag{5.30}$$

The number of photons absorbed per unit time is:

$$\frac{n \cdot \alpha(\lambda)}{t} = \frac{n(1 - \rho(\lambda) - \tau(\lambda))}{t}; \tag{5.31}$$

where $\alpha$ is the absorbance, $\rho$ is the reflectance, and $\tau$ represents the transmittance. These are fractions between 0 and 1 and therefore with no dimensions. The sum of these three coefficients is equal to one. For this reason:

$$i = \frac{n \cdot [1 - \rho(\lambda) - \tau(\lambda)] \cdot q(\lambda) \cdot e}{t}; \tag{5.32}$$

where $q(\lambda)$ is number of charges/absorbed photon, and $e = 1.6022 \times 10^{-19}$ C, the charge of the electron. We obtain finally:

$$R(\lambda) = \frac{[1 - \rho(\lambda) - \tau(\lambda)] \cdot q(\lambda) \cdot \lambda}{\frac{h \cdot c_0}{e}} \left(\mathrm{A\,W^{-1}}\right) \tag{5.33}$$

where $c_0 = 2.9979 \times 10^{14}\ \frac{\mu m}{s}$.

The thermodynamic efficiency of an energy converter machine driven by radiative energy is often expressed by empirical formulations, for instance (Ksenzhek and Volkov 1998):

$$\eta = \frac{W}{E_s} = 1 - \frac{4}{3}\left(\frac{T_c}{T_s}\right) + \frac{1}{3}\left(\frac{T_c}{T_s}\right)^4; \tag{5.34}$$

where s refers to the source of the radiation while c is referring to the energy converter (the organism using that radiation), and $T$ is the temperature. The previous equation shows a straight dependence of the efficiency on the temperatures of the sources and the converter.

## 5.4 The Photosynthetic Active Radiation

Plants absorb photons which are above a particular frequency (Aaslyng et al. 1999), and this could be for instance referred as $430 \times 10^{-12}$ Hz with an energy of 1.77 eV (electronvolt) (Archer and Barber 2004). However, there is also a maximum at which plants absorb energy and this value is 750 THz. The photon energy is expressed also as:

$$\epsilon_0 = \frac{h\nu_0}{e_0} \tag{5.35}$$

in this case:

$h = 6.626 \times 10^{-34}$ J s is the Planck constant.
$\nu_0$ is the threshold frequency: $430 \times 10^{12}$ Hz.
$e_0$ is the electron charge, $1.602 \times 10^{-19}$ C.

As a matter of fact, the value of 1.77 eV is obtained if we do the following calculation:

$$h \cdot 430 \times \frac{10^{12}(\text{Hz})}{\text{charge of electron}} = 1.77\ \text{eV} \tag{5.36}$$

It is good to stress here that only a fractional amount of the intensity of solar radiation is utilized as photosynthetic active radiation, PAR (Knyazikhin et al. 1998). To give a more practical example, the PAR can be compared to the activation energy needed for chemical reactions to take place; more simply, a comparison can be made between two molecules and two stones: Let us assume that we want to break the stones or at least one of them by means of using their kinetic energy, and the stones will not break unless we give to them a sufficient amount of energy (in this case kinetic energy). The same is happening for chemical reactions where the energy in that case is determined by the temperature, and, as a matter of fact, there is a direct correlation between the kinetic energy and the temperature for molecules. This is no exception when photons are considered in this case.

The photosynthetic process can be also simplified to take into account for the main steps required to convert the incoming energy to work. In Fig. 5.2, a simplified version of what can be denominated as "the light-harvesting engine" is given.

In Fig. 5.2, $E_s$ is the energy coming from the source (sun for instance), $E_a$ is the energy absorbed, $E_e$ is the energy required to constitute an excited state, $E_r$ is the part reflected, $E_q$ is the part lost under the form of heat, $M$ is the imaginary engine, and $W$ is the work obtained. $T$ represents the temperature at the various stages.

The four stages for the conversion of energy into work are as follows. According to the literature (Hall and Rao 1999) 47% of the incoming radiation from the sun is lost because of being outside the PAR range. This can be expressed as:

$$\eta_1 = \frac{\int_{v_0}^{v_1} I(v)\mathrm{d}v}{\int_0^{\infty} I(v)\mathrm{d}v} \tag{5.37}$$

where $I(v)$ is the intensity of the incoming radiation as a function of the frequency, while $v_0$ and $v_1$ are the frequencies related to the PAR. This will give 53% as the fraction of energy in the PAR ($100 - 47 = 53$). Now, about 30% of this 53% is lost to incomplete absorption (Hall and Rao 1999). This will leave another 15.9% from the 53% giving the previously mentioned 37%.

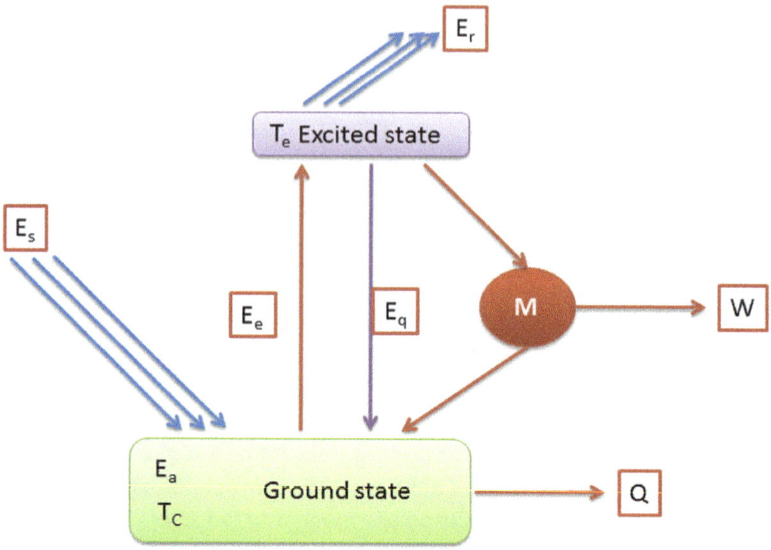


**Fig. 5.2** Light-harvesting engine, simplified representation

$$\eta_2 = \frac{E_a}{E_u} \tag{5.38}$$

where $E_u$ is the energy which can be utilized and given previously. Considering that 24% of this 37% is lost due to inefficiency in creating the excited state (Hall and Rao 1999), this will leave another 8.9% from the original 100% and therefore (37 − 8.9 = 28.2). The efficiency in generating the excited state will be then:

$$\eta_3 = \frac{E_e}{E_a} = \frac{28.2}{37} = 0.76 \tag{5.39}$$

where $E_a$ is the absorbed energy and $E_e$ is the one referred to the excited state. This could be also expressed as:

$$\eta_3 = 1 - \frac{T_e}{T_s} \cdot r; \tag{5.40}$$

where $r$ is a coefficient related to a dilution factor (Ksenzhek and Volkov 1998). The utilization of the excited state of the working body can be expressed as:

$$\eta_4 = 1 - \left(\frac{T_c}{T_e}\right)^{0.5}; \tag{5.41}$$

where c refers to the converter and e refers to the excited state. As a summary: $\eta_1$ is the fraction of photosynthetic active radiation for different temperatures; $\eta_2$ is the

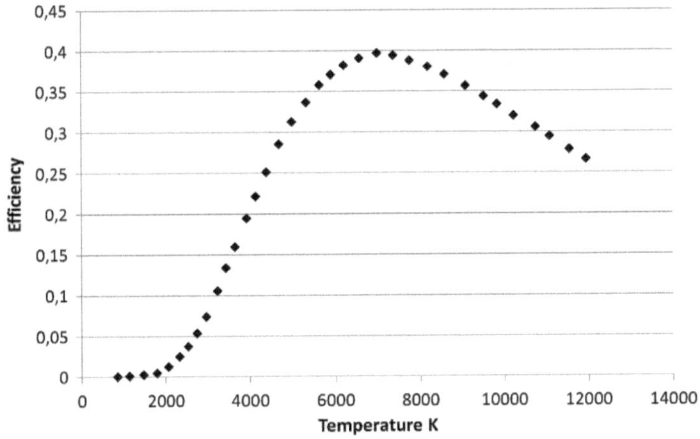


**Fig. 5.3** Efficiency of light absorption as a function of the temperature

efficiency for the absorption of this energy; $\eta_3$ is the part of energy transferred in the excited state; $\eta_4$ is the fraction of energy in the threshold level.

From Fig. 5.3, it is possible to see that the highest absorption efficiency is situated around 7000 K (Ksenzhek and Volkov 1998); however, it is known that the sun is at a temperature of 5762 K at the surface (Albarrán-Zavala and Angulo-Brown 2007) which suggests that the sun could have been diminished its temperature at the surface since the evolution of the plants.

It should be mentioned here that there are several values given in the literature for the overall efficiency calculated for the energy conversion in the process of photosynthesis and there are diverse methods. Here, it is reported one definition given by Ksenzhek and Volkov (1998).

$$\eta_{\mathrm{PS}} = \frac{\text{rate of enthalpy storage } \left(\mathrm{W\,m^{-2}}\right) \text{ in plants as D-glucose}}{\text{solar irradiance } \left(\mathrm{W\,m^{-2}}\right) \text{ acting on the plant}} \tag{5.42}$$

which can be also written as:

$$\eta_{\mathrm{PS}} = \frac{\int_{\lambda\mathrm{min}}^{\lambda\mathrm{max}} \frac{J_{\lambda}^{\mathrm{S}} \alpha_{\lambda} \Phi_{\lambda}}{N_{\mathrm{A}}} \Delta H \,\mathrm{d}\lambda}{\int_0^{\infty} E_{\lambda}^{\mathrm{S}} \,\mathrm{d}\lambda} \tag{5.43}$$

where

$N_{\mathrm{A}}$ is the Avogadro number;
$J_{\lambda}^{\mathrm{S}}$ is the solar photon flux $\left(\frac{\mathrm{photons}}{\mathrm{m^2\,s\,nm}}\right)$, at a wavelength between λ-min and λ-max;
$E_{\lambda}^{\mathrm{S}}$ is the solar irradiance [W/(m$^2$ nm)];
$\alpha_{\lambda}$ is the fraction of light absorbed by the plant (absorbance);

$\Phi_\lambda$ is the quantum yield for a consumption of $CO_2$, and ideal value is 0.125 photons/(moles of $O_2$ produced);

$\Delta H$ is the enthalpy of the photosynthesis (overall reaction).

In the literature, thermodynamic models for the photosynthetic process are available. In case we consider the overall reaction taking place to form glucose molecules (Albarrán-Zavala and Angulo-Brown 2007):

$$6CO_2(g) + 6H_2O(liq) \rightarrow C_6H_{12}O_6(s) + 6O_2; \tag{5.44}$$

data for the standard enthalpy of formation could be retrieved from databases (National Institute of Standards and Technology 2018). Considering standard conditions, the enthalpies of formation for the $CO_2$, $H_2O$, and glucose are respectively −393.52 kJ/mol, −285.820 kJ/mol, and −1273.3 kJ/mol. Summing up according to:

$$E^0_{\mathrm{Ph}} = \Delta H^0_{\mathrm{Glu}} + 6\Delta H^0_{\mathrm{ox}} - 6\Delta H^0_{\mathrm{CO_2}} - 6\Delta H^0_{\mathrm{w}}; \tag{5.45}$$

where Glu stands for Glucose, ox is the oxygen, and w is the water, this calculation will give that the energy required to form one mole of glucose would be $E^0_{\mathrm{Ph}} =$ 2802.74 kJ/mol. Additional data are given in terms of standard internal energy at a particular wavelength and in terms of Gibbs free energy (Albarrán-Zavala and Angulo-Brown 2007).

$$\Delta U^0 = 10555.287\ \frac{\mathrm{kJ}}{\mathrm{mol}}\ \text{at}\ \lambda = 680\ \mathrm{nm}$$
$$\Delta G^0 = 2880.31\ \frac{\mathrm{kJ}}{\mathrm{mol}}\ \text{(sugar)}$$

## References

Aaslyng, J. M., Rosenqvist, E., & Høgh-Schmidt, K. (1999). A sensor for microclimatic measurement of photosynthetically active radiation in a plant canopy. *Agricultural and Forest Meteorology, 96*(4), 189–197. https://doi.org/10.1016/S0168-1923(99)00057-X.

Albarrán-Zavala, E., & Angulo-Brown, F. (2007). A simple thermodynamic analysis of photosynthesis. *Entropy, 9*(4), 152–168. https://doi.org/10.3390/e9040152.

Archer, M. D., & Barber, J. (2004). *Molecular to global photosynthesis*. London: Imperial College Press.

Bonanno, A., Schlattl, H., & Paternò, L. (2002). The age of the sun and the relativistic corrections in the EOS. *Astronomy & Astrophysics, 390*(3), 1115–1118. https://doi.org/10.1051/0004-6361:20020749.

Bridgman, P. W. (1963). *Dimensional analysis*. London: Yale University Press.

DeWitt, D. P., & Nutter, G. (1988). *Theory and practice of radiation thermometry*. Wiley.

Hall, D. O., & Rao, K. (1999). *Photosynthesis*. Cambridge University Press.

Incropera, F. P., Bergman, T. L., & Lavine, A. S. (2006). *Introduction to heat transfer* (5th Edition). Wiley.

Knyazikhin, Y., Martonchik, J. V., Myneni, R. B., Diner, D. J., & Running, S. W. (1998). Synergistic algorithm for estimating vegetation canopy leaf area index and fraction of absorbed photosynthetically active radiation from MODIS and MISR data. *Journal of Geophysical Research: Atmospheres, 103*(D24), 32257–32275. https://doi.org/10.1029/98JD02462.

Ksenzhek, O. S., & Volkov, A. G. (1998). *Plant energetics*. San Diego, California: Academic Press.

Liddle, A. (2015). *An introduction to modern cosmology* (3rd ed.). Wiley.

National Institute of Standards and Technology. (2018). *NIST chemistry webbook*. Retrieved October 3, 2018, from https://webbook.nist.gov/chemistry/.

Odenwald, S. (2018). IMAGE Education Center. *Imager for magnetopause to aurora global exploration*. US: NASA National Aeronautics and Space Administration. Retrieved from https://image.gsfc.nasa.gov/poetry/ask/a11354.html.

Richtmyer, F. K., Kennard, E. K., & Cooper, J. N. (1969). *Introduction to modern physics* (6th ed.). McGraw-Hill.

Tijms, H. C. (2003). *A first course in stochastic models*. Chichester, West Sussex, England: Wiley.

Williams, D. R. (2013). *Sun fact sheet*. NASA Goddard Space Flight Center.

# Chapter 6
# Some Types of Analyses Conducted on Biomass

*The creatures instead ..., the creatures are divided in creatures of love and creatures of freedom.*
Adapted translation from the book: Così parlò Bellavista. Napoli, amore e libertà. Mondadori, 2010.
Luciano De Crescenzo.

## 6.1 Determination of the Total Nitrogen Content, the Kjeldahl Method

One of the major problems encountered when microorganisms are utilized for biofuel production is the determination of how much microorganisms are actually contained within the slurry at a determinate time. To be noticed is that there is a difference between the overall amount of biomass contained in the reactor and the overall amount of biomass that has to do with the actual microbes. For this reason, different methods have been developed for this purpose. One classification of these methods involves the measurement of cell mass with instrumental analysis by:

- dry weight technique;
- measurement of the total nitrogen content;
- measurement of turbidity.

Figure 6.1 shows a qualitative description of the turbidity measurement by using for instance laser diffraction technique; in this case, the more is the obscuration measured, the higher is the number of cells in the slurry. However, this kind of analysis is not the most accurate even if it gives a good indication of the amount of cells present in the slurry sample.

Other examples of techniques used to measure the amount of cells present in the biochemical reactor are given as follows:

- measurement of cell number (counting);
- weighing;
- separation of organisms from the medium;



C. De Blasio, *Fundamentals of Biofuels Engineering and Technology*,
Green Energy and Technology, https://doi.org/10.1007/978-3-030-11599-9_6

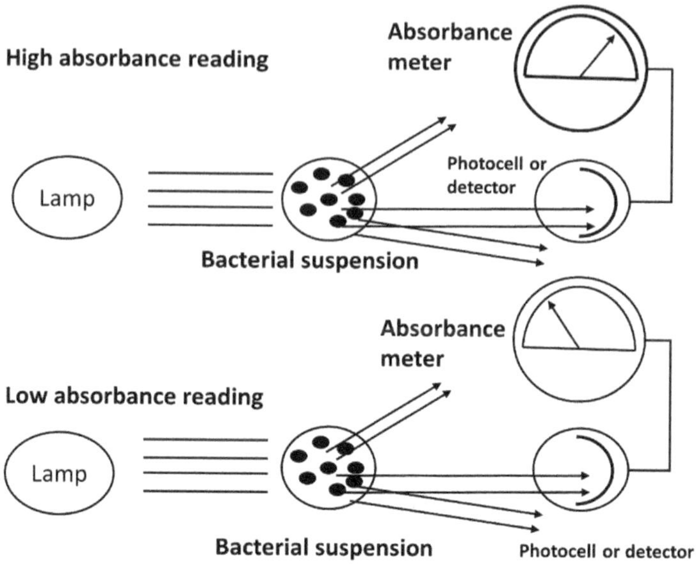


**Fig. 6.1** A qualitative description of the turbidity measurement by using laser/light diffraction technique

- drying of biomass;
- light absorbance measurements.

In this manuscript, the evaluation of cell mass by measuring one cell component will be taken into account. In this case, the component will be the nitrogen, N, and that will give an approximate amount of protein from which the number of cells or their concentration can be derived. It is known for instance that 60–65% of the dry weight of bacteria is protein, or more in detail, there are $3 \times 10^6 \frac{\text{proteins}}{\mu m^3}$ (the volume refers to the cell volume) (Milo 2013). Additionally, it is known that the amount of nitrogen is 8–12% of the dry cell weight (Bradley and Nichols 1918).

The main methods for the determination of the protein content are:

- Biuret [in this case we have a titration, change in color (Keppy and Allen 2018)];
- Lowry (titration, change in color) (Mæhre et al. 2018);
- Kjeldahl (measurement of total nitrogen by chemical treatment, there is release of ammonia which is then titrated).

In this book, the Kjeldahl method (Kjeldahl 1883) will be explained since it is a quite suitable method to measure the total organic nitrogen content of the biomass sample. The main reactions of the method are listed as follows:

$$\text{Organic N} + H_2SO_4 \rightarrow (NH_4)_2SO_4 + H_2O + CO_2 + \text{by products} \tag{6.1}$$

$$(NH_4)_2SO_4 + 2NaOH \rightarrow 2NH_3 \uparrow + Na_2SO_4 + 2H_2O \tag{6.2}$$

$$2NH_3 + H_2SO_4 \rightarrow (NH_4)_2SO_4 + \text{Excess}\,(H_2SO_4) \quad (6.3)$$

The organic nitrogen is first digested with an excess of sulfuric acid, Eq. (6.1), while ammonium sulfate is produced; then, the ammonium sulfate is reacting with sodium hydroxide, Eq. (6.2), to release all the nitrogen as ammonia which is condensed to an additional flask bottle. The ammonia is then reacting with a known excess amount of sulfuric acid which will form ammonium sulfate, Eq. (6.3). The excess sulfuric acid left is then titrated with sodium hydroxide, Eq. (6.4).

$$\text{Excess}\,(H_2SO_4) + 2NaOH \rightarrow Na_2SO_4 + 2H_2O \quad (6.4)$$

Figure 6.2 is a simplified illustration of the process.

The wt% of nitrogen is then calculated from the formula:

$$\text{wt\% N} = \frac{[L(H_2SO_4) \cdot N(H_2SO_4) - L(NaOH) \cdot N(NaOH)] \cdot 14.0067}{\text{sample gram}} \times 100 \quad (6.5)$$

where

L liters;
N normality.

To be remembered is that the molecular weight of N = 14.0067 g/mol.

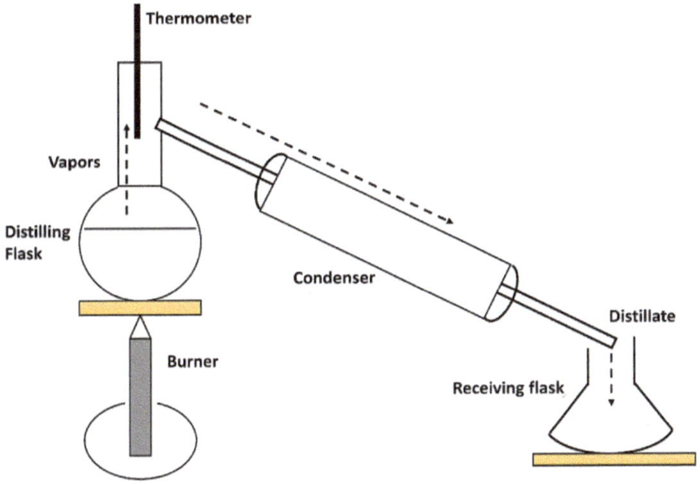


**Fig. 6.2** Simplified illustration of the Kjeldahl method

**Table 6.1** Protein factors for selected substances

| Protein factor | Protein sources |
|---|---|
| 6.38 | Milk and dairy |
| 6.25 | Grains |
| 5.95 | Rice |
| 5.70 | Wheat flour |

The wt% of N is finally multiplied by the protein factor giving the amount of protein in weight percentage. The so-called protein factor is calculated for different substances (Mariotti et al. 2008), and some examples are given in Table 6.1.

The Kjeldahl method can be executed also by using HCL; in this case, the procedure follows the steps:

1. Nitrogenous organic compound + concentrated $H_2SO_4 \rightarrow (NH_4)_2SO_4$ (ammonium sulfate)
2. $(NH_4)_2SO_4 + 2NaOH \rightarrow 2NH_3 + Na_2SO_4 + 2H_2O$
3. $NH_3 + HCl\ (0.5\ N) \rightarrow NH_4^{+}Cl^{-} + HCl$ (this quantity is the left back)
4. The left back HCl is titrated with standard NaOH (0.5 N).

The percentage of N is then estimated by the simple relation:

$$\%\ \text{of N} = \frac{\text{weight of N in grams}}{\text{weight of substance in grams}} \times 100 \tag{6.6}$$

An additional procedure exists (Martín et al. 2017). In this procedure, the previous step No. 3 of the HCL procedure is done by using boric acid.

$$NH_3 + H_3BO_3 \rightarrow NH_4^{+}H_2BO_3^{-}\ \text{(ammonium borate)} \tag{6.7}$$

Ammonium borate can be directly titrated by standard acid using methyl red as indicator.

The advantages with this last method are:

- We could prevent the loss of ammonia by volatilization.
- Boric acid is an acid which is too weak to interfere with further titration of ammonium borate.
- There is no need of back titration as in the case of reaction with HCl.

## 6.2 Determination of the Higher Heating Value

The determination of the gross calorific value for a biomass sample is of crucial importance when the biomass itself will be used as fuel for power plants. Let us assume that we have a biomass-fired power plant, and there are diverse sources from where we could buy the fuel. The first step, in this case, is to assess the fuel itself, and the first kind of analysis to be performed would be the evaluation of the gross calorific value. This is done by using the so-called calorimetric bomb. The understanding of how this measure is made will help the reader to avoid mistakes done commonly in the calculation of the overall efficiency of a power plant. To be noticed is that, in case the lower heating value would be taken into consideration for the inlet energy, the overall efficiency of the power plant could be over 100% if also a condenser is used. The efficiency of a power plant cannot be over 100% as obvious and also by referring to the common laws of the thermodynamics.

The functioning of the calorimetric bomb is quite simple (these steps will be described more in detail later):

1. The biomass is inserted in a chamber.
2. The chamber is closed and inserted in a volume of water.
3. The biomass is combusted completely with excess of oxygen.
4. The heat produced is transferred to the known amount of water.

Figure 6.3 shows a photograph of common laboratory equipment used at Åbo Akademi University, Faculty of Science and Engineering. The equipment is a Parr 1672 calorimetric bomb which includes a thermometer.

The amount of energy released by the biomass sample during its combustion and the condensation of the produced water is not straightly related to the specific heat of water and its variation in temperature. As a matter of fact, diverse factors should be taken into consideration and some parameters are related to the particular instrument used. In this case, the released heat is calculated according to Eq. (6.8)

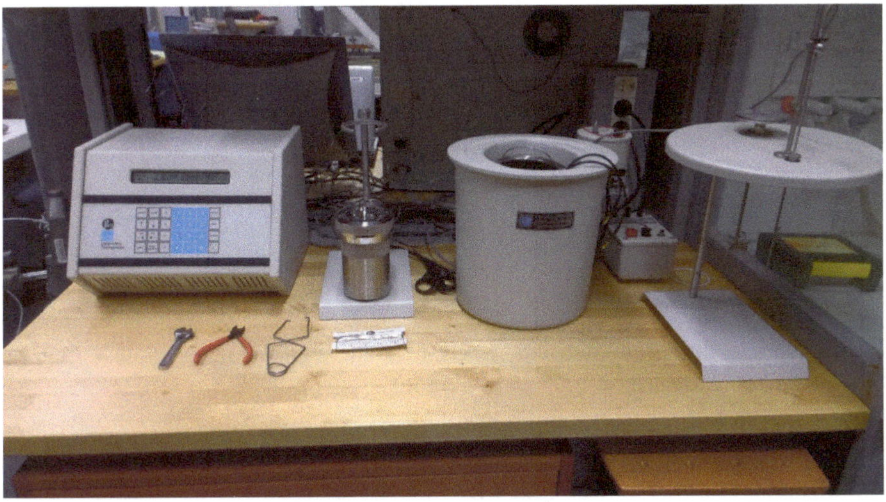

**Fig. 6.3** A laboratory setup for calorimetric measurements

$$\Delta H = m_{H_2O} \cdot cp_{H_2O} \cdot \Delta T + K \cdot \Delta T \tag{6.8}$$

where $m_{H_2O}$ is the mass of water which is present outside the calorimetric bomb and $K$ is an instrument-related constant. Remember that the value of $cp$, the specific heat of water, is not constant with the temperature; however, the amount of water is quite large compared with the mass of the fuel that we want to analyze; for this reason, the water temperature will be raised only of some degrees. If the temperature is not changing much, then it is a good approximation to keep it constant. In any case, there is the machine-related constant helping us.

The main parts of the calorimeter are:

- the bomb cylinder with sample holder;
- cooling water vessel;
- stirring device;
- ignition system;
- temperature measurement system (temperature increase measurement system).

The following factors affect the measurement results:

- the stirring device operation;
- the heat loss through external walls;
- the increase in temperature of metal components of the system (we take into consideration of this in the water equivalent coefficient);
- the presence of other combustible materials within the cylinder.

The higher heating value for the considered instrument is evaluated by:

$$\text{HHV} = \frac{(m_w + K)c_w \cdot \Delta t - \sum q_i}{m} \tag{6.9}$$

where

- $m_w$—mass of water in the vessel;
- $K = 0.441$ kg—water equivalent value of the calorimeter;
- $c_w$—water specific heat capacity;
- $\Delta t$—increase in water temperature throughout the experiment;
- $\Sigma q_i$—sum of correction factors for additional thermal effects;
- $m$—weight of the combusted fuel.

To be specific, some heat is required to combust the wire and thread which are used during the starting of the process according to:

$$\sum q_i = \text{HHV}_{wi} m_{wi} + \text{HHV}_t m_t \tag{6.10}$$

where

- $m_{wi}$—mass of burned wire;
- $m_t$—mass of combusted thread;

- $HHV_{wi} = 6700$ kJ/kg—higher heating value of wire;
- $HHV_{t} = 14{,}700$ kJ/kg—higher heating value of thread.

Specific heats, enthalpies, specific volumes, internal energies, and entropy values are tabulated for several compounds; a good reference for obtaining such data is for instance the Handbook of Chemical Engineers (Green and Perry 2008); additional data concerning water can be found in The IAPWS Formulation 1995 for the Thermodynamic Properties of Ordinary Water Substance for General and Scientific Use (Wagner and Pruß 2002).

Data for the specific heat can be obtained from enthalpy data, and the specific heat at 25 °C, for instance, can just be approximated by the enthalpy value at 25 °C minus the enthalpy at 24 °C. It is possible to derive the *cp* from these data since it will be possible to plot them, and with a suitable program, we can evaluate a relation for $cp = f(T)$.

The relation between the mass, the specific heat, and the temperature is given as a consequence:

$$\Delta H = mc_p\Delta T \tag{6.11}$$

where $m$ is in kg, $cp$ in (kJ/kg K), and $T$ in Kelvin, and $\Delta H$ is then the variation in enthalpy expressed in kJ.

Dividing by the weight of the sample, we have produced enthalpy per unit weight of sample. The lower heating value, LHV, is evaluated from the HHV as:

$$\text{LHV} = \text{HHV} - 2441.8\left(\frac{\text{kJ}}{\text{kg}}\right)(9\cdot\text{H} + \text{moisture fraction}) \tag{6.12}$$

In Eq. (6.12) the value 2441.8 (kJ/kg) refers to the heat of vaporization of water, while H is the fraction of hydrogen in the fuel sample expressed in weight.

We take into account the hydrogen which is present in the fuel since when the hydrogen reacts with oxygen it releases a big quantity of energy and forms water. Now, the water is not at the liquid state when the sample is combusted, but it is present in the form of vapor, and for this reason, some extra calories are required to keep this state. When the combustion is ended, the vapor will give back this amount of energy while it is condensing.

## 6.3 A Procedure for the HHV Calculation Based on a Particular Instrument

With reference to Fig. 6.4, the main parts of the overall calorimeter are listed as follows: thermometer (1), thermometer bracket (2), thermometer support washer (3), thermometer lens (4), thermometer support rod (5), motor (6), motor pulley (7), stirrer drive belt (8), stirrer pulley (9), stirrer bearing assembly (10), ignition wires (11), stirrer shaft and impeller (12), bucket (13), calorimeter jacket and cover (14), and oxygen combustion bomb or calorimetric embodiment (15). From now on, the part of the calorimeter where the combustion of the sample takes place will be denominated as calorimetric embodiment.

The main components constituting the calorimetric embodiment are shown in Fig. 6.5.

The assembly of the embodiment itself and the arrangements of the biomass sample within the calorimeter is made according to following steps:

a. The first step is to weight the capsule (where the sample will be positioned) and set the tare (Fig. 6.6).
b. Then, the sample should be put in the capsule and a note of the weight on the display should be taken (hint: it is a good idea to let the value stabilize since this measure should be quite accurate).
c. The capsule should be arranged in the ring of the calorimeter embodiment. This ring is pending from the top of the embodiment as shown in Fig. 6.7.
d. The next step is to cut 10 cm length of the ignition wire, using the ruler already drawn on the reel (Fig. 6.8).
e. The following operation is not that easy to perform and requires a certain manual capacity. The goal here is to combust the sample pellet, and, in order to do this, we must heat up the sample in some ways. This is done by means of the resistance wire which has to be in contact with the sample. At this point, the extremities of the wire should be inserted in holes which are positioned in top-lateral part

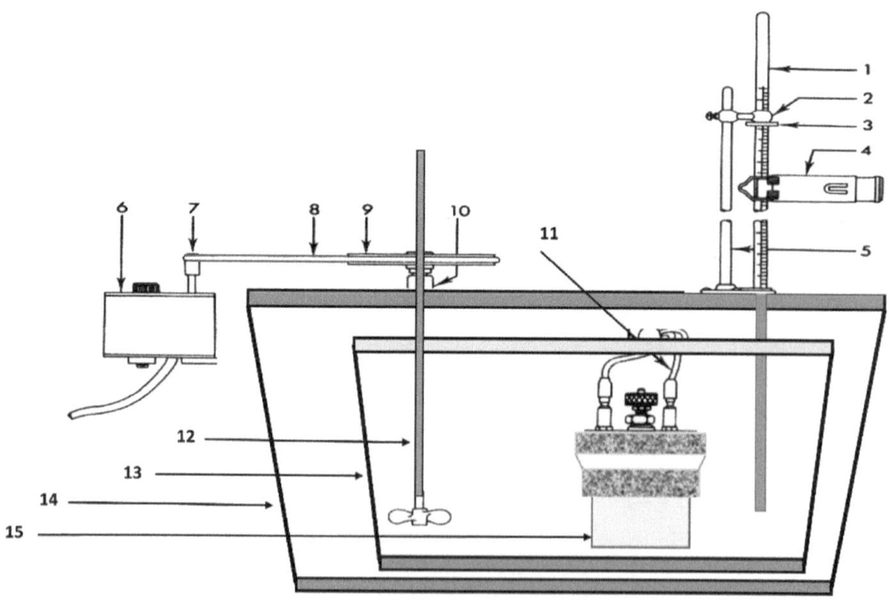


**Fig. 6.4** Main parts of the overall calorimeter

**Fig. 6.5** Main components of the calorimetric embodiment

**Fig. 6.6** Scale used for sample weight measurement

of the calorimetric embodiment (see Fig. 6.9) and, as mentioned previously, the wire should be shaped in such a way that it must be in contact with the pelletized fuel, but without touching the capsule (otherwise you will create a short circuit which will burn only the wire).

f. The next step is the closure of the calorimetric bomb. To do this, it has to be kept in mind that if there is some extra pressure into the container, the vessel could not close properly. Therefore, we should unscrew the knurled knob which is situated on the top of the vessel (this valve must be open to properly close the bomb). The top part of the embodiment should be carefully inserted on the cylinder and closed. The knob should be closed too (see Figs. 6.10 and 6.12).
g. At this point, the embodiment shown in Fig. 6.10 should be pressurized with oxygen. The concept here is that the wire will be brought to very high temperature and become incandescent. When these conditions are achieved, the oxygen will be combusted powerfully. The oxygen for instance could have a pressure of 30 bar, and in order to do this, a oxygen bottle should be connected to the top of the embodiment through the suitable one-way valve. Of course, the pressure is read by means of a pressure meter as the one given in Fig. 6.11.
h. At this stage, the calorimetric embodiment is ready (Fig. 6.12).

**Fig. 6.7** Positioning of the capsule inside the embodiment ring

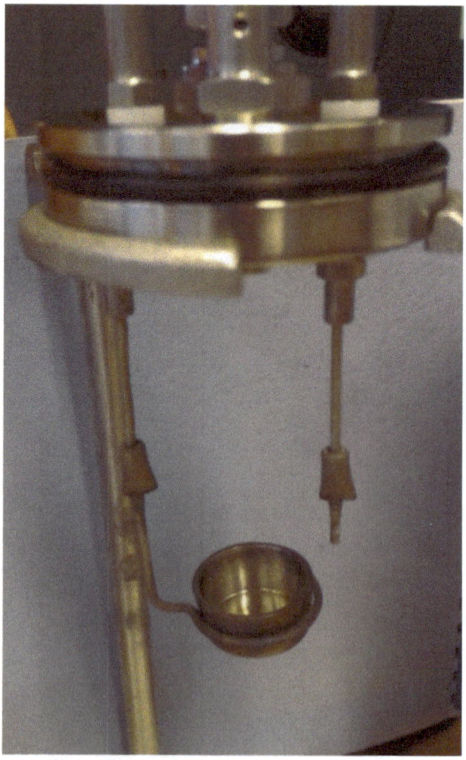

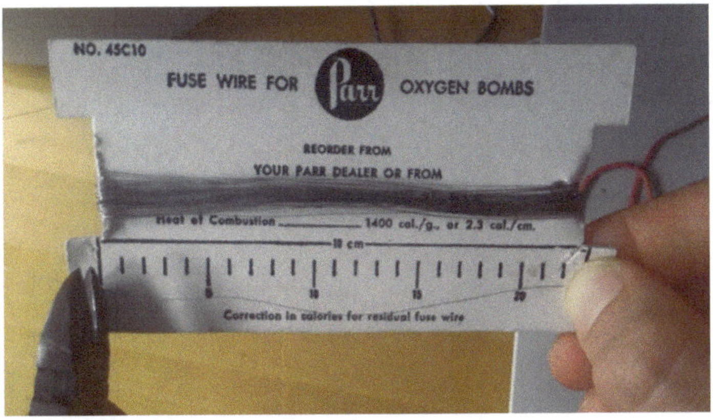


**Fig. 6.8** Cutting the ignition wire

**Fig. 6.9** Positioning of the ignition wire, to be noticed that the wire has to be in this kind of shape but of course has to contact the sample

Figure 6.13 shows the section where the calorimetric embodiment is placed. In this section, the water will "collect" the calories contained within the combusted biomass sample and its temperature will be raised accordingly. In this device, the rise in water temperature is observed, registered, and subsequently processed by the digital thermometer.

Continuing with the procedure for the utilization of the instrument, the following steps should be performed:

**Fig. 6.10** Oxygen filling of the calorimetric embodiment

i. Figure 6.14 shows the container where the calorimetric embodiment is placed and where water is filled. In this case, the procedure is to fill the container with exactly 2 L of distilled water. Of course, a tare should be taken on the empty container before filling it with water.
j. This container should be placed within the calorimetric jacket, and usually, there are some signs where the container should be fitted. This signs are also at the bottom of the jacket (Fig. 6.15).
k. The calorimetric embodiment shown in Fig. 6.16 should be, at this stage, inserted into the container filled with water. This operation should be done with care by using a lifting handle to be attached to two suitable holes at the top sides of the embodiment itself. The embodiment should be transported carefully to not disturb the sample and its arrangements. The embodiment is placed at the bottom of the water contained where suitable signs are present to allow for its correct positioning. The two ignition electrodes should be plugged into the terminal connections on the top of the calorimetric embodiment. Attention should be paid to not remove any water from the bucket during the whole procedure (Fig. 6.17).

When the sample is combusted into the calorimetric embodiment, the temperature of the water is measured. However, it is also true that the temperature should be uniform within the medium in order to have an accurate measure. For this reason, stirring is required. The procedure continues with the following actions:

l. The jacket should be covered and the cover should be positioned in the right way so that we have the stirring in the water. The stirring is done by means of a belt (see Fig. 6.18), and this belt is connected to an electric motor with a switch.

**Fig. 6.11** Oxygen bottle with pressure meters

Data acquisition and elaboration are performed by the data logger, Fig. 6.19, which eventually provides with the higher heating value of the sample of interest and corrects it from the influence of nitrogen and sulfur occasionally present in the fuel and from the amount of heat generated by the ignition wire.

In this case, the procedure continues as follows, to be noticed that this procedure refers to the particular instrument used:

m. If the assembly has been done in the correct way and depending on the instrument used, on the display of the data logger we could have a message of "READY", or we could confirm that we have finished the procedure.
n. The value of the sample mass previously weighted should be input.
o. At this point, the measurement has started and no actions are required. If everything has been assembled correctly, the process can take about 15 min before outputting the HHV value. During this time, the sample will pass through three "periods": the *preperiod*, during which the mixing makes the water temperature in the water container homogeneous; the *rise period* starts when the sample is

Fig. 6.12 A top view of the calorimetric embodiment closed and ready

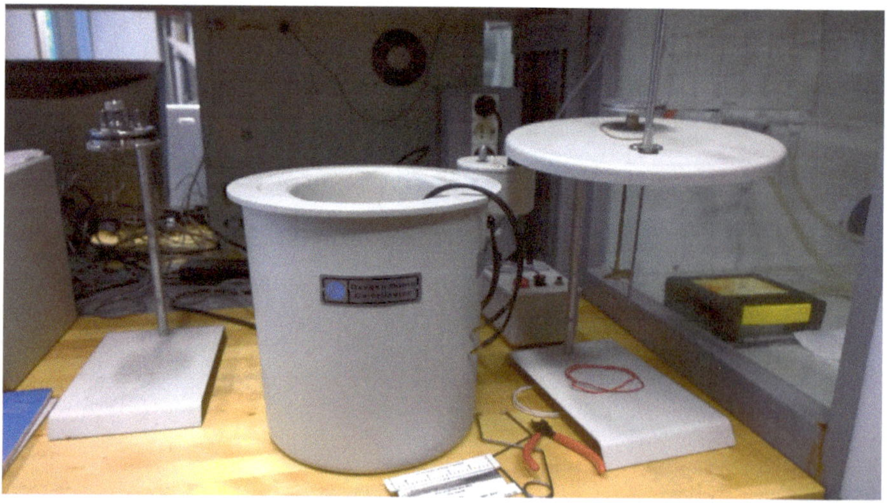

Fig. 6.13 Calorimeter jacket and cover

**Fig. 6.14** Water container

**Fig. 6.15** Positioning of the water container

**Fig. 6.16** Handling of the calorimetric embodiment by suitable handle

combusted; the *post-period* starts when the temperature stops rising and will carry on till the temperature becomes constant.

p. The machine will advise with a message or sound when the measurement is ended, and the HHV value will be displayed.

The calorimeter could be programmed to perform the calculation of the gross calorific value in accordance to Eq. (6.13).

$$\mathrm{HHV} = \frac{W \cdot T - e_1 - e_2 - e_3}{m} \tag{6.13}$$

where

HHV Higher heating value (J/kg);
$W$ the energy equivalent of the calorimeter, the amount of energy to rise 1 °C the calorimeter $T$ (J/°C);
$e_1$ the heat produced by burning the nitrogen entrapped in the bomb to form nitric acid (J);

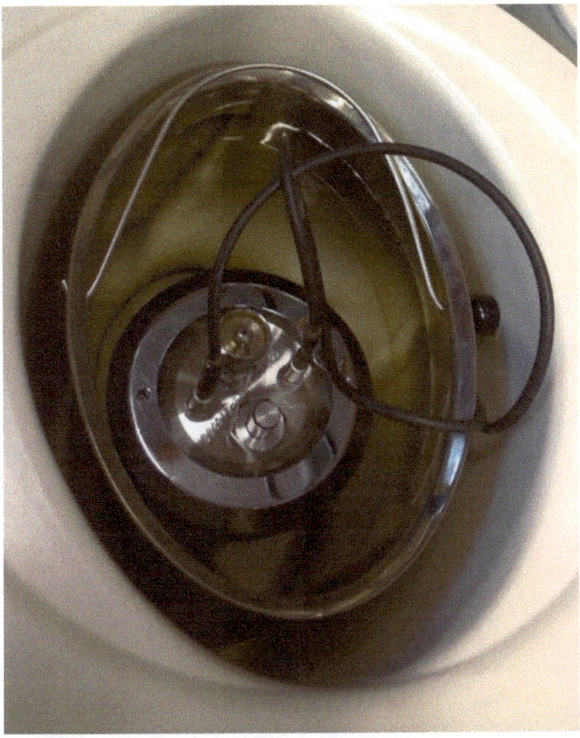

**Fig. 6.17** Positioning of the calorimetric embodiment inside the water container and connection of the cables

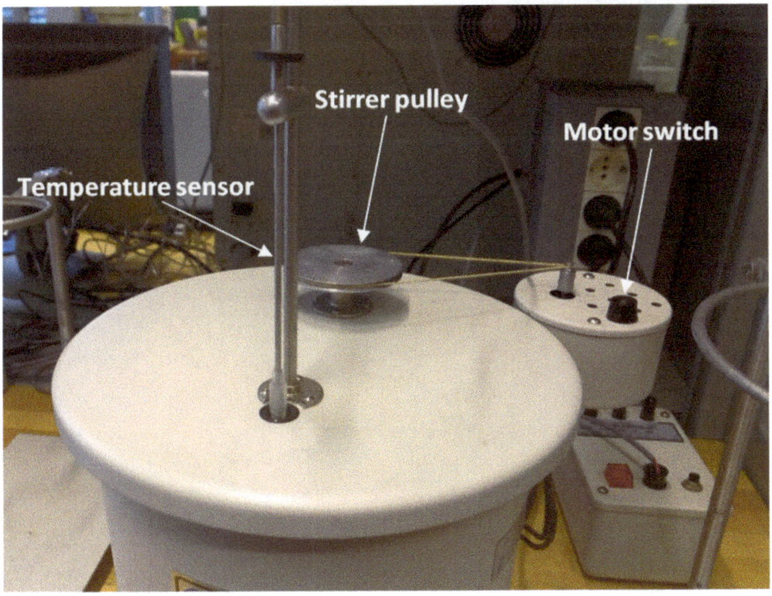


**Fig. 6.18** Positioning of the jacket cover and the elastic belt

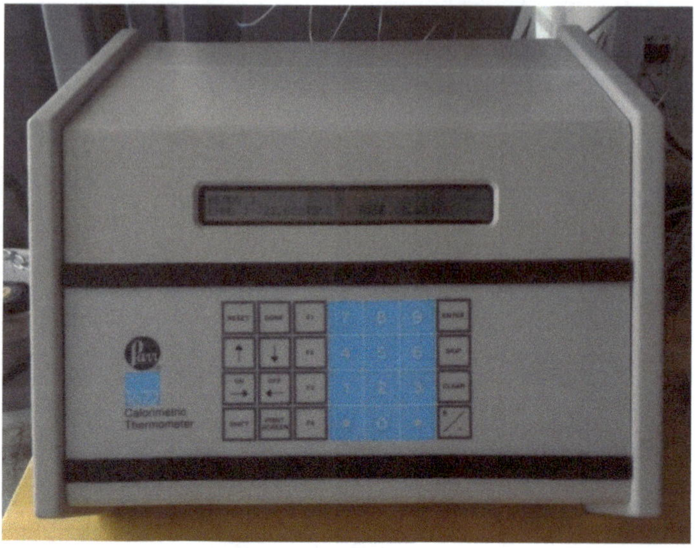


**Fig. 6.19** Data logger

$e_2$ the extra heat produced by burning sulfur to sulfur trioxides and forming sulfuric acid instead of sulfur dioxide (J);
$e_3$ the heat produced by the burning fuse wire (J);
$m$ the mass of the sample (g).

The influence of the fuse could be estimated by considering the length of the left-over wire. We assume in this case that the disappeared part was actually combusted, and the following calculation steps will be done:

$$E_{tot} = HHV_{meas} \cdot m_{sample} \tag{6.14}$$

$$E_{fuse} = HHV_{fuse} \cdot l_{burnt\ fuse} \cdot 4.186 \times 10^{-6} \tag{6.15}$$

$$E_{sample} = E_{tot} - E_{fuse} \tag{6.16}$$

$$HHV_{sample} = \frac{E_{sample}}{m_{sample}} \tag{6.17}$$

where
The term $4.186 \times 10^{-6}$ is a conversion factor.

$m_{sample}$ is the mass of the sample (kg);
$l_{burnt\ fuse}$ is the length of the burnt wire (cm);
$E_{tot}$ is the total energy developed by the combustion (MJ);

$E_{fuse}$ is the energy developed by the combustion of the only fuse wire (MJ);
$E_{sample}$ is the energy developed by the combustion of the only sample (MJ);
$HHV_{meas}$ is the measured higher heating value, that is, the amount of energy generated by the combustion of all substances inside the bomb, related to the mass of sample ($MJ_{tot}/kg_{sample}$);
$HHV_{fuse}$ is the higher heating value of the nickel/chromium wire as indicated on the reel. In this case, this is equal to 2.3 ($cal_{fuse}/cm$);
$HHV_{sample}$ is the higher heating value of the sample only ($MJ_{sample}/kg_{sample}$).

The chlorine content determination is done in a similar way to the determination of the higher heating value. The method takes into account the combustion of a test sample in an oxygen bomb calorimeter (same as the HHV determination), and then:

- The solution is rinsed with distilled water.
- A solution which contains chloride ions is obtained.
- The solution is reacting with silver nitrate to form silver chloride and then recovered using a precipitation technique.

The reaction is expressed as follows:

$$Cl^- + Ag^+(aq) \rightarrow AgCl\,(s) \tag{6.18}$$

## References

Bradley, H. C., & Nichols, M. S. (1918). Nitrogen content of bacterial cells. *Journal of Biological Chemistry, 33*, 525. Retrieved from http://www.jbc.org/content/33/3/525.full.pdf.

Green, D. W., & Perry, R. H. (2008). *Perry's chemical engineers' handbook* (8th ed.). McGraw-Hill. Retrieved from https://doi.org/10.1036/0071422943.

Keppy, N. K., & Allen, M. W. (2018). *The biuret method for the determination of total protein using an evolution array 8-position cell changer* (pp. 2). Thermo Fisher Scientific.

Kjeldahl, J. (1883). Neue Methode zur Bestimmung des Stickstoffs in organischen Körpern. *Zeitschrift für Analytische Chemie, 22*(1), 366–383.

Mæhre, H. K., Dalheim, L., Edvinsen, G. K., Elvevoll, E. O., & Jensen, I. -J. (2018). Protein determination—Method matters. *Foods, 7*(1). https://doi.org/10.3390/foods7010005.

Mariotti, F., Tomé, D., & Mirand, P. P. (2008). Converting nitrogen into protein—Beyond 6.25 and Jones' factors. *Critical Reviews in Food Science and Nutrition, 48*(2), 177–184. https://doi.org/10.1080/10408390701279749.

Martín, J., Fernandez Sarria, L., & Asuero, A. G. (2017). The Kjeldahl titrimetric finish: On the ammonia titration trapping in boric acid. In *Advances in titration techniques*. https://doi.org/10.5772/intechopen.68826.

Milo, R. (2013). What is the total number of protein molecules per cell volume? A call to rethink some published values. *BioEssays, 35*(12), 1050–1055. https://doi.org/10.1002/bies.201300066.

Wagner, W., & Pruß, A. (2002). The IAPWS formulation 1995 for the thermodynamic properties of ordinary water substance for general and scientific use. *Journal of Physical and Chemical Reference Data, 31*(2), 387–535. https://doi.org/10.1063/1.1461829.

# Chapter 7
# Thermogravimetric Analysis (TGA)

*When Alexander the Great (the king of the Greek kingdom of Macedon) asked to Diogenes of Sinope: "what I can do for you?"*
*Diogene answered: "Stand aside so I can get some sun".*
Free translation from: Epictetus, The Discourses, Book 3, 108 AD.

## 7.1 Basic Principles of Thermogravimetric Analysis

Thermogravimetric analysis (TGA) is often used within studies of thermal decomposition of biomass. For instance, with this method we could study the pyrolysis of substances at diverse conditions and heating rates. The instrumental measure in some cases can give the so-called differential thermal gravimetric (DTG) thermogram from which assumptions are done on the reaction rates and the rates of the diverse steps in the thermal treatment. To give one example, the pyrolysis of some organic substances and wastes like manures is studied by this method (Chong et al. 2019) or, for instance, the decomposition of micro-algae (Xu et al. 2017).

The thermogravimetric method is obviously belonging to the so-called thermal analysis, and this is because the sample is treated at diverse temperatures and with different atmospheres. The method allows for a simultaneous measurement of temperature, time, and mass of a sample in a controlled dynamic atmosphere (Fig. 7.1).

The foundation on the basis of these kinds of measures is the variation of weight of the sample as a consequence of the thermal treatment. Because of the process, some mass will be lost in terms of volatiles or decomposition and this weight loss is measured normally by means of a microbalance. The sample is inserted in a "sample pan" or container which is specifically designed for this kind of measure. Concerning the temperature conditions and its measurement, the temperature variation is set according to a customized temperature program which may include isothermal and ramp steps with different heating rates, while the temperature is measured with thermocouples which are in contact with the sample container (Fig. 7.2).

The pan is usually positioned on a sample holder, which is connected with the mass sensitive element (the microbalance). The TG system (sample, sample holder,



C. De Blasio, *Fundamentals of Biofuels Engineering and Technology*,
Green Energy and Technology, https://doi.org/10.1007/978-3-030-11599-9_7

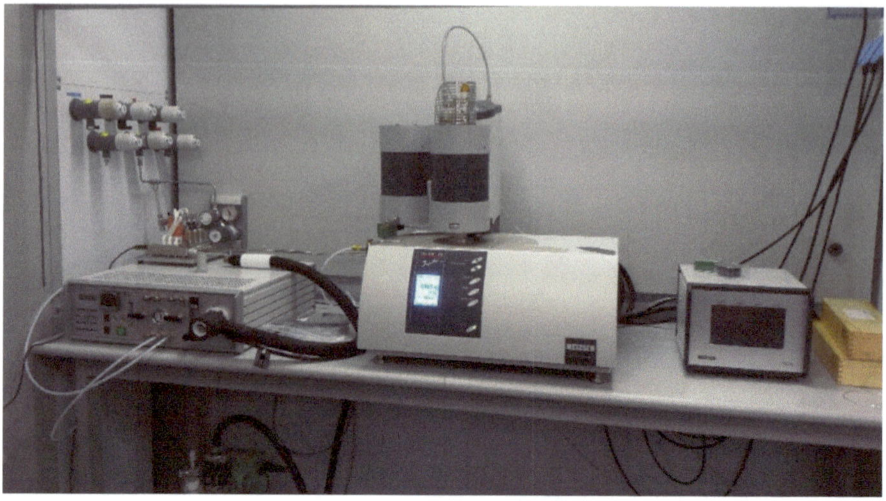

**Fig. 7.1** Example of thermogravimetric analyzer. Curtesy of Prof. Antonio Galvagno and Mauro Prestipino at University of Messina, Italy

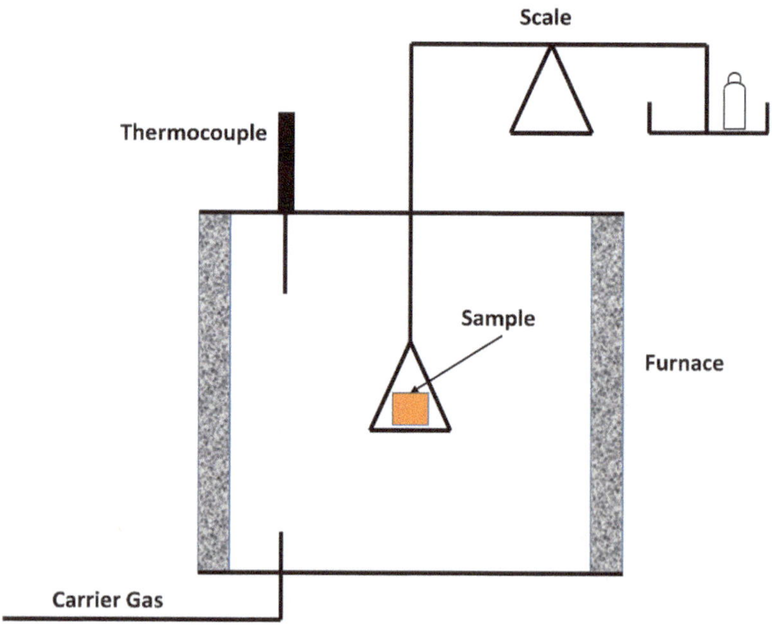


**Fig. 7.2** Simple representation of the thermogravimetric method

pan) is heated using an electric furnace, which in particular cases can reach up to 2000 °C, depending on the specific materials and design of the furnace and other components. The sample pans that are used in a TGA device can be of different shapes and materials. They should be able to safely store the sample, to not react with the sample, and suitable for the desired temperature. The most common sample pans are made of alumina, platinum, and aluminum.

## 7.2 TGA Classification and Types of Equipment

The TGA devices can be grouped in three main types, depending on the design of the sample holder-microbalance system, as shown in Fig. 7.3: cantilever (a), hang down (b), and top-loading column (c).

The *hang down* and the *top-loading* systems are different with respect to the position of the scale. In the first case, the sample is hanged and the scale is on top of it, and in the second case, the scale is at the bottom. Both methods take into account the gravity force. The devices with two sample holders are used for differential thermal analysis with reference materials, as discussed later in this section.

During a TGA, the wide temperature variation at which the gas may be subjected involves a variation in the purge gas density that cannot be neglected. This effect is denominated as buoyancy effect and can be observed as an apparent mass variation, whose magnitude depends on the kind of the TG apparatus. As a matter of fact, this phenomenon depends on the volume of the sample and the density of the gas which is forming the gas atmosphere into the chamber. In the numerical processing of the TGA curves, this phenomenon is often hidden or not taken into consideration (Auroux 2013).

For these reasons, some of the trials require a blank test (i.e., with the empty sample pan) for each TGA program, in order to make the correction curve that will be subtracted from the sample's test. A TGA program is the combination of the different temperature variation steps, characterized by specific heating rates and gas

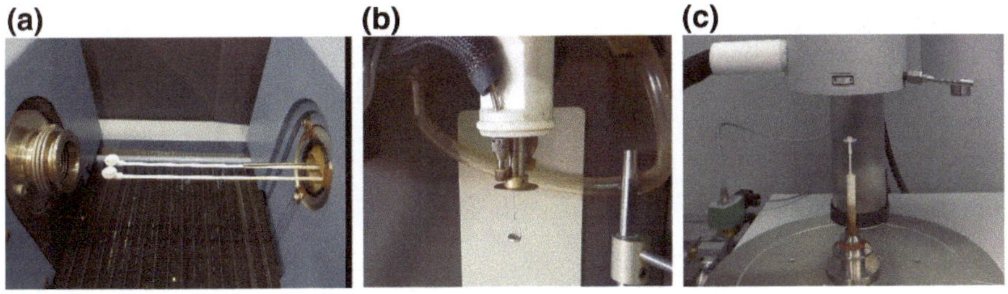


**Fig. 7.3** Three of the most common TGA configurations: cantilever (**a**), hang down (**b**), and top-loading column (**c**)

flow. When a variation occurs from a test to another, a new blank test should be performed.

This method of analysis can be coupled with other methods like differential thermal analysis (DTA) and differential scanning calorimetry (DSC). This is usually done when the composition of the sample is not known (Van Humbeeck 1998).

## 7.3 Procedure and Sample Analysis

The TGA apparatus is usually loaded with a sample mass in the range 2–20 mg, depending on the density of the sample, the sample pan size, and the purpose of the analysis. Concerning the physical properties of the sample, these can affect the TG curves. For instance, in the case of an exothermic reaction, higher mass loading may lead to higher onset temperature because of the higher heat released by the sample, which involves a sudden local temperature increase. Hence, considering the thermal inertia of the system and the fact that the thermocouple is positioned in close contact with the sample, the detected temperature may apparently result higher than the one observed during tests with lower mass. The magnitude of this effect is correlated to the physical properties of the sample, the sample pan, and the amount of heat released or absorbed during the sample transformation. A similar effect is noticeable with different sample's sizes and shapes. Powders, flakes, and blocks have different effects on heat diffusion, which can lead to an apparent retard or anticipation of the onset temperature or time. Furthermore, in the case of powders, significant differences in the particle size between two runs may affect the repeatability of the results. To give one example, the results of thermogravimetric analysis has been reported to be highly dependent on the sample size for particular samples (Stawski 2009), while also the gas flow and the geometry of the pan can affect the results.

## 7.4 Positive and Negative Mass Variation

The thermogravimetric analysis is used for characterizing the thermal behavior of materials during thermal treatment in either inert or reactive atmosphere. In a reactive atmosphere, it is also possible to study the reactions that occur between the sample and the purge gas if it involves mass variation (increase or decrease). Hence, it is possible to classify the thermal processes that occur in a TGA device according to the positive or negative variation of mass as follows:

Positive mass variation:

- adsorption;
- oxidation;
- reduction.

Negative mass variation:

- desorption;
- thermal decomposition (with volatiles formation);
- oxidation (combustion);
- vaporization;
- sublimation.

From the list above, another classification can be made between thermal and thermochemical processes, depending on the capacity of the purge gas to react with the sample in the temperature range of the test.

## 7.5 Reporting TGA Results

As mentioned above, TGA results are plotted as mass versus temperature or time. Another kind of representation of the results is the derivative thermogravimetry (DTG), which is the first derivative of the TGA curve. Hence, it describes the mass change rate. DTG is also used to detect more subtle effects or when kinetic parameters should be investigated (Gallagher 1998). Since the TGA signal is reported in mg or % (100% is the initial mass at time zero), the DTG is usually plotted as mg/min or %/min. Figure 7.4 shows the plot of both TGA and DTG curves obtained with a thermogravimetric test of a biomass sample under nitrogen atmosphere and constant heating rate up to 750 °C. The TGA curve reported in figure describes the thermal

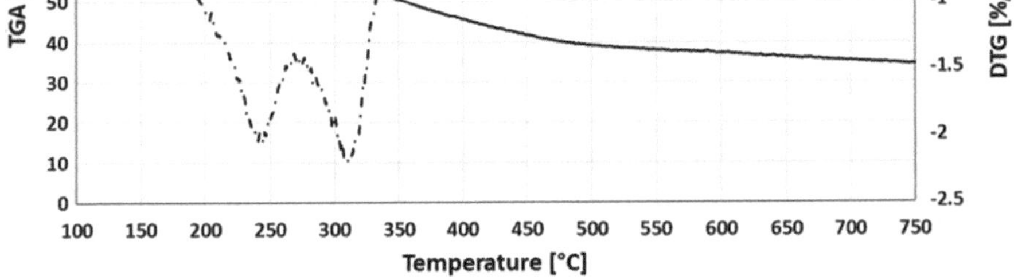


**Fig. 7.4** TGA (solid line) and DTG (dashed line) signals of biomass conversion in inert atmosphere

decomposition of biomass following different steps that occur at a different temperature. From a TGA plot, a constant slope characterizes the different steps of thermal decomposition. In the case of Fig. 7.4, these steps are hardly detectable from the different slopes of the curve, because they occur in overlapped temperature ranges. The DTG plot helps the detection of transformations that overlap or are associated with poor mass loss. Indeed, the peaks of DTG indicate a specific transformation, and the minimum values coincide with the maximum rate of mass loss of the related decomposition step.

## 7.6 Differential Thermal Analysis and Differential Scanning Calorimetry

The TGA can be combined with differential scanning calorimetry (DSC) or differential thermal analysis (DTA) in order to carry out a complete analysis of the thermal behavior of materials with one equipment. Indeed, in addition to the mass variation, it is possible to provide quantitative and qualitative information on the endothermic and exothermic phenomena. Both DTA and DSC are based on the comparison of the different thermal behavior between the reference material and the sample (Pope and Judd 1997).

The DTA measures the temperature difference between the reference and the sample when they are subjected to the same heat flux. As it is for the TGA, the differential temperature is plotted against time or temperature. Any change in the sample that involves heat release (exothermic process) also involves the detection of higher temperature in the sample than the reference. The opposite happens when the changes in the sample cause heat absorption (endothermic process). An important observation for the correct interpretation of the DTA and DSC data is that the differences in heat capacity and thermal conductivity between the sample and the reference may cause variation of the differential temperature as well, even if no transformations are occurring. It follows that these thermal analytical techniques allow measuring the heat capacity of materials. Furthermore, they allow for the detection of those transitions that do not involve heat release or adsorption but imply the variation of thermal properties of the sample, as in the case of some solid-phase changes.

The main difference between the DSC and DTA lies in the method used for calculating the different behavior between the inert reference material and the sample. Indeed, the DSC records the energy that is necessary to provide to the sample or the reference in order to determine a zero temperature difference between them. A DSC device has two individual heaters, one for the sample and one for the reference, and two individual temperature sensors. This method is also called the power compensation principle. Some devices are claimed as DSC even if they have a single heat source. In this case, they should be defined as false DSC, because the principle is closer to a DTA since the energy needed for keeping zero temperature is calculated and not directly recorded by the heating unit, which is shared by both sample and

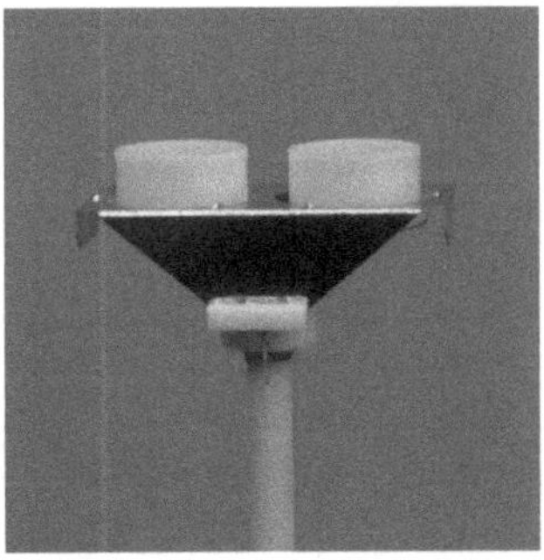

**Fig. 7.5** TGA/DTA sample holder of a top-loading column TGA/DTA device

reference. The latter method is also called heat flow DSC. It, therefore, follows that, in many cases, combined TGA-DSC devices do not have real DSC. In principle, the two DTA and DSC are equivalent for qualitative analysis, with the difference that DSC gives better performance at a low heating rate.

Furthermore, because of the materials used for accurate measurement, most of power compensation DSC devices can run in an inert atmosphere and at a lower temperature than DTA (600–750 °C). However, for accurate quantitative analysis power compensation DSC should be recommended. Figure 7.5 shows the TG/DTA sample holder of a top-loading sample holder with both reference and sample pans.

The study of biomass thermal behavior through TGA provides information about its composition and reactivity. Indeed, several characteristics can be determined, such as moisture content, fixed carbon, volatile matter, and ash (Basu 2013), making TGA a reliable and rapid method for performing the proximate analysis of biomass. Some authors developed methodologies for biomass characterization based on thermogravimetry, which allows calculating the content of the main pseudo-components of biomass, i.e., hemicellulose, cellulose, and lignin, as well as the proximate analysis, with a single thermogravimetric test (Saldarriaga et al. 2015). These methods combine TGA, deconvolution of the DTG signals, and empirical correlations.

## 7.7 TGA and the Study of the Kinetics of Diverse Thermal Processes

The correlation between the mass variation and time in TGA signals allows studying the kinetic of thermochemical conversion under different atmospheres, which means it is possible to investigate the kinetics of torrefaction, pyrolysis, gasification, and combustion. The knowledge of the kinetic behavior of biomass conversion is of primary importance in order to optimize the design of the reactors and the operating conditions (residence time, temperature, pressure, oxidant medium flows).

The thermogravimetric experiments for carrying out the kinetic studies can be performed according to isothermal, linear, or stepwise linear heating programs. With such analysis, it is possible to determine the evolution of conversion rates as a function of the conversion's level, temperature, or time. Also, it is possible to extrapolate the kinetic parameters of biomass conversion by performing TGA at different partial pressures (only in case of reactive purge gas and by fixing the temperature) and different temperatures (fixing the partial pressure of the reactive gas), for isothermal runs, or different heating rates for non-isothermal runs. The details of the different approaches that can be applied to calculating the kinetic parameters are widely described in the specific literature (Vyazovkin et al. 2011). Some authors suggested alternative methods that allow computing the activating energy or reaction order with a single run by repeatedly shifting the temperature or the gas partial pressure between two predetermined levels, respectively (Zimbardi 2000). Additional research works showed the possibility to use some empirical correlations between the conversion models and the inorganics embodied in biomass, as well as between the kinetic constant and the inorganics, in order to predict the conversion profiles and kinetics of thermogravimetric processes under specific reacting environment (Dupont et al. 2011; Prestipino et al. 2018).

## 7.8 TGA Coupled with Mass Spectrometry, Gas Chromatography, and Infrared Spectroscopy

The study of the thermochemical behavior of biomass through TGA can be improved by coupling the TG devices with the analysis of the evolved gas. This upgrade can enhance the outcomes of the tests by providing the variation of the gas evolution from the sample as a function of time and temperature, correlated to the corresponding mass variation. The most common techniques for the evolution gas analysis coupled with TGA are mass spectrometry (MS) and infrared spectroscopy (IR). In some cases, gas chromatography combined with mass spectroscopy (GC/MS) is also used for a more comprehensive characterization (Cai et al. 2018). TG-FTIR and GC/MS can be used together to study the mechanisms involved in the thermal treatment of the sample; for instance, they can be coupled to study the mechanisms of pyrolysis (Xu et al. 2018). These solutions are usually said to be hyphenated, and they can be even used when we want to know the composition of an aqueous sample (Garavaglia 2012). Hyphenated techniques are applied to study diverse kinds of samples, and for different purposes, they usually include two or more analytical techniques which are coupled together online (Niessen 2017).

The kinetic studies of pyrolysis and torrefaction are usually carried out in nitrogen, argon, or helium flow in order to simulate endothermic processes in the absence of oxygen. With the most common TG devices, the maximum heating rate for reliable results is 50 K/min (Cai et al. 2018), even though most of the manufacturers state that the devices can work up to 100 K/min. This poses some limitations in scaling up the results of the TGA pyrolysis to real-scale processes, which are often performed under fast and flash conditions (Basu 2013).

During thermochemical gasification, biomass undergoes to various steps, such as drying, pyrolysis, gasification, and partial combustion, which are characterized of different temperatures. After pyrolysis, the solid biomass is converted into char. The latter is the actual solid reactant that is subjected to the partial oxidation and the different heterogeneous reactions with the gasification mediums and the products of combustion. Hence, the kinetics of gasification and combustion is studied using the char of parent biomass. The gasification step is also the slowest one, which involves its reactions being the controlling ones and the most important in the design of the whole gasification process reactors. For the kinetics of combustion, different oxygen partial pressures are adopted, while in the study of gasification kinetics, char is subject to a reacting environment consisting of carbon dioxide or steam at different partial pressures, using an inert gas as the complement. When steam is used as gasification medium in combination with air in a real gasification process, the thermogravimetric characterization with steam is of primary importance to understand the kinetic behavior. Indeed, in steam or air-steam gasification, the heterogeneous steam–carbon reaction is the most relevant one. The reaction between carbon dioxide and carbon, with the formation of carbon monoxide (reverse Boudouard), should always be studied, primarily when the air is used as gasification medium since it is the most limiting one, even in the case of air-steam gasification. However, at temperatures lower than 750 °C and high steam partial pressures, the kinetics of steam-char is still the most relevant for the design of the reactor and the proper modeling of the process.

The importance of studying the conversion kinetics of biomass is evidenced in Fig. 7.6. The figure shows the isothermal runs of different residual chars in a steam environment with a TG device, where the conversion ($X$) is plotted as a function of time. In Fig. 7.6, we have results for the olive pomace char (OPChar), citrus peel char (CPChar), and grape pomace char (GPChar).

The conversion expresses the fraction of mass loss compared to the final theoretical mass of complete conversion ($X = 1$), as expressed in Eq. (7.1)

$$X = \frac{m_{\mathrm{i}} - m_t}{m_{\mathrm{i}} - m_{\mathrm{f}}} \tag{7.1}$$

where $m_{\mathrm{i}}$, $m_t$, and $m_{\mathrm{f}}$ are the initial mass, the mass at time $t$, and the theoretical final mass, respectively.

From Fig. 7.6, it is possible to observe that using different parent biomass may lead to very different degrees of conversion at a fixed time $t$. Indeed OPChar achieves only 55% of conversion after 1000 s, at 750 °C and 50 kPa of steam partial pressure, while CPChar and GPChar are almost entirely converted. The different behavior is attributed to the different composition of the inorganics, which affect the kinetic of steam gasification of chars (Dupont et al. 2011; Prestipino et al. 2018). It follows that performing a thermogravimetric run at proper conditions it is a rapid and straightforward method that allows understanding the affinity of different feedstock to be converted at the same conditions, in addition to the possibility of providing preliminary information on the proper process parameters.

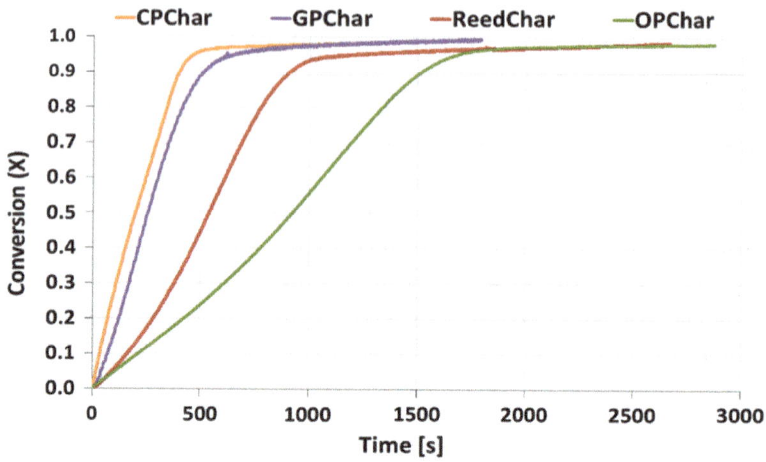


**Fig. 7.6** Conversion versus time curves of different residual chars tested through a TGA in 50 kPa steam atmosphere at 750 °C and isothermal conditions. Data from Prestipino et al. (2018)

## 7.9 Methods and Experimental Conditions Suggested

When performing TGA for studies of kinetics related to gasification and combustion of chars, some precautions should be taken into account in order to avoid diffusion limitation and keeping kinetic controlled conditions. To this regards, the temperature should be <900 °C and the particle size <0.1 mm (Di Blasi 2009). As particle size increases, the diffusion of gas reactants into the char is limited, which leads to the diffusion-controlled regimes. The heat transfer plays an essential role in the transition between kinetic and diffusion-controlled regimes. Indeed, higher temperature involves higher heat and mass transfer, moving away from the kinetic regime. Although using TGA it is possible to get very accurate data for the kinetics of biomass conversion, the difference between the particles shape and size in the laboratory analysis and those in a real-scale reactors poses the problem of addressing the mass and heat transfer of real-scale biomass particles. This limitation can be partially overcome by using macro-TGA devices that allow for testing biomass samples in the mass range 30–200 g where heat and mass transfer limitations can be taken into account. In some cases, the reactor itself can be used as a macro-TG if it is continuously weighted (Onsree and Tippayawong 2017). The drawbacks of using macro-TGA are the loss of resolution and precision in the study of the real reactions' kinetics.

## References

Auroux, A. (Ed.). (2013). *Calorimetry and thermal methods in catalysis*. Berlin, Heidelberg: Springer. Retrieved from www.springer.com/gp/book/9783642119538.

Basu, P. (2013). *Biomass gasification, pyrolysis and torrefaction—Practical design and theory* (2nd ed.). Oxford, UK: Academic Press. Retrieved from https://doi.org/10.1016/c2011-0-07564-6.

Cai, J., Xu, D., Dong, Z., Yu, X., Yang, Y., Banks, S. W., et al. (2018). Processing thermogravimetric analysis data for isoconversional kinetic analysis of lignocellulosic biomass pyrolysis: Case study of corn stalk. *Renewable and Sustainable Energy Reviews, 82,* 2705–2715. https://doi.org/10.1016/j.rser.2017.09.113.

Chong, C. T., Mong, G. R., Ng, J. -H., Chong, W. W. F., Ani, F. N., Lam, S. S., et al. (2019). Pyrolysis characteristics and kinetic studies of horse manure using thermogravimetric analysis. *Energy Conversion and Management, 180,* 1260–1267. https://doi.org/10.1016/j.enconman.2018.11.071.

Di Blasi, C. (2009). Combustion and gasification rates of lignocellulosic chars. *Progress in Energy and Combustion Science, 35*(2), 121–140. https://doi.org/10.1016/j.pecs.2008.08.001.

Dupont, C., Nocquet, T., Da Costa, J. A., & Verne-Tournon, C. (2011). Kinetic modelling of steam gasification of various woody biomass chars: Influence of inorganic elements. *Bioresource Technology, 102*(20), 9743–9748. https://doi.org/10.1016/j.biortech.2011.07.016.

Gallagher, P. K. (1998). Thermogravimetry and thermomagnetometry, Chap. 4. In M. E. Brown (Ed.), *Handbook of thermal analysis and calorimetry* (Vol. 1, pp. 225–278). Elsevier Science B.V. https://doi.org/10.1016/S1573-4374(98)80007-1.

Garavaglia, M. (2012). *Analysis of an unknown aqueous sample by TG-IR-GC/MS*. PerkinElmer. Retrieved from http://photos.labwrench.com/equipmentManuals/14602-5645.pdf.

Niessen, W. M. A. (2017). Hyphenated techniques, applications of in mass spectrometry. In J. C. Lindon, G. E. Tranter, & D. W. Koppenaal (Eds.), *Encyclopedia of spectroscopy and spectrometry* (3rd ed., pp. 174–180). Oxford: Academic Press. https://doi.org/10.1016/B978-0-12-409547-2.05240-9.

Onsree, T., & Tippayawong, N. (2017). Application of Gaussian smoothing technique in evaluation of biomass pyrolysis kinetics in macro-TGA. *Energy Procedia, 138,* 778–783. https://doi.org/10.1016/j.egypro.2017.10.059.

Pope, M. I., & Judd, M. D. (1997). *Differential thermal analysis—A guide to the technique and its applications*. Heyden.

Prestipino, M., Galvagno, A., Karlström, O., & Brink, A. (2018). Energy conversion of agricultural biomass char: Steam gasification kinetics. *Energy, 161,* 1055–1063. https://doi.org/10.1016/j.energy.2018.07.205.

Saldarriaga, J. F., Aguado, R., Pablos, A., Amutio, M., Olazar, M., & Bilbao, J. (2015). Fast characterization of biomass fuels by thermogravimetric analysis (TGA). *Fuel, 140,* 744–751. https://doi.org/10.1016/j.fuel.2014.10.024.

Stawski, D. (2009). The effect of sample weight in thermogravimetric analysis of low viscosity polypropylene in air atmosphere. *Polymer Testing, 28*(2), 223–225. https://doi.org/10.1016/j.polymertesting.2008.12.003.

Van Humbeeck, J. (1998). Simultaneous thermal analysis, Chap. 11. In M. E. Brown (Ed.), *Handbook of thermal analysis and calorimetry* (Vol. 1, pp. 497–508). Elsevier Science B.V. https://doi.org/10.1016/S1573-4374(98)80014-9.

Vyazovkin, S., Burnham, A. K., Criado, J. M., Pérez-Maqueda, L. A., Popescu, C., & Sbirrazzuoli, N. (2011). ICTAC kinetics committee recommendations for performing kinetic computations on thermal analysis data. *Thermochimica Acta, 520*(1), 1–19. https://doi.org/10.1016/j.tca.2011.03.034.

Xu, F., Wang, B., Yang, D., Ming, X., Jiang, Y., Hao, J., et al. (2018). TG-FTIR and Py-GC/MS study on pyrolysis mechanism and products distribution of waste bicycle tire. *Energy Conversion and Management, 175,* 288–297. https://doi.org/10.1016/j.enconman.2018.09.013.

Xu, Q., Ling, C., & Li, J. (2017). Microalgae decomposition in $CO_2$ atmospheres by thermogravimetric analysis. *Energy Procedia, 123,* 381–386. https://doi.org/10.1016/j.egypro.2017.07.273.
Zimbardi, F. (2000). Evaluation of reaction order and activation energy of char combustion by shift technique. *Combustion Science and Technology, 156*(1), 251–269. https://doi.org/10.1080/00102200008947305.

# Chapter 8
# Chromatography

*We consider a body to be in motion in case it changes its relative position with respect to a group of bodies which is considered to be at rest. But as all bodies, even the ones which appear in the most perfect rest, may be in motion.*

Free translation from Mecanique Celeste, Pierre Simon Marquis de Laplace. 1799–1825.

## 8.1 Principles of Chromatography

The basic principles of chromatography are given here with a particular focus on gas chromatography. Gas and liquid chromatography have many things in common, and they are widely described in the literature (Rouessac and Rouessac 2007); however, in this section the focus is given on the main parameters of the derived chromatogram for pedagogical reasons. Gas chromatography is used for analytical purposes, and most commonly refers to the concentration measurement of gas mixtures. The gas to be analyzed is sent to a chromatographic column which is constituted by a stationary phase and a mobile phase. The substance behaves differently depending on the different affinities of each substance in the mixture with the phases (Fig. 8.1).

The chromatographic instrument is simply described to be constituted by an oven which should be thermostable. Within the oven, there is the chromatographic column. The column is formed commonly by a thin glass capillary tube which can be some meters long. In the tube, a thin layer of the fixed phase is fixed on the internal walls. The substance forming the fixed layer on the tubes of the column should have a certain degree of affinity with the gas mixture, and this is because in this column we want to separate the single components before arriving to the sensor section. There is a mobile carrier which is used to transport the mixture into the capillary column. This gas usually is inert, and hydrogen, $H_2$; nitrogen, $N_2$; and helium, He are used (Rouessac and Rouessac 2007; Skoog et al. 2006). The sample is inserted through an injector. Figure 8.2 shows a simple drawing with the components of the chromatographic column.

The basic functioning principle is described as follows: There is a certain affinity between the components of the gas mixture to be analyzed and the stationary phase



C. De Blasio, *Fundamentals of Biofuels Engineering and Technology*,
Green Energy and Technology, https://doi.org/10.1007/978-3-030-11599-9_8

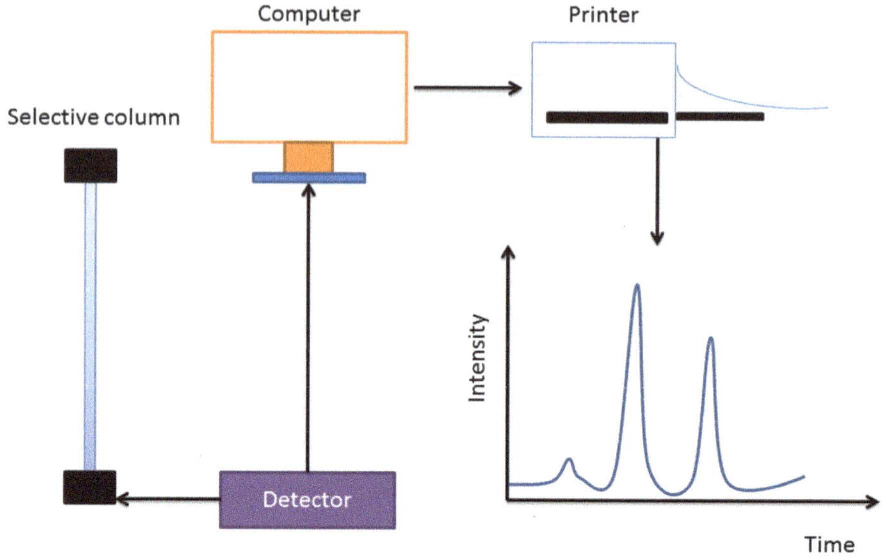


**Fig. 8.1** Simple representation of the mains steps in chromatography

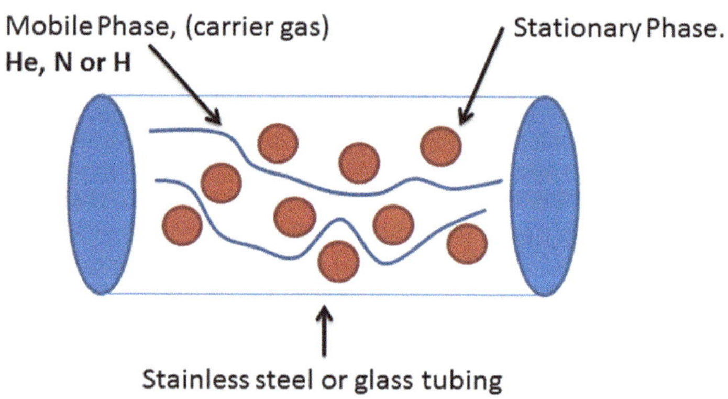


**Fig. 8.2** Draft of the chromatographic column

of the column. Because of this, some elements will travel faster than others into the column. The consequences will be that each of the components will be separated among each other. Because the column will be long enough (otherwise there is some serious problems with the design of the apparatus), each of the components will across the final detector one at the time. In this way, only when all the molecules of the faster element have crossed that section, the second element will arrive and so forth the other elements. This is why, as we will see later, each column has a particular resolution, and the value of the resolution has to be large enough. As an example, a detector can convert difference in absorbance in diverse current signals (mA) and plot them against time. This is the so-called chromatogram.

Concerning the different kinds of detectors used in chromatography, some of them will be listed here. One of the most used, especially for gas chromatography, is the thermal conductivity detector (TCD). This type of detectors is based on the principle of heat conductivity, and the central part of the detector is a Wheatstone bridge circuit. In this system, the sample gas which is coming from the column has a different conductivity with respect to the carrier gas. Now, let us assume that we have a hot filament where current is passing; the filament will be at diverse temperatures if the gas flowing on it will be of different nature and this is because they have different conductivity. If the temperature of the filament is higher, this will increase or decrease the electrical resistance of the filament and therefore, the voltage will be also different. The sensor gives a continuous comparison of the voltage produced by the carrier gas and the voltage produced by the sample gas. This is because two resistances are used in parallel. The concentration of the sample gas is directly proportional to the voltage output.

The sensitivity of these sensors is directly proportional to the current flowing into the filament, and it is increased by higher temperature differences between the filament and the case which encloses the filament itself. However, it has to be noticed that the temperature of the filament cannot be increased over certain limits to not burn the wire itself and for this reason the sensitivity of this kind of detectors cannot be of very high values. The sensitivity is inversely proportional to the flow of gas.

The flame ionization detector (FID) is the most used within gas chromatography; the principle on the basis of this detector is that the sample is combusted in a flame produced by hydrogen and air. During the combustion, in case we have molecules where carbon is present, we have the formation of CH radicals, which in turn can produce $CHO^+$ ions as follows:

$$CH + O \rightarrow CHO^+ + e^- \tag{8.1}$$

Now, because a difference in voltage of some hundreds volt is applied between two electrodes where the electrons are, a current is formed and this current can be recorded (Stauffer et al. 2008b). The current produced is directly proportional to the formyl cation produced. This method is used for organic compounds, but it is not really suitable for compounds like $O_2$, $CO_2$, $H_2O$, and $NH_3$.

The refractive index detector (RI) is used for instance in high-pressure liquid chromatography, HPLC, and they generate a signal which is based on the difference between the RI of the carrier mobile phase and the RI of the sample mixed with the mobile phase. The sample cell and the reference cell are separated; when the sample cell contains only the carrier phase, the incoming beam is focused on a photodetector (which can be a photo-resistor), and when the sample is present, the beam will be refracted to an angle. The more the concentration of the sample compound, the more the beam is refracted. A simple description of the functioning principles of RI measurements is given in Fig. 8.3.

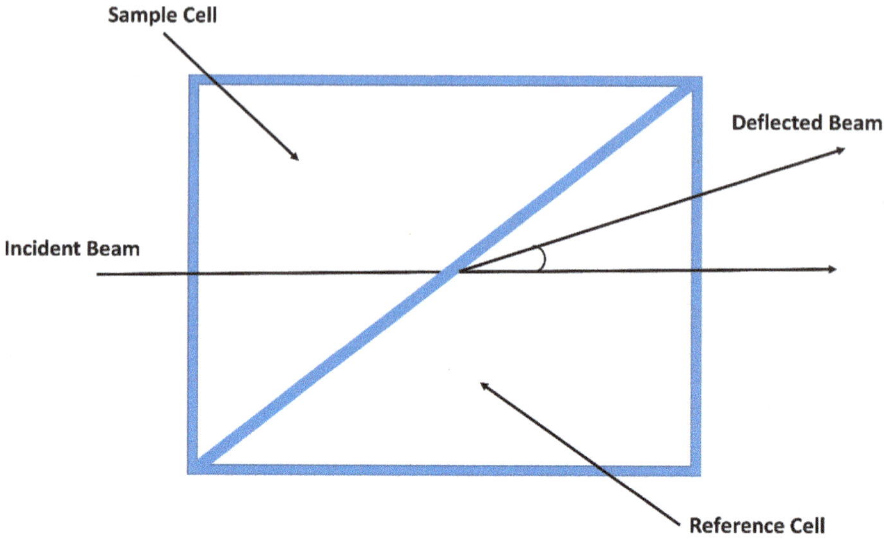


**Fig. 8.3** Functioning principle of refractive index detectors

HPLC is used for studying a variety of liquid products, and it can be used to analyze the composition of fatty acid methyl esters (FAME) within the production of biodiesel (Syed 2017).

Ultraviolet detectors, also denominated as ultraviolet photoelectric detectors, are also widely used in chromatography. Here, UV or visible radiation in the range of 190–800 nm of wavelength is used; the light is then absorbed by the sample within the flow cell, and chromophoric groups are then detected by this method (Baiocchi et al. 2002). A simple representation of the functioning principles of this method is given in Fig. 8.4.

The UV light is directed through a filter and a slit to a first mirror from where the light is sent to a diffraction grating which divides the light into different wavelengths and frequencies. This is the key step in this process; the diffracted light is then sent to a second mirror and divided into two beams; one is going to a reference diode, and the other one is going to the flow cell with the sample. The light from the two diodes will produce a signal which is recorded and compared at all times. In this case, absorbance-time plots are obtained and peak areas are directly proportional to the sample concentration.

There are mainly three types of UV detectors:

- fixed-wavelength filter-photometers;
- variable-wavelength spectrophotometers;
- photodiode-array detectors.

The positive aspects that have made chromatography used broadly can be summarized as follows (Skoog et al. 2006):

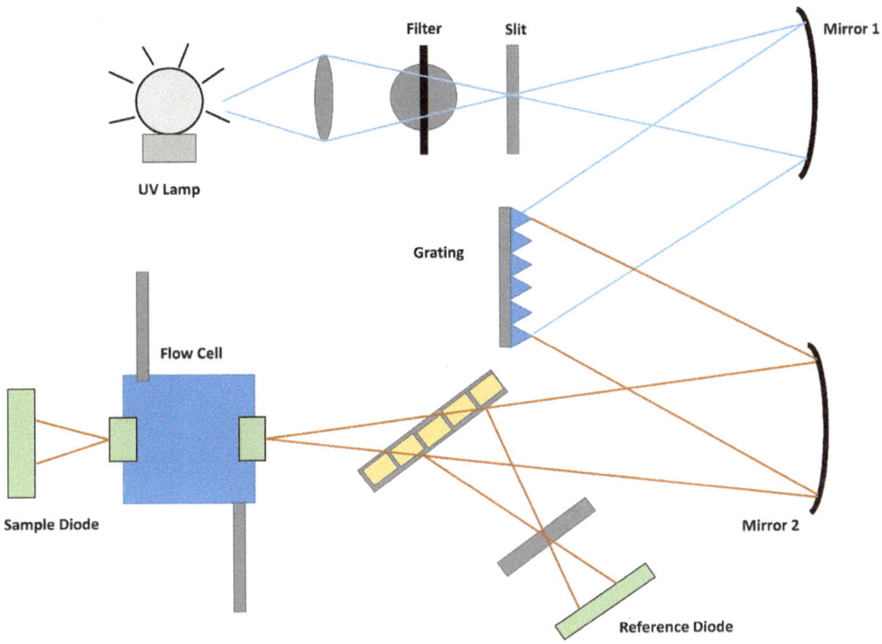


**Fig. 8.4** Basic principles of the ultraviolet detectors

- very short analysis times (one measure can take 15 min for instance);
- possibility of separating, working in suitable conditions any mixture of substances;
- possibility of carrying out analysis in series, since the same column can be regenerated continuously by the carrier gas;
- high sensitivity (amount of substance analyzed can be in the range of $10^{-5}$–$10^{-12}$ g).

Figure 8.5 shows the basic schematic of a gas chromatograph. The cylinder contains the gases that constitute the mobile phase (this will be the carrier); this is purified by the usage of some traps before being inserted into the column. The chromatographic column containing the stationary phase is placed inside a chamber which is thermostabilized appropriately in order to keep the various constituents in the gaseous phase of the mixture to be separated (Guiochon and Guillemin 1988). The mixture of gases to be analyzed is introduced into the column through the injector. This can be injected in the gaseous phase if the mixture is a gas, or dissolved in a suitable solvent, if the mixture to be analyzed is liquid or solid. To give one example, if the sample gases are collected in a gas sampling bag with septum and valve options, then the gas could be taken directly with a syringe and injected directly into the chromatographic column.

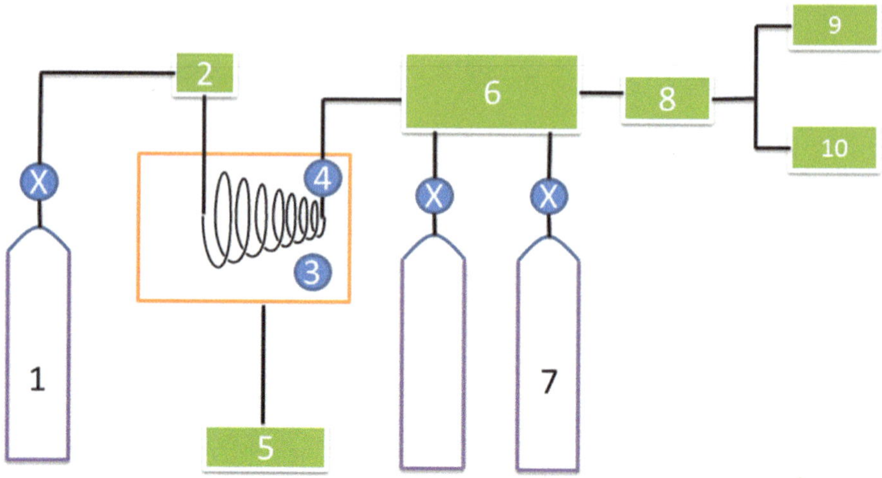


**Fig. 8.5** System adopted for chromatographic measurements

The parts shown in Fig. 8.5 are:

1. gas carrier tank;
2. injector block;
3. room columns;
4. chromatographic column;
5. electronic control system of the temperature;
6. detector;
7. pure gas tanks in service of the detector;
8. electronic processing of the detector signal;
9. recorder;
10. dedicated computer integrator or PC connected to the gas chromatograph, with software for GC.

An elementary representation of the main steps involved during the measurement of gas/liquid concentration in chromatography is shown in Fig. 8.6. The mechanisms of chromatographic separation are based on adsorption, distribution, ion exchange, exclusion, and affinity. Figure 8.6 shows the sequence of these steps:

The main steps shown in Fig. 8.6 are described as follows:

(a) the necessary parts of this measurement are column, SP—stationary phase, MP—mobile phase, and sample;
(b) introduction of the sample;
(c) start of elution;
(d) recovery of the products following the separation.

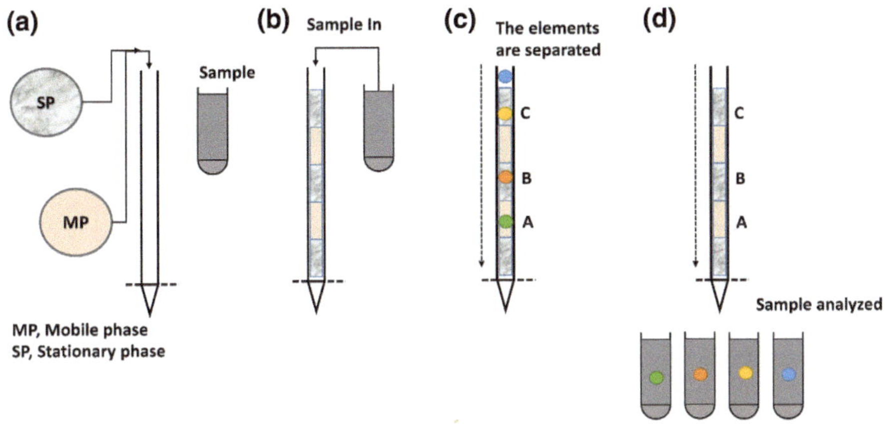


**Fig. 8.6** Main steps involved during the measurement of gas/liquid concentration in chromatography. Adapted from Rouessac and Rouessac (2007)

## 8.2 The Chromatogram

As mentioned previously, at the end of the column is placed a detector which is able to highlight the various substances emerging at different times. They emit a signal, and the intensity of this signal is proportional to their concentration. The signal is recorded by an instrument that will thus provide a chromatogram as shown in Fig. 8.7.

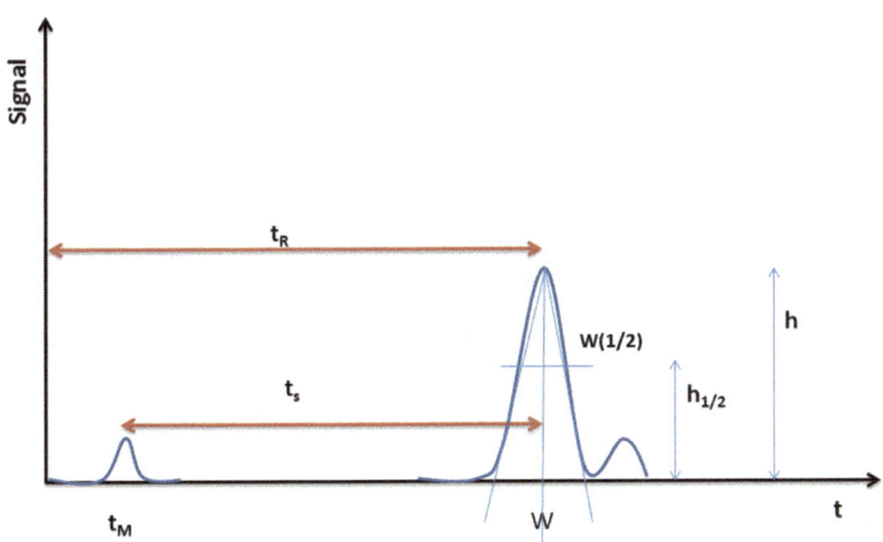


**Fig. 8.7** Example of chromatogram with main parameters

The most important parameters in a chromatogram are as follows. In reference to Fig. 8.7, some of them are shown as:

- $w$ is the peak amplitude.
- $w_{1/2}$ is the amplitude of half height (in $h/2$); in practice, it is the amplitude of the curve where the height is half of the maximum.
- $h$ is the peak height.
- $t_R$ is the retention time. This parameter is of outmost importance, and it is the time needed for the sample to exit from the column. To be noticed that this is approximately the point at which 50% of the total quantity has passed the detector. In other words, this would be the value of the abscissa at the point where the concentration is maximum. The retention time is dependent on the nature of the component.
- $t_M$ is the so-called dead time. This is the time required for the carrier to across all the column; the carrier is suppose to not spend time on the stationary phase and therefore this is the shortest time a component will spend in the column.
- $Q_C$ is the carrier flow; this is the flow rate of the carrier gas through the column.
- $V_R$ is the retention volume. This is the volume of the carrier gas used in the transport of the sample from the starting point (when we inject the sample into the column) to the exit of the column. This is calculated by:

$$V_R = t_R Q_C \tag{8.2}$$

- $V_M$ is the so-called dead volume. This is the volume of carrier gas that passes through the column within the dead time.
- $V_S$ is the corrected retention volume, and this is, on the other hand, the actual volume of carrier necessary for the transport of a certain component. The corrected retention volume is calculated accordingly to:

$$V_S = V_R - V_M \tag{8.3}$$

- $j$ is the compressibility factor. This value is related to the pressure of the gas inside the chromatographic column. One simple way to assess the compressibility factor of a real gas is to evaluate it from a simple relation like (Acrivos 1988):

$$Z = \frac{PV}{nRT} \tag{8.4}$$

And for ideal gases, the value of $Z$ would be then equal to 1.

The net volume, $V_{NET}$, is calculated according to Eq. (8.5). This is given by the actual volume of carrier, which is first corrected with a suitable factor to give $V'_M$ and then multiplied by the compressibility of the gas in the column;

$$V_{\text{NET}} = V'_{\text{M}} \cdot j \tag{8.5}$$

It is useful to examine in detail all the factors involved in the physics of the system.

## 8.3 Distribution Constants

During the chromatography process, there is equilibrium between the solute in the stationary and mobile phases (Dettmer-Wilde and Engewald 2014):

$$A_{\text{mobile}} \Leftrightarrow A_{\text{stationary}} \tag{8.6}$$

The equilibrium constant $K_{\text{C}}$ for the distribution of species $A$ between the two phases is called the distribution constant. This is defined as:

$$K_{\text{C}} = \frac{(a_A)_{\text{S}}}{(a_A)_{\text{M}}} \tag{8.7}$$

where

- $(a_A)_{\text{S}}$ is the activity of solute $A$ in the stationary phase.
- $(a_A)_{\text{M}}$ is the activity of solute $A$ in the mobile phase.

Under some conditions which can be low concentrations or when, for instance, non-ionic species are involved, the activity coefficients are almost equal to 1; then it is possible to write:

$$K_{\text{C}} = \frac{C_{\text{S}}}{C_{\text{M}}} = \frac{\frac{n_{\text{S}}}{V_{\text{S}}}}{\frac{n_{\text{M}}}{V_{\text{M}}}} \tag{8.8}$$

where

- $C_{\text{S}}$ is the molar analytical concentration of the solute in the stationary phase.
- $C_{\text{M}}$ is the molar analytical concentration of the solute in the mobile phase.
- $n_{\text{S}}$ is the number of moles of analyte in the stationary phase.
- $n_{\text{M}}$ is the number of moles of analyte in the mobile phase.

$K_{\text{C}}$ is constant over a large range of solute concentrations. Because of this, it is possible to assume that $C_{\text{S}}$ is directly proportional to $C_{\text{M}}$. This type of chromatography is called linear chromatography and provides the results as symmetric Gaussian-type peaks and retention times independently of the quantity of analyte in input.

## 8.4 Retention Time

This quantity is a function of $K_C$, and it is measurable. For a better understanding, it is good to observe Fig. 8.7 where the small peak on the left represents a specie that is not retained by the column. This is happening often since the sample to be analyzed or the mobile phase contains a specie (or more) which is not retained. When one or more species are not retained, this could help in the identification of the other species present in the sample. The larger peak on the right is that of a component, and the time required to reach the detector after the sample injection is called retention time, $t_R$, as seen previously. The specie has been retained because it spends a time $t_S$ in the stationary phase. Then the retention time can be defined as (Rouessac and Rouessac 2007; Skoog et al. 2006):

$$t_R = t_S + t_M \tag{8.9}$$

In addition to the previous discussion, it is also possible to define the following parameters:

- $v$: average linear rate of solute migration through the column (cm/s);

$$v = \frac{L}{t_R} \tag{8.10}$$

- $u$: average linear velocity of the mobile-phase molecules (cm/s);

$$u = \frac{L}{t_M} \tag{8.11}$$

where $L$ is the length of the column packing.

## 8.5 Volumetric Flow Rate

In chromatography, the mobile-phase flow is usually characterized by the volume flow rate $F$ at the column outlet (cm$^3$/min). For an open tubular column, we have:

$$F = u_0 A = u_0 \pi r^2 \tag{8.12}$$

where

- $u_0$, linear velocity at the column outlet;
- $A$, cross-sectional area of the tube.

When a stationary phase is present, for instance in the case of packed columns, the volume of the entire conduct is not available to the gas flow and therefore, a correction term must be introduced as follows:

$$F = u_0 \pi r^2 \varepsilon \tag{8.13}$$

where $\varepsilon$ is the column porosity, a fraction of total column volume available to the liquid.

## 8.6 Retention Time and Distribution Constant

Consider the average linear rate of solute migration through the column, $v$, Eq. (8.10), and the average linear velocity of the mobile-phase molecules, $u$, Eq. (8.11). These two quantities are related as follows:

$$v = uf = u\frac{t_M}{t_R} \tag{8.14}$$

where $f$ is the fraction of time the solute spends in the mobile phase.

This fraction is also equal to the average number of mole of solute in the mobile phase divided by the total number of moles of solute in the column at any instant:

$$v = u \cdot \frac{\text{moles of solute in mobile phase}}{\text{total moles of solute}} \tag{8.15}$$

Furthermore, we consider that:

moles of solute in mobile phase $= C_M\, V_M$
moles of solute in stationary phase $= C_S\, V_S$

And therefore:

$$v = u\frac{C_M V_M}{C_M V_M + C_S V_S} = u\frac{1}{1 + \frac{C_S V_S}{C_M V_M}} \tag{8.16}$$

Substituting in this equation, the equation of the distribution constants is obtained:

$$v = u\frac{1}{1 + \frac{K_C V_S}{V_M}} \tag{8.17}$$

## 8.7 Retention Factor

The retention factor, $k$, is an experimental quantity widely used to compare the migration rates of solutes in columns. This quantity is very important because $k$ does not depend on the geometry or on volumetric flow rate of the column. The retention factor $k_A$ for a solute $A$ is defined as:

$$k_A = \frac{K_A V_S}{V_M} \tag{8.18}$$

where $K_A$ is the distribution constant for the solute $A$ view previously.

Substituting this equation with the one related to the rate of migration of a solute, Eq. (8.17), the following expression is obtained:

$$v = u \frac{1}{1 + k_A} \tag{8.19}$$

To demonstrate how $k_A$ can be derived from a chromatogram, it is shown here how it is possible to apply some substitutions. Giving that:

$$\frac{L}{t_R} = \frac{L}{t_M} \frac{1}{1 + k_A} \tag{8.20}$$

and rearranging, we have

$$k_A = \frac{t_R - t_M}{t_M} = \frac{t_S}{t_M} \tag{8.21}$$

In Eq. (8.21), $t_S$ is the corrected retention time, which is the actual retention time of a certain component in the column. Note that $t_R$ and $t_M$ are readily obtained from the chromatography. In case we have values of $k$ much less than unity, this means that the solute passes through the column at a time close to the value of the dead time. On the other hand, when the retention factor is greater than 20, the elution times become excessively long. Then the ideal separation is carried out in conditions where the retention factor solutes in a mixture are comprised between 1 and 10. Optimization of the separation therefore occurs by optimizing the values of the retention factor for the components of interest.

## 8.8 Parameters Characterizing a Chromatographic Column

There are different parameters by which we are able to characterize a chromatographic column. Among the most important, there are selectivity, efficiency, and resolution. The main references used in the following derivations are in this case (Rouessac and Rouessac 2007; Skoog et al. 2006; Stauffer et al. 2008a).

### 8.8.1 Selectivity

The selectivity of a column expresses the capability of a certain column to give well-separated peaks of components. It can be expressed in relation to the factors of separation, for example, let us consider two generic components *A* and *B*, according to previous mathematical relations existing between the various quantities, the following parameter, $\alpha$ (the selectivity), can be obtained:

$$\alpha = \frac{K_B}{K_A} = \frac{k_B}{k_A} = \frac{(t_R)_B - t_M}{(t_R)_A - t_M} \tag{8.22}$$

$\alpha$ should be as far as possible diverse from one. According to this definition, $\alpha$ is always greater than unity. The last equality permits the determination of $\alpha$ from an experimental chromatogram.

### 8.8.2 Efficiency

The following correlated terms are widely used as quantitative measures of chromatographic column efficiency:

- Plate height, *H*;
- Plate count or number of theoretical plates, *N*.

The two are related in this way:

$$N = \frac{L}{H} \tag{8.23}$$

where *L* is the length of the column packing (cm).

This efficiency increases if the number of plates, *N*, grows and if the plate height *H* is lesser (Stauffer et al. 2008a). The efficiency of a column depends on the column type and phases.

Considering a chromatographic column, there will be equilibrium between the liquid phase and the vapor phase if we are taking into consideration liquids. On the other hand, there will be also a certain equilibrium between the rate at which molecules are in the mobile phase and on the fixed phase. To be noticed that when a component passes the detector, it is derived by logical reasoning that the values of its concentration with time will have a Gaussian-shaped curve.

The Gaussian curve is described by its standard deviation $\sigma$ or its variance $\sigma^2$. It is convenient to define the efficiency of a column in terms of variance per unit length of column. For this reason, by definition, the plate height $H$ is given by:

$$H = \frac{\sigma^2}{L} \tag{8.24}$$

Figure 8.8 shows a column having a packing $L$ in length. In the same figure, there is a curve showing the distribution of the number of molecules (or its concentration) along the length of the column when the component is crossing the detector. This is a Gaussian curve, and the position of $L + \sigma$ and $L - \sigma$ is given.

It is possible to notice that if $\sigma$ is in $cm^2$, then $H$ represents a linear distance in cm. As a matter of fact, the plate height can be imagined as the length of the column that contains a fraction of the analyte that lies between $L$ and $L - \sigma$. Since by statistical derivation (Montgomery 2003) the area under a normal curve which is situated between $\pm\sigma$ is approximately 68% of the total area, the plate height contains about 34% of the component analyzed.

The variance of the solute peak, $\tau^2$, expressed in seconds ($s^2$) is distinguished from $\sigma^2$ which is expressed in length units ($cm^2$). The relation given between these two standard deviations $\tau$ and $\sigma$ is:

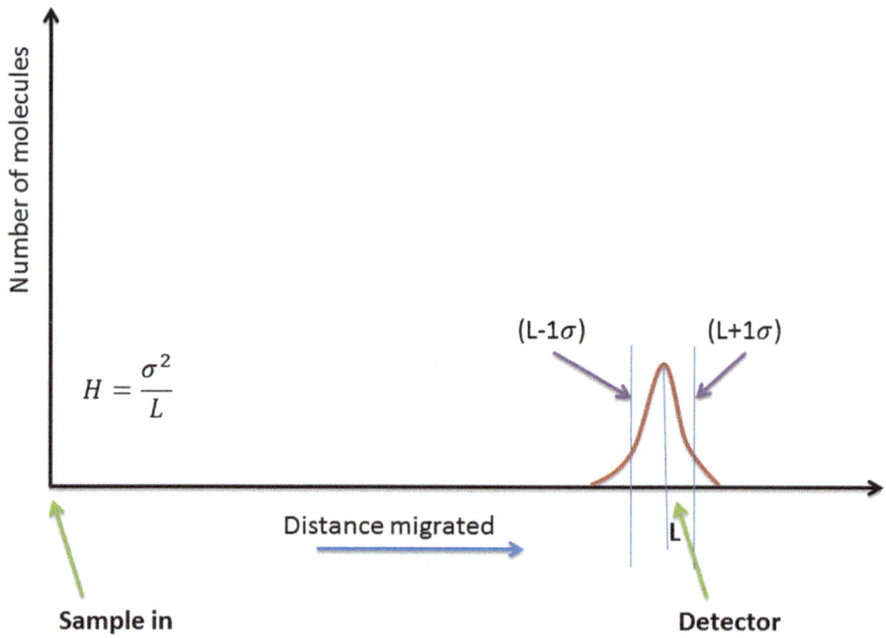


**Fig. 8.8** Typical response as a function of distance

$$\tau = \frac{\sigma}{\frac{L}{t_R}} = \frac{\sigma}{v} \tag{8.25}$$

Figure 8.9 also shows a method for approximating these two standard deviations in an experimental plot. According to the procedure proposed by Skoog and collaborators (2006), if we take the tangents at the inflection points on both sides of the peak in order to form a triangle with the baseline of the chromatogram, then this area subtended by the triangle is about 96% of the total area under the peak in case the peak is Gaussian curve. Therefore, one can conclude that these intercepts are at approximately ($\pm 2\tau$) from the maximum point of the curve, and it is also given that the amplitude of the base of the triangle $W$ is equal to $4\tau$.

Substituting $w = 4\tau$ into Eq. (8.25) and rearranging, it is possible to find:

$$\sigma = \frac{LW}{4t_R} \tag{8.26}$$

Equation (8.24) will be then:

$$H = \frac{LW^2}{16t_R^2} \tag{8.27}$$

From which:

$$N = 16\left(\frac{t_R}{W}\right)^2 \tag{8.28}$$

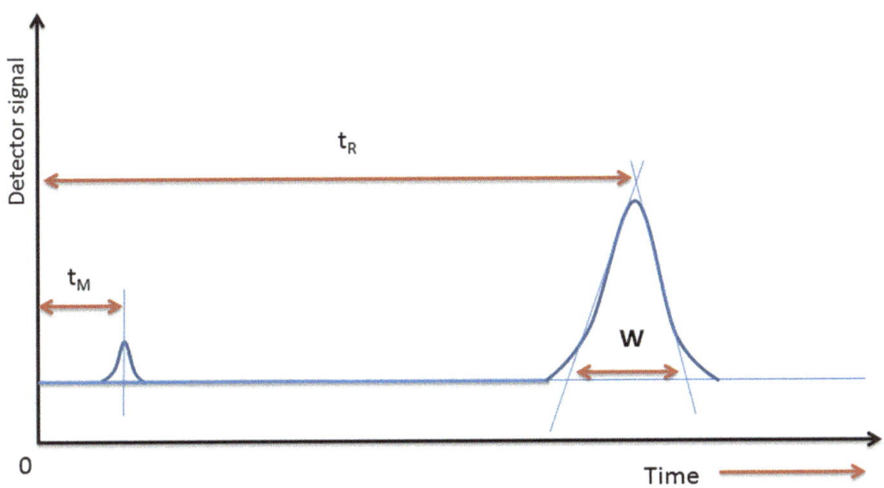


**Fig. 8.9** Typical Gaussian curves for a component into a chromatographic column

It is possible also to use the empirical relations given previously to connect the height equivalent to some basic parameters. For instance, the Van Deemter equation (van Deemter et al. 1995) connects the quantity $H$ with the flow $Q_C$ of the carrier gas.

$$H = A + \frac{B}{Q_C} + CQ_C \tag{8.29}$$

This is the equation of a hyperbola which is not equilateral. In Eq. (8.29), we have that the term $A$ is the micro-eddy diffusion coefficient. This coefficient is a function of the particles diameters average in the filling. It is mentioned (Skoog et al. 2006) that the coefficient tends to zero in the capillary columns in which the regime is not swirling flows.

The term $B$ is the molecular diffusion coefficient. This term is a function of the diffusion coefficients of gases. It is minimized by using low temperatures, high pressures, and gas high molecular weight.

The term $C$ is the inertia ratio to the passage of molecules from the liquid phase to the gas phase. This value is depending on the thickness of the stationary phase (for instance, if the stationary phase is liquid) and its viscosity. It is minimized at high temperature.

From Fig. 8.10, it can be seen that to minimize the height $H$ of a theoretical plate, it is necessary to use the optimal flow values between the values $Q_1$ and $Q_2$.

The coefficients $A$, $B$, and $C$ are characteristic of a certain column and are responsible for enlargement of the slots and therefore for enlargement of the peaks.

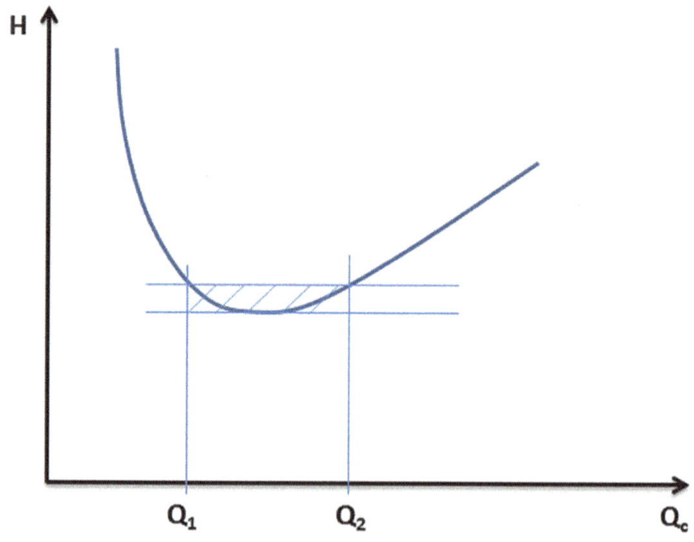


**Fig. 8.10** Relation between flow and column height

A similar equation can be written also with regard to the temperature; there is also an optimum temperature $T$ that minimizes the value of $H$ improving the efficiency of a given column.

$$H = A + \frac{B}{T} + CT \tag{8.30}$$

## 8.9 Resolution

The resolution represents one of the most important parameters given for a chromatographic column. As a matter of fact, each column behaves in a different way in case the same mixture should be analyzed. Each column has a different resolution. The resolution gives a certain value, and this value indicates how far two peaks are within the chromatogram. Analytically, the resolution of each column is defined in this way (Fig. 8.11):

$$R_s = \frac{\Delta Z}{\frac{W_A}{2} + \frac{W_B}{2}} = \frac{2\Delta Z}{W_A + W_B} = \frac{2\left[(t_R)_B - (t_R)_A\right]}{W_A + W_B} \tag{8.31}$$

It is given that for a resolution of 1.5, the resolution value is the minimal to have a good understanding of the concentrations for each of the components of the mixture.

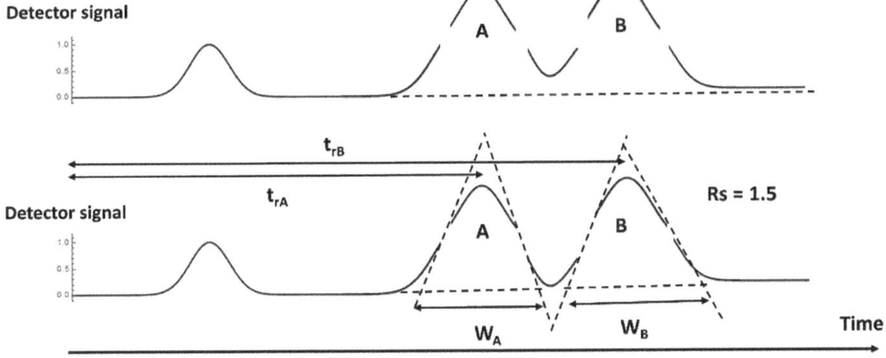


**Fig. 8.11** Main parameters and resolution of a chromatogram. Adapted from Skoog et al. (2006)

On the other hand, a resolution of 0.75 does not give a complete separation of $A$ and $B$. With a resolution of 1.0, the peak $A$ contains approximately 4% of $B$ and therefore the peak $B$ contains about 4% of $A$. Increasing the number of plates and increasing also the length of the column give a better resolution of course.

## 8.10 Parameters Influencing the Resolution

There are mathematical relationships that link the resolution of a column and the retention factors $k_A$ and $k_B$ for two solutes, the selectivity factor $\alpha$, and the number of plates $N$. Under the hypothesis that two solutes $A$ and $B$ have retention times close enough between them, it is possible to write:

$$W_A = W_B \approx W \tag{8.32}$$

Rewriting Eq. (8.31) which is related to the column resolution, it is possible to find:

$$R_s = \frac{(t_R)_B - (t_R)_A}{W} \tag{8.33}$$

We rearrange this by substituting

$$N = 16\left(\frac{t_R}{W}\right)^2; \tag{8.34}$$

and we obtain:

$$R_s = \frac{(t_R)_B - (t_R)_A}{(t_R)_B}\frac{\sqrt{N}}{4} \tag{8.35}$$

In case we use the substitution:

$$k_A = \frac{t_R - t_M}{t_M} = \frac{t'_R}{t_M}; \tag{8.36}$$

the resolution is obtained as:

$$R_s = \frac{k_B - k_A}{1 + k_B}\frac{\sqrt{N}}{4} \tag{8.37}$$

In addition, we could also use the definition of selectivity factor:

$$\alpha = \frac{k_B}{k_A}; \tag{8.38}$$

and obtain

$$R_s = \left(\frac{\alpha - 1}{\alpha}\right)\left(\frac{k_B}{1 + k_B}\right)\frac{\sqrt{N}}{4} \tag{8.39}$$

In Eq. (8.39), $k_B$ is the retention factor related to the slower-moving species and $\alpha$ represents the selectivity factor. The equation can be also rewritten in the following form to give the number of plates needed to realize a given resolution:

$$N = 16R_s^2\left(\frac{\alpha}{\alpha - 1}\right)^2\left(\frac{1 + k_B}{k_B}\right)^2 \tag{8.40}$$

Since occasionally it is needed for certain species to divide among each other, it would be necessary to have a model which takes into account this variable.

During chromatography, we want to achieve the highest possible resolution in a short length of the column or residence time. Nevertheless, it is necessary to optimize the two parameters since they go in opposite directions.

The element which is the slowest in the column represents the limiting step, and this will determine the length of the column. It is possible to recall from what was written previously:

$$v = \frac{L}{(t_R)_B}, \quad N = \frac{L}{H}, \quad v = u\frac{1}{1 + k_A} \tag{8.41}$$

From these equations, we have:

$$(t_R)_B = \frac{NH(1 + k_B)}{u}; \tag{8.42}$$

where $(t_R)_B$ is the time required to bring the peak for $B$ component to the end of the column when the velocity of the mobile phase is $u$.

Putting together the following two equations, Eqs. (8.43) and (8.44):

$$(t_R)_B = \frac{NH(1 + k_B)}{u}; \tag{8.43}$$

$$N = 16R_s^2\left(\frac{\alpha}{\alpha - 1}\right)^2\left(\frac{1 + k_B}{k_B}\right)^2; \tag{8.44}$$

Equation (8.45) is obtained:

$$(t_R)_B = \frac{16R_s^2}{u}\left(\frac{\alpha}{\alpha - 1}\right)^2\frac{(1 + k_B)^3}{(k_B)^2} \tag{8.45}$$

Since $H$ is a function of $u$ and $\alpha$ is a function of $k$, such an equation is difficult to solve and use except when doing comparisons and simplifications.

## 8.11 Qualitative Analysis

A qualitative analysis of chromatography is performed by studying different conditions and physical parameters used. For instance, we would like to study the mixture at hand by using different temperatures or different fixed phases. The quality of the measurements will be a function of these parameters.

In addition to chromatography, there are different technologies providing a greater quantity of information with highest level of accuracy compared to chromatography. These could be IR, nuclear magnetic resonance, or mass spectrometry. Nevertheless, chromatography is a widely used technique for identify the presence or not of components of mixtures.

Chromatography is an essential step in qualitative analyses and can be considered the first step to do in order to have preliminary information on the sample and continue with the further investigation. As a matter of fact, with complex samples it is difficult to obtain the wanted information by using only one method.

As a summary, even if this technique is not the most accurate among the available methods, it can give very valuable information. For instance, we can see if a compound is present in the sample or it is entirely absent from the mixture. Chromatography is a very good tool when we are analyzing gas products coming for instance from gasification of biomass to produce syngas (De Blasio et al. 2019).

## 8.12 Elemental Analysis

The elemental composition of a sample in terms of, for instance, oxygen, nitrogen, hydrogen, and sulfur is usually measured by elemental analysis. The principle on the basis of these types of analyses is that the sample is heated to very high temperatures and, at these temperatures the sample is oxidized or undergoing to a thermal degradation where known compounds are formed. By means of specifically designed columns, the produced gas is further treated to remove excess compounds end/or producing particular reactions.

Some types of elemental analyzers utilize very high temperatures where even the molecular bonds of the molecules forming the sample are then broken. Therefore, a gas consisting of only the elements is formed.

The temperatures used for transforming the sample into its elements can be of thousands of degrees (for instance, 1500–3000 °C), and the power used to do this can be of some kilowatts (for instance, 8–9 kW). The kinds of detectors used for the analyses can be based on infrared detection (one example could be a solid-state detector) and thermal conductivity (TCD). Recall that some of the detectors used in chromatography are similar. The analysis allows for detecting concentrations of few ppm, and therefore, it is quite accurate. Figure 8.12 shows a schematic for the CHNS and oxygen elemental analysis. This example refers to a Thermo Fisher Scientific Flash series instrument.

The maximum weight of the sample is normally specified by the particular machine used for the analysis; the sample is weighted accurately and inserted in a suitable sampling device. The machine will automatically allocate the sample in the space where it will be processed thermally. In the case the CHNS amount should be derived, the sample is usually positioned in a tin capsule. After the sample is positioned in place, oxygen will be automatically pumped to allow for the combustion of the sample; this is taking place at around 1800 °C in this example.

The products of combustion are sent to a copper oxide column where we have mainly $N_2$, nitrogen oxides, $CO_2$, water, $SO_2$, and $SO_3$. Copper oxide is used here to help the oxidation process. The excess of oxygen (the one which is reacting with the carrier gas and the one which is oxidizing further the $SO_2$) is removed into a column where electrolytic copper is placed. In this section, we have the reduction of nitrogen oxides to nitrogen and the excess oxygen is also trapped here (Gentile et al. 2013). In this place, we have mainly $N_2$, $CO_2$, $SO_2$, and water. These compounds are then sent to the TCD detector through a CHNS column which is maintained at a specific temperature depending on the machine used. The results will be shown in a very similar way that was used for the gas chromatography.

In the case of oxygen analysis, the sample should be enclosed within a silver capsule and introduced into the sampling unit. Also in this occasion, the machine will

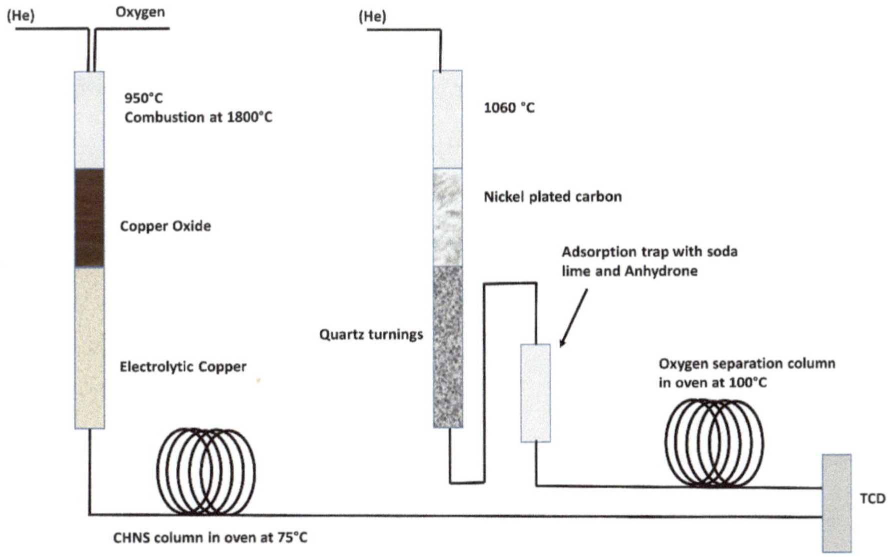


**Fig. 8.12** Schematics of a CHNS/O analyzer functioning

automatically send the sample to the chamber where it will be processed thermally at high temperatures. Temperatures between 1000 and 1100 °C are used in this example to convert the sample into gas products which are treated into a nickel-plated carbon column. Here, mainly CO and $N_2$ are formed at these temperatures (Berezkin 1983). The components are sent to the detectors after passing an adsorption trap with soda lime and anhydrone and an oxygen separation column which is maintained at 100 °C. By absorbing the oxidized molecules, soda lime helps to reduce further the products from the thermal processing and release all the oxygen to be measured (Weimann et al. 1997).

In the case of elemental analysis, the calibration of the machine is done with particular compounds of known composition and, like for the gas chromatography, the machine will derive a diagram showing different peaks. Each peak will be related to a different component, and the concentrations of C, H, N, S, and oxygen can be then retrieved. Since the sample to be inserted into the analyzer is a small solid quantity, the reference materials have to be also solid. In case liquids should be tested, the liquid should be then absorbed into an inert powder and then inserted into the sampling section. Examples of quantitative content determination in GC and EA will be shown in Sect. 9.5.

## References


Acrivos, J. (1988). Physical chemistry, third edition (Levine, Ira N.). *Journal of Chemical Education, 65*(12), A335. https://doi.org/10.1021/ed065pA335.3.

Baiocchi, C., Brussino, M. C., Pramauro, E., Prevot, A. B., Palmisano, L., & Marcì, G. (2002). Characterization of methyl orange and its photocatalytic degradation products by HPLC/UV–VIS diode array and atmospheric pressure ionization quadrupole ion trap mass spectrometry. *International Journal of Mass Spectrometry, 214*(2), 247–256. https://doi.org/10.1016/S1387-3806(01)00590-5.

Berezkin, V. G. (1983). *Chemical methods in gas chromatography*. Elsevier Science Publishers.

De Blasio, C., De Gisi, S., Molino, A., Simonetti, M., Santarelli, M., & Björklund-Sänkiaho, M. (2019). Concerning operational aspects in supercritical water gasification of kraft black liquor. *Renewable Energy, 130,* 891–901. https://doi.org/10.1016/j.renene.2018.07.004.

Dettmer-Wilde, K., & Engewald, W. (2014). *Practical gas chromatography. A comprehensive reference*. Springer.

Gentile, N., Rossi, M. J., Delémont, O., & Siegwolf, R. T. W. (2013). δ15 N measurement of organic and inorganic substances by EA-IRMS: A speciation-dependent procedure. *Analytical and Bioanalytical Chemistry, 405*(1), 159–176. https://doi.org/10.1007/s00216-012-6471-z.

Guiochon, G., & Guillemin, C. L. (Eds.). (1988). Methodology gas chromatographic instrumentation. *Journal of Chromatography Library, 42*, 319–391. https://doi.org/10.1016/S0301-4770(08)70081-2.

Montgomery, D. C. (2003). *Applied statistics and probability for engineers* (3rd ed.). New York, USA: Wiley.

Rouessac, F., & Rouessac, A. (2007). *Chemical analysis: Modern instrumentation methods and techniques* (2nd ed.). Wiley.

Skoog, D., Holler, F. J., & Crouch, S. R. (2006). *Principles of instrumental analysis* (6th ed.). Brooks Cole.

Stauffer, E., Dolan, J. A., & Newman, R. (2008a). Gas chromatography and gas chromatography—-Mass spectrometry, Chap. 8. In E. Stauffer, J. A. Dolan, & R. Newman (Eds.), *Fire debris analysis* (pp. 235–293). Burlington: Academic Press. https://doi.org/10.1016/B978-012663971-1.50012-9.

Stauffer, E., Dolan, J. A., & Newman, R. (2008b). Detection of ignitable liquid residues at fire scenes. In E. Stauffer, J. A. Dolan, & R. Newman (Eds.), *Fire debris analysis* (pp. 131–161). Burlington: Academic Press. https://doi.org/10.1016/B978-012663971-1.50009-9.

Syed, M. B. (2017). Analysis of biodiesel by high performance liquid chromatography using refractive index detector. *MethodsX, 4,* 256–259. https://doi.org/10.1016/j.mex.2017.07.002.

van Deemter, J. J., Zuiderweg, F. J., & Klinkenberg, A. (1995). Longitudinal diffusion and resistance to mass transfer as causes of nonideality in chromatography. *Chemical Engineering Science, 50*(24), 3869–3882. https://doi.org/10.1016/0009-2509(96)81813-6.

Weimann, J., Hagenah, J. U., & Motsch, J. (1997). Reduction in nitrogen dioxide concentration by soda lime preparations during simulated nitric oxide inhalation. *British Journal of Anaesthesia, 79*(5), 641–644. https://doi.org/10.1093/bja/79.5.641.

# Chapter 9
# Examples of Quantitative Content Determination in Chromatography and Elemental Analysis

## 9.1 Instrumentation

The quantitative analysis of results obtained from chromatography refers to the evaluation of the actual concentration of each component of the mixture. The first step is the evaluation of the peak area, which usually is given by the instrument used. An example of chromatography analyses done on gas products derived from supercritical water gasification of Biomass (De Blasio et al. 2015) is given here.

The instrument used in this case is the gas chromatograph (GC) Clarus 500 (Fig. 9.1). The detector utilized in this case is a thermal conductivity detector (TCD).

The gas chromatograph returns as output a plot, which consists of several peaks (Fig. 9.2). This is what we have seen previously to be the chromatogram.

Each peak corresponds to a compound among those measurable by the machine. More specifically, each of them has a different retention time which makes it recognizable to the operator. The area of these peaks, through certain coefficients, is proportional to the volumetric concentration of the element in the sample.

For a correct analysis, it is useful to calculate these coefficients related to the measure of each compound through the preliminary study of a known substance. This is known as the calibration of the machine. After calibrating the instrument, it will be possible to evaluate the composition of a sample whose composition is not known.

In few words, since we know that a compound is detected at a particular residence time and that the concentration of that compound is directly proportional to the area under the curve, then, by the utilization of a suitable coefficient of proportionality (coefficient or function), it will be possible to estimate accurately the concentration of that component in the sample tested.



C. De Blasio, *Fundamentals of Biofuels Engineering and Technology*,
Green Energy and Technology, https://doi.org/10.1007/978-3-030-11599-9_9

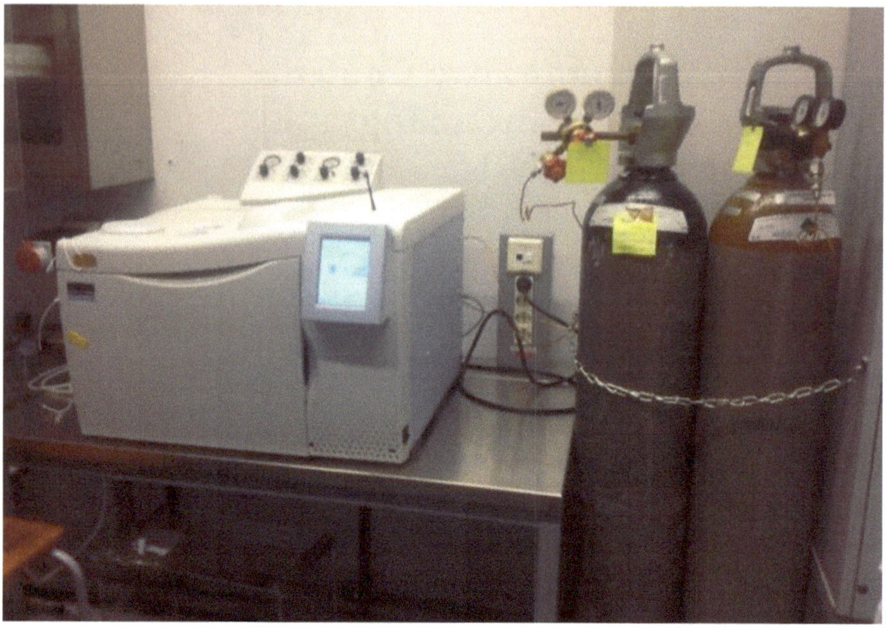

**Fig. 9.1** A photo of the gas chromatography instrumentation

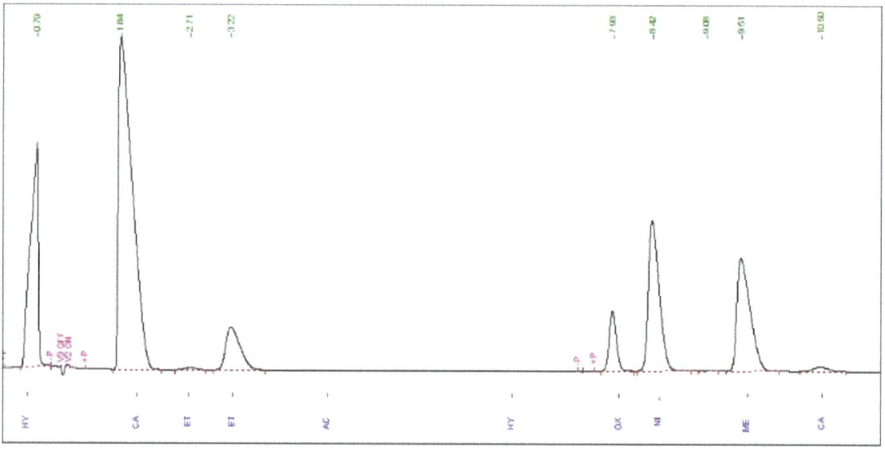

**Fig. 9.2** Example of output of the chromatographic analysis. As shown by the instrument

## 9.2 Preparation and Introduction of the Sample

The samples to be introduced into the chromatograph are obtained from the collection of the products of gasification or, in the case of calibration, from bottles of known composition. In the first case, this operation is done by using the gas bags provided with a membrane that allows for the injection of a needle without loss of gas (gas sampling bags with valve and septum). In the second case, it is possible to release the gas directly into a syringe. This is visible from Fig. 9.3.

Figure 9.4 shows the injection of the gas into the gas chromatograph. When the sample is inserted into the instrument, a small amount of air could be aspirated too or for instance, some air could have been still present into the chromatograph chamber where the sample gas is momentarily stored during the sampling step before

**Fig. 9.3** Collection of samples of known composition

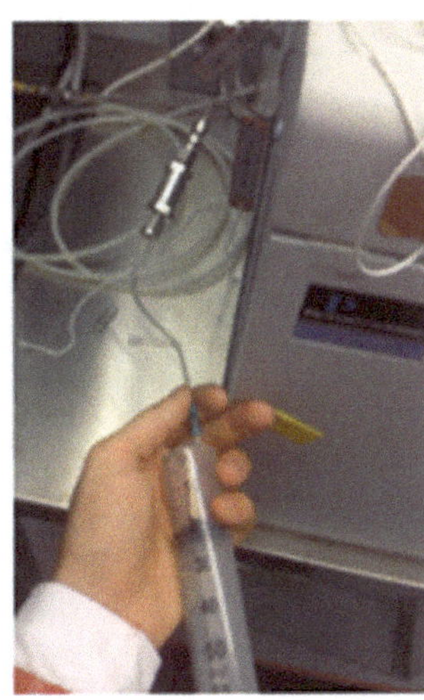

**Fig. 9.4** Injection of the gas into the gas chromatograph

measuring. For this reason, in the analysis of the final composition of the products, the oxygen and the nitrogen forming the air can be usually subtracted.

## 9.3 Calculation of the Calibration Coefficients

To calculate the calibration constants (the coefficient of proportionality), diverse substances with known composition can be taken into consideration. For this particular example, these substances are summarized in Table 9.1.

The change in the volume of sample introduced should not cause any change in the area of the peaks of the individual compounds.

For the calculation of the coefficients, it is necessary to divide the volume fraction of the individual compound by the peak area resulting from the chromatogram related to the same compound. This is done during the calibration process where the volume fractions are known.

$$K_{\text{compound}} = \frac{\text{Volume fraction}_{\text{compound}}}{\text{Peak area}_{\text{compound}}} \tag{9.1}$$

To be noticed that this is a simplification of the procedure, since not necessarily the volume areas and the known volumetric fractions have a linear dependence. The accurate procedure would be to measure the areas for several different known compositions and then determine the proportionality coefficient which can be a number of a function.

**Table 9.1** Substances with known composition used

| | |
|---|---|
| Bottle 1 | $H_2$ 100% Mol |
| Bottle 2 | CO 5% Mol<br>Air BAL |
| Bottle 3 | $CH_4$ 60% Mol<br>$CO_2$ BAL |
| Bottle 4 | $H_2$ 10% Mol<br>$N_2$ BAL |
| Bottle 5 | $H_2S$ 5000 ppm<br>$N_2$ BAL |
| Bottle 6 | $C_2H_4$ 100% |
| Bottle 7 | $CO_2$ 1%<br>$C_2H_4$ 0.099%<br>$C_2H_6$ 0.972%<br>$CH_4$ 1.02%<br>He BAL |

*BAL* stands for balanced

### 9.3.1 Calibration Constant for Hydrogen

The procedures which follow take into account the given Eq. (9.1). To find the correct calibration constant for hydrogen gas, sampling bottles with different compositions have been used in this example. The data and the results are visible in Table 9.2.

For the sensitivity of the machine, diluted samples are more reliable compared to those concentrated (bottle 4 compared to 1). This is also visible from the asymmetry of the Gaussian curve relative to the concentration 100% $H_2$. Indeed, the Gaussian of the sample 4 shows a trend very symmetrical and therefore more precise.

### 9.3.2 Calibration Constant for Carbon Monoxide

To find the correct calibration constant of carbon monoxide, also in this case, a calibration bottle has been used. The data and the results are visible in Table 9.3.

### 9.3.3 Calibration Constant for Methane

In this case, the data and the results are visible in Table 9.4.

**Table 9.2** Calculation of the calibration constant for hydrogen

| | Sample volume (mL) | Volume fraction (%) | Peak area | $K$ |
|---|---|---|---|---|
| *Bottle 1* | | | | |
| Test 1 | 11 | 100 | 4,020,820.16 | $2.49 \times 10^{-5}$ |
| Test 2 | 10 | 100 | 4,057,932.57 | $2.46 \times 10^{-5}$ |
| Test 3 | 10 | 100 | 4,232,846.05 | $2.36 \times 10^{-5}$ |
| *Bottle 4* | | | | |
| Test 1 | 10 | 10 | 422,134.15 | $2.36 \times 10^{-5}$ |
| Test 2 | 10 | 10 | 441,125.63 | $2.27 \times 10^{-5}$ |
| Test 3 | 9 | 10 | 443,963.95 | $2.25 \times 10^{-5}$ |

**Table 9.3** Calculation of the calibration constant of carbon monoxide

| Bottle 2 | Sample volume (mL) | Volume fraction (%) | Peak area | $K$ |
|---|---|---|---|---|
| Test 1 | 15 | 5 | 422,441.18 | $1.18 \times 10^{-5}$ |
| Test 2 | 10 | 5 | 419,887.24 | $1.19 \times 10^{-5}$ |
| Test 3 | 10 | 5 | 416,974.83 | $1.20 \times 10^{-5}$ |

**Table 9.4** Calculation of the calibration constant of methane

| Bottle 3 | Sample volume (mL) | Volume fraction (%) | Peak area | $K$ |
|---|---|---|---|---|
| Test 1 | 10 | 60 | 3,941,285.96 | $1.52 \times 10^{-5}$ |
| Test 2 | 10 | 60 | 3,922,941.64 | $1.53 \times 10^{-5}$ |
| Test 3 | 15 | 60 | 3,911,516.61 | $1.53 \times 10^{-5}$ |

### *9.3.4 Calibration Constant for Carbon Dioxide*

The data and the results for $CO_2$ are visible in Table 9.5.

### *9.3.5 Calibration Constant for Hydrogen Sulfide*

For $H_2S$, the data and the results are visible in Table 9.6.

### *9.3.6 Calibration Constant for Nitrogen*

For nitrogen, data and results are visible in Table 9.7.

**Table 9.5** Calculation of the calibration constant of carbon dioxide

| | Sample volume (mL) | Volume fraction (%) | Peak area | $K$ |
|---|---|---|---|---|
| *Bottle 3* | | | | |
| Test 1 | 10 | 40 | 3,443,657.61 | $1.16 \times 10^{-5}$ |
| Test 2 | 10 | 40 | 3,412,977.81 | $1.17 \times 10^{-5}$ |
| Test 3 | 15 | 40 | 3,398,244.43 | $1.18 \times 10^{-5}$ |
| *Bottle 7* | | | | |
| Test 1 | 20 | 1 | 99,606.84 | $1.00 \times 10^{-5}$ |
| Test 2 | 11 | 1 | 97,003.13 | $1.03 \times 10^{-5}$ |
| Test 3 | 9 | 1 | 96,241.89 | $1.04 \times 10^{-5}$ |

**Table 9.6** Calculation of the calibration constant of hydrogen sulfide

| Bottle 5 | Sample volume (mL) | Volume fraction (%) | Peak area | $K$ |
|---|---|---|---|---|
| Test 1 | 13 | 0.5 | 32,427.53 | $1.54 \times 10^{-5}$ |
| Test 2 | 10 | 0.5 | 26,015.42 | $1.92 \times 10^{-5}$ |
| Test 3 | 10 | 0.5 | 25,667.64 | $1.95 \times 10^{-5}$ |

**Table 9.7** Calculation of the calibration constant of nitrogen

| | Sample volume (mL) | Volume fraction (%) | Peak area | *K* |
|---|---|---|---|---|
| *Bottle 4* | | | | |
| Test 1 | 10 | 90 | 7,529,354.09 | $1.20 \times 10^{-5}$ |
| Test 2 | 10 | 90 | 7,469,558.98 | $1.21 \times 10^{-5}$ |
| Test 3 | 9 | 90 | 7,429,618.32 | $1.21 \times 10^{-5}$ |
| *Bottle 5* | | | | |
| Test 1 | 13 | 99.5 | 8,339,365.39 | $1.19 \times 10^{-5}$ |
| Test 2 | 10 | 99.5 | 8,342,276.79 | $1.19 \times 10^{-5}$ |
| Test 3 | 10 | 99.5 | 8,337,046.56 | $1.19 \times 10^{-5}$ |

### 9.3.7 Calibration Constant for Oxygen

For oxygen, the data and the results are visible in Table 9.8.

### 9.3.8 Calibration Constant for Ethylene

For ethylene, data and results are shown in Table 9.9.

### 9.3.9 Calibration Constant for Ethane

For ethane, data and results are shown in Table 9.10.

**Table 9.8** Calculation of the calibration constant of oxygen

| Bottle 2 | Sample volume (mL) | Volume fraction (%) | Peak area | *K* |
|---|---|---|---|---|
| Test 1 | 10 | 19.95 | 1,597,280.74 | $1.25 \times 10^{-5}$ |
| Test 2 | 10 | 19.95 | 1,597,499.02 | $1.25 \times 10^{-5}$ |
| Test 3 | 15 | 19.95 | 1,598,675.99 | $1.25 \times 10^{-5}$ |

**Table 9.9** Calculation of the calibration constant of ethylene

| Bottle 6 | Sample volume (mL) | Volume fraction (%) | Peak area | *K* |
|---|---|---|---|---|
| Test 1 | 10 | 100 | 9,871,250.99 | $1.01 \times 10^{-5}$ |
| Test 2 | 10 | 100 | 9,702,196.38 | $1.03 \times 10^{-5}$ |
| Test 3 | 20 | 100 | 9,600,094.28 | $1.04 \times 10^{-5}$ |

**Table 9.10** Calculation of the calibration constant of ethane

| Bottle 7 | Sample volume (mL) | Volume fraction (%) | Peak area | $K$ |
|---|---|---|---|---|
| Test 1 | 20 | 0.972 | 105,370.12 | $9.23 \times 10^{-6}$ |
| Test 2 | 11 | 0.972 | 102,544.15 | $9.48 \times 10^{-6}$ |
| Test 3 | 9 | 0.972 | 101,783.81 | $9.55 \times 10^{-6}$ |

**Table 9.11** Summary of the results

| | Retention time | $K$ |
|---|---|---|
| $H_2$ | 0.73–0.96 | $2.29 \times 10^{-5}$ |
| CO | 10.38–10.40 | $1.19 \times 10^{-5}$ |
| $CH_4$ | 9.28 | $1.53 \times 10^{-5}$ |
| $CO_2$ | 1.83–2.08 | $1.10 \times 10^{-5}$ |
| $H_2S$ | 6.64–6.69 | $1.80 \times 10^{-5}$ |
| $N_2$ | 8.17–8.19 | $1.20 \times 10^{-5}$ |
| $O_2$ | 7.82–7.83 | $1.25 \times 10^{-5}$ |
| $C_2H_4$ | 2.06–2.08 | $1.03 \times 10^{-5}$ |
| $C_2H_6$ | 3.30 | $9.42 \times 10^{-6}$ |

### 9.3.10 *Summary of the Results*

Table 9.11 shows a summary of the results obtained for each compound and also the typical retention times of each of them; the retention time allows to recognize the compound separated when the substances have of unknown composition.

## 9.4 An Example of Results from Elemental Analysis

### 9.4.1 *Instrumentation*

In order to assess the composition of certain liquid and solid samples, feeds and products, it is necessary an elemental analysis; for this purpose, a typical instrument used is the elemental analyzer (EA). An example will be given here (like for the case of GC). In the following text, a series of laboratory results will be given. The EA used is shown in Fig. 9.5. In particular, this instrument is a Flash EA1100 from Thermo Quest.

The EA returns as output a plot consists of several peaks (Fig. 9.6).

Each peak corresponds to an element among those measurable by the machine, more specifically according to the following order N–C–H–S. The peak areas, through certain coefficients, are proportional to the concentration of the element in the sample.

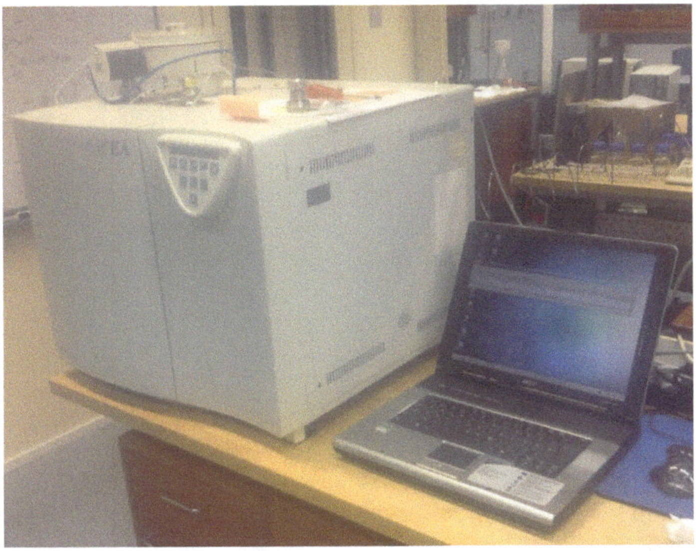

Fig. 9.5 Elemental analyzer

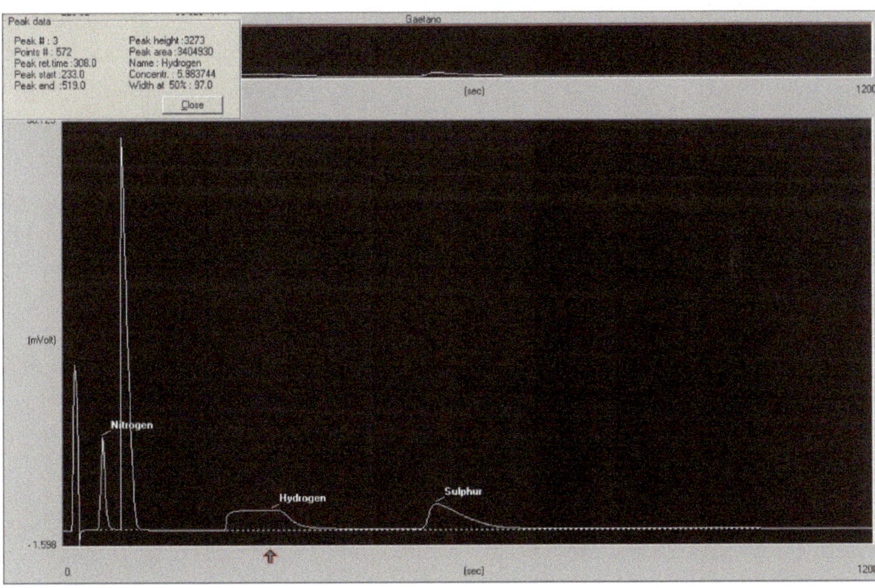


Fig. 9.6 Example of output of the elemental analyzer

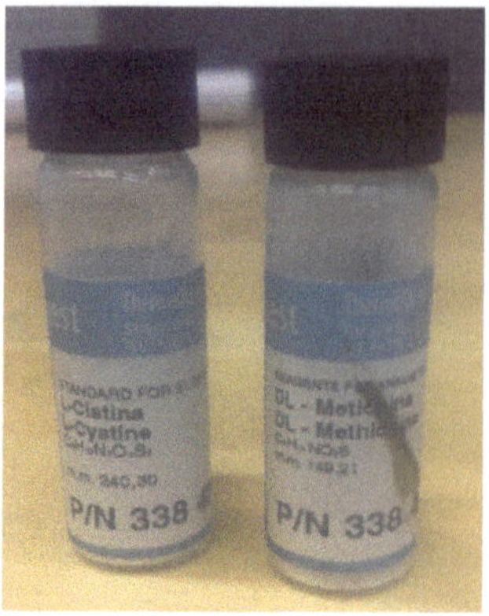


**Fig. 9.7** Reference substances

Also in this case, it is necessary to calibrate the machine through the preliminary study of known substances. After this procedure, it will be possible to evaluate the composition of a sample whose composition is not known.

In the same way as for the GC, in this case, a proportionality exists between the weight of the element in the sample and its peak area. Here the proportionality coefficient is evaluated by dividing the weight of the individual elements by the area shown on the output plot. Also the area obviously is related to the element considered.

$$K_{\text{element}} = \frac{\text{Weight}_{\text{element}}}{\text{Peak area}_{\text{element}}} \tag{9.2}$$

The analysis processes are similar to the case of GC, while only the final calculation changes; it is pointed out that in this case, we assume that a linear proportionality exists between the weight of the element and its measured area. However, this is an approximation since the proportionality coefficient could be also represented by a suitable function.

### *9.4.2 Calculation of the Calibration Coefficients*

To calculate the calibration constants, two substances with known composition have been taken into consideration. These substances are L-cystine and DL-methionine. Figure 9.7 shows the two reference samples.

Table 9.12 shows the weight percent of the elements in the two reference substances.

The following are examples of real measures and the determination of calibration coefficients for the reference substances. The tests are normally repeated to have averaged values and more accurate evaluation of the coefficients. Measures are as follows:

**Table 9.12** Elemental composition of the reference substances

| | L-Cystine(w%) | DL-Methionine (w%) |
|---|---|---|
| C | 29.99 | 40.25 |
| H | 5.03 | 7.43 |
| N | 11.66 | 9.39 |
| S | 26.69 | 21.49 |
| O | 26.63 | 21.45 |

**Table 9.13** Calibration coefficients

| | Weight (mg) | Area (mVs) | *K* |
|---|---|---|---|
| C | 1.889 | 7,979,878 | $2.367 \times 10^{-7}$ |
| H | 0.317 | 4,034,818 | $7.857 \times 10^{-8}$ |
| N | 0.735 | 1,395,280 | $5.268 \times 10^{-7}$ |
| S | 1.682 | 3,193,019 | $5.268 \times 10^{-7}$ |

Test n. 1: 6.30 mg of L-cystine (Table 9.13).
Test n. 2: 5.98 mg of L-cystine (Table 9.14).
Test n. 3: 5.24 mg of L-cystine (Table 9.15).
Test n. 4: 6.20 mg of DL-methionine (Table 9.16).
Test n. 5: 4.86 mg of DL-methionine (Table 9.17).

### *9.4.3 Summary of the Results*

The calibration coefficients can be derived from tests performed by averaging the results. The results obtained are reported in Table 9.18.

**Table 9.14** Calibration coefficients

| | Weight (mg) | Area (mVs) | *K* |
|---|---|---|---|
| C | 1.793 | 7,609,669 | $2.356 \times 10^{-7}$ |
| H | 0.301 | 3,892,033 | $7.734 \times 10^{-8}$ |
| N | 0.697 | 1,277,633 | $5.455 \times 10^{-7}$ |
| S | 1.596 | 3,040,440 | $5.249 \times 10^{-7}$ |

**Table 9.15** Calibration coefficients

| | Weight (mg) | Area (mVs) | *K* |
|---|---|---|---|
| C | 1.572 | 6,745,124 | $2.331 \times 10^{-7}$ |
| H | 0.264 | 3,404,930 | $7.754 \times 10^{-8}$ |
| N | 0.611 | 1,091,720 | $5.597 \times 10^{-7}$ |
| S | 1.399 | 2,676,692 | $5.227 \times 10^{-7}$ |

**Table 9.16** Calibration coefficients

| | Weight (mg) | Area (mVs) | *K* |
|---|---|---|---|
| C | 2.496 | 10,669,690 | $2.339 \times 10^{-7}$ |
| H | 0.461 | 6,102,209 | $7.555 \times 10^{-8}$ |
| N | 0.582 | 1,050,285 | $5.541 \times 10^{-7}$ |
| S | 1.332 | 2,583,730 | $5.155 \times 10^{-7}$ |

**Table 9.17** Calibration coefficients

| | Weight (mg) | Area (mVs) | *K* |
|---|---|---|---|
| C | 1.956 | 8,390,162 | $2.331 \times 10^{-7}$ |
| H | 0.361 | 4,756,623 | $7.589 \times 10^{-8}$ |
| N | 0.456 | 808,920 | $5.637 \times 10^{-7}$ |
| S | 1.044 | 2,060,116 | $5.068 \times 10^{-7}$ |

**Table 9.18** Summary of the results

| | Test 1 Cystine | Test 2 Cystine | Test 3 Cystine | Test 4 Methionine | Test 5 Methionine | Average results |
|---|---|---|---|---|---|---|
| $K_C$ | $2.37 \times 10^{-7}$ | $2.36 \times 10^{-7}$ | $2.33 \times 10^{-7}$ | $2.34 \times 10^{-7}$ | $2.33 \times 10^{-7}$ | $2.35 \times 10^{-7}$ |
| $K_H$ | $7.86 \times 10^{-8}$ | $7.73 \times 10^{-8}$ | $7.75 \times 10^{-8}$ | $7.56 \times 10^{-8}$ | $7.59 \times 10^{-8}$ | $7.70 \times 10^{-8}$ |
| $K_N$ | $5.27 \times 10^{-7}$ | $5.46 \times 10^{-7}$ | $5.60 \times 10^{-7}$ | $5.54 \times 10^{-7}$ | $5.64 \times 10^{-7}$ | $5.50 \times 10^{-7}$ |
| $K_S$ | $5.27 \times 10^{-7}$ | $5.25 \times 10^{-7}$ | $5.23 \times 10^{-7}$ | $5.16 \times 10^{-7}$ | $5.07 \times 10^{-7}$ | $5.19 \times 10^{-7}$ |

It is also important to notice that the element with the most variation is hydrogen. This is easily understood from the fact that the elongated shape of its peak is more affected by the error compared, for example, with the uniform peak of carbon.

In addition, the values obtained are very good approximations because the results of the various tests are not very different and a slight variation of them does not cause a significant error in the measurement of the concentration of the elements.

## 9.5 Analysis of Unknown Substances

### GC

Once the calibration constants of the GC are obtained, the volume concentration of a compound in the sample is calculated by:

$$\text{Volume fraction}_{\text{compound}} = \text{K}_{\text{compound}} \cdot \text{Peak area}_{\text{compound}} \tag{9.3}$$

**EA**

The weight of the elements of a sample of unknown composition is easily achieved with the help of the EA. In this case, it is necessary to multiply the peak area of the element for its constant (the proportionality coefficients obtained previously during the calibration).

$$\text{Weight}_{\text{element}} = \text{K}_{\text{element}} \cdot \text{Peak area}_{\text{element}} \tag{9.4}$$

For the analysis of liquids, it is a good practice to use a fixed inert and perform the analysis in the same way previously viewed. The inert material used in this study is Chromosorb (reagent for elemental analysis).

## Reference

De Blasio, C., Lucca, G., Özdenkci, K., Mulas, M., Lundqvist, K., Koskinen, J., et al. (2015). A study on supercritical water gasification of black liquor conducted in stainless steel and nickel-chromium-molybdenum reactors. *Journal of Chemical Technology and Biotechnology, 91*(10), 2664–2678. https://doi.org/10.1002/jctb.4871.

# Chapter 10
# Some Considerations and Statistical Derivations for the Concentration Profile and Gaussian Curve

*The time to be lost is never lost.*
A suggestion which is almost always valid.

In the previous discussion on the principles of chromatography, it was mentioned that it is convenient to express the chromatograms' curves with Gaussian functions. It is obvious that this assumption is reasonable and it reflects very well the reality. In fact, in case one element is flowing into a conduct after being injected in it, it will spread the concentration along with the flowing direction. This is because some molecules of that "packet" are slower and some others are faster. The spread will be in some way related to the diffusivity of that element in the particular carrier.

A Gaussian function is expressed as follows:

$$y = \frac{1}{\sigma\sqrt{2\pi}} \cdot \exp\left[-\frac{(x-\mu)^2}{2\sigma^2}\right] \quad (10.1)$$

where $\mu$ is denominated as the variance or the expected value (or mean value) of the function, while $\sigma$ is the standard deviation of the function and $\sigma^2$ is its variance. Equation (10.1) is a probability density function which is related to the measures in experiments and it is really much applied in science and engineering. Therefore, in the following discussion, we will demonstrate some of the features of this curve.

The Gaussian function is characterized by a symmetrical curve. For the standard normal distribution (if we consider the axes origin placed in the middle of the curve, the maximum is at $x = 0$, $y = 03999$), the curve has two inflection points at $x = \pm 1$, and for these points, the ordinate value is 0.242 (this is 60.6% of the maximum value). We know from literature that the mean of this probability density function should be $\mu$ (Montgomery 2003).



C. De Blasio, *Fundamentals of Biofuels Engineering and Technology*,
Green Energy and Technology, https://doi.org/10.1007/978-3-030-11599-9_10

It will be demonstrated here that the mathematical derivation of the mean value of a function (integral value) will give $\mu$. The mean value of the function $y = f(x)$ is:

$$\int_{-\infty}^{\infty} xf(x)\mathrm{d}x = \mu \tag{10.2}$$

In the resolution of Eq. (10.2), we solve for the indefinite integral and then we consider the boundaries. We substitute:

$$x - \mu = t, \rightarrow x = t + \mu \text{ and } \mathrm{d}x = \mathrm{d}t \tag{10.3}$$

for this reason, there is no change in *dx*, and *dx* is just substituted with *dt*. After this, we substitute:

$$\frac{(x-\mu)^2}{2\sigma^2} = y^2 \rightarrow \frac{t^2}{2\sigma^2} = y^2 \rightarrow t^2 = y^2 \cdot 2\sigma^2 \rightarrow t = \sqrt{y^2 \cdot 2\sigma^2} \tag{10.4}$$

In Eq. (10.4), the arrow $\rightarrow$ means "which gives". From Eq. (10.4) we have

$$t = \sqrt{2} \cdot \sigma \rightarrow y\mathrm{d}t = \sqrt{2} \cdot \sigma \cdot \mathrm{d}y \tag{10.5}$$

For this reason, substituting in our probability density function Eq. (10.1) we have:

$$f(x)\mathrm{d}x = \frac{1}{\sigma\sqrt{2\pi}} \cdot \exp[-y^2] \cdot \sqrt{2} \cdot \sigma \cdot \mathrm{d}y \tag{10.6}$$

From:

$$(x-\mu)^2 = y^2 \cdot 2\sigma^2 \rightarrow x = \sqrt{y^2 \cdot 2\sigma^2} + \mu \rightarrow x = \sqrt{2} \cdot \sigma \cdot y + \mu; \tag{10.7}$$

and therefore:

$$\int xf(x)\mathrm{d}x = \left(\sqrt{2} \cdot \sigma \cdot y + \mu\right) \cdot \frac{1}{\sqrt{\pi}} \cdot \exp[-y^2]\mathrm{d}y \tag{10.8}$$

We remember the method of integration per parts:

$$\begin{aligned} D[f(x) \cdot g(x)] &= f'(x) \cdot g(x) + g'(x) \cdot f(x) \\ &\Rightarrow \int D[f(x) \cdot g(x)] \\ &= \int f'(x) \cdot g(x) + \int g'(x) \cdot f(x) \end{aligned} \tag{10.9}$$

where $f(x)$ and $g(x)$ are two diverse functions of the variable $x$. And as a consequence:

$$\int f'(x) \cdot g(x) = [f(x) \cdot g(x)] - \int g'(x) \cdot f(x) \tag{10.10}$$

Integrating per parts Eq. (10.8):

$$\int \left(\sqrt{2} \cdot \sigma \cdot y + \mu\right) \cdot \frac{1}{\sqrt{\pi}} \cdot \exp[-y^2]\mathrm{d}y = \frac{\sqrt{\pi}}{\sqrt{\pi} \cdot 2} \cdot \left(\sqrt{2} \cdot \sigma \cdot y + \mu\right) - \int \frac{\sqrt{\pi}}{2} \cdot \frac{\sqrt{2} \cdot \sigma}{\sqrt{\pi}} \mathrm{d}y \tag{10.11}$$

and finally we obtain:

$$\frac{\sqrt{2}}{2} \cdot \sigma \cdot y + \frac{\mu}{2} - \frac{\sqrt{2}}{2} \cdot \sigma \cdot y = \frac{\mu}{2} \tag{10.12}$$

and this result has to be multiplied by 2 because we have integrated only between zero and infinite. In this way, Eq. (10.2) is proven.

## Reference

Montgomery, D. C. (2003). *Applied statistics and probability for engineers* (3rd ed.). New York, USA: Wiley.

# Part II
# Notions of Biochemical Reactors Design and Biofuels Production

# Chapter 11
# Introduction to Part 2

*The most terrible of misfortunes, the death, means nothing for us. Since when we are living, there is no death and when the death is there, we don't exist anymore.*
Epicurus, Letter to Menoeceus, Third century B.C.

Liquid fuels are particularly suitable for our needs. They contain large amount of energy in terms of calorific value, they are easily transported and most of all, and the power delivered by this kind of fuels is constantly the same, no matter the amount of the fuel present in the tank of your car. This is a quite important aspect of liquid fuels which is not usually mentioned. If we think about batteries, indeed there are also positive aspects with their usage, and however, it is also to be realized that they are heavy and the weight of the batteries is carried around at all times no matter if there is no power anymore left in them. Additionally, the power delivered by fully charged batteries is different from when the battery is almost empty. As a matter of fact, the differential voltage is indeed correlated to the discharge of the battery (Kato et al. 2018). This can be soon realized after buying a simple electric scooter and using it a bit.

Liquid fuels can be produced by using biomass feedstock, and therefore, this would make them renewable energy sources. The kinds of feedstock that can be used are also different; for instance, algae can be utilized for producing biodiesel (Khan et al. 2017), and also cellulose can be utilized to produce ethanol (Singh et al. 2018).

In order to reduce oil dependency, one of the strategies proposed usually is to utilize biomass for the production of fuels and also for the production of diverse chemicals and materials (Klass 1998). To support this sentence, we see from the following graph (Fig. 11.1) that the use of domestic wood can be increased sustainably in the future.

An industrial plant where operations are done on biomass to produce energy, fuels, chemicals, or materials is usually defined as a biorefinery. The goal in a biorefinery is that all the fractions and sidestreams are efficiently converted into products and that these products could be reused and sold on the market; this is directly related to



C. De Blasio, *Fundamentals of Biofuels Engineering and Technology*,
Green Energy and Technology, https://doi.org/10.1007/978-3-030-11599-9_11

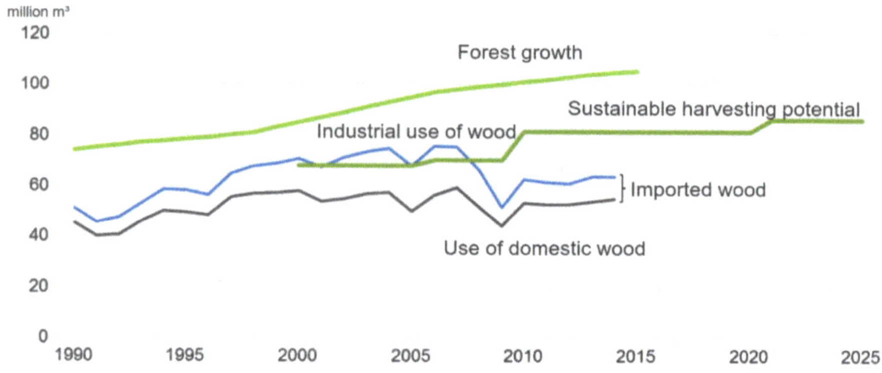


**Fig. 11.1** Forest growth plotted with the sustainable harvesting potential and industrial use of wood. Metsä Group. With permission, the figure is based on Natural Resource Institute Finland 2015 (Luke)

the sustainability of the activities carried on in a biorefinery, and this is the basis for the development of a bio-economy (Palmeros Parada et al. 2017). In this way, the waste is minimized, and the production should be more sustainable. The definition given by the International Energy Agency, IEA, for a biorefinery is: "Biorefinery is the sustainable processing of biomass into a spectrum of marketable products and energy" (Liguori and Faraco 2016). Or, as stated by Valmet "Biorefining means creating more value from biomass" (Valmet 2014). Biorefineries can be classified, for example, based on the used feedstock or technology and this kind of processes are utilizing biomass in a more sustainable way and also in a more productive way. As a matter of fact, three aspects should be considered when biorefinery is considered (Pandey et al. 2015). The first point is that the process should be economically attractive, and this means that all the products have a market, and they can be sold at competitive price. The profitability of the process can be improved by increasing the efficiency of the biomass utilization. The biorefinery gives also less pollution and less emission, and therefore, this is the meaning for sustainability. In the literature, novel biorefinery processes are given (Özdenkçi et al. 2017), and possible options could be also represented by the integration of technologies like supercritical water gasification of biomass and hydrothermal liquefaction (De Blasio et al. 2015) even if some kinds of feedstock would give problems when this kind of technologies are adopted (De Blasio et al. 2019).

Biofuels can be produced from diverse kinds of feedstock, and this would be the basis for the classification of biofuels in first generation, second generation, and third generation. For instance, the first-generation biofuels are produced from vegetable oil, sugars, starch, or animal fats; the second generation is produced from feedstock which is not related to food crops, while the third-generation biofuels are those produced from algae.

Concerning the first-generation biofuels, there are concerns that this kind of biofuels' production is not even ethical since producing crops not for food purposes but

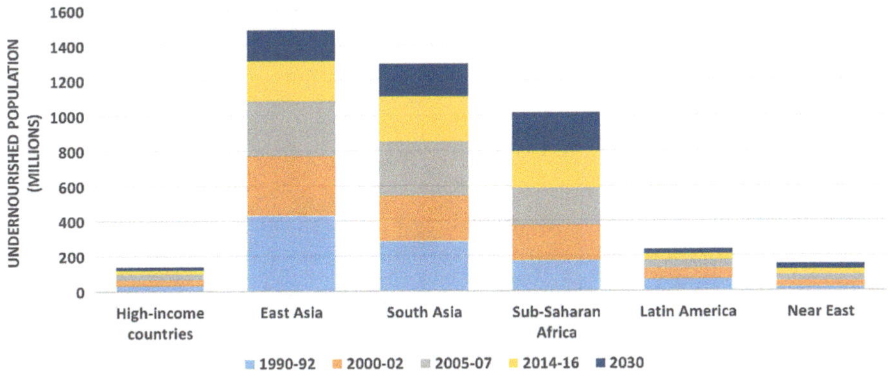


**Fig. 11.2** Undernourished population in the world by region. Data from Statista (2018a)

for driving cars is not the best idea thinking that in the world there are millions of children actually dying because of food scarcity.

In fact, the children dying every year for this reason are millions, and they represent almost 50% of all the children's deaths (UNICEF, World Health Organization, and The World Bank 2018).

Because of the food we do not actually need (the one in excess), we are actually occupying precious land which could be utilized for other purposes. This is because we consume extra energy in our activities, for instance, walking.

A simple example will be given here. In the literature, it is possible to find estimations of the consumption of energy of a human being at running or walking, but many of this data are not congruent, and therefore, in this case some assumptions will be done.

Let us assume here that the consumption of energy at walking is 0.456 kcal for each kilogram of carried extra weight and for each kilometer of walking distance. We assume also that a person normally walks only 3 km per day. Another assumption is that in a very nice season the production of potatoes could be 22.7 kg for square meter, and this is very high over-estimation since this normally does not happen. Now, in the world there are 7.53 billion people (Statista 2018b), and unfortunately, 796 million people are undernourished as can be derived from Fig. 11.2.

Subtracting the undernourished population from the total population, we assume here that half of the remaining fraction has at least 3 kg of overweight, and this would be 3.37 billion people. For the energy content of potatoes, it was chosen here the value of 0.81 kcal/kg. Also in this case, there are many different values given in the literature. The potato peel, for instance, has been reported to have energy content very similar to coal (Uzun et al. 2017). In short, it would be needed half hectare just to provide the energy required to transport some extra weight.

It is clear that the utilization of crops for replacing oil decreases the land area available for food production. Therefore, this would be a non-sustainable solution.

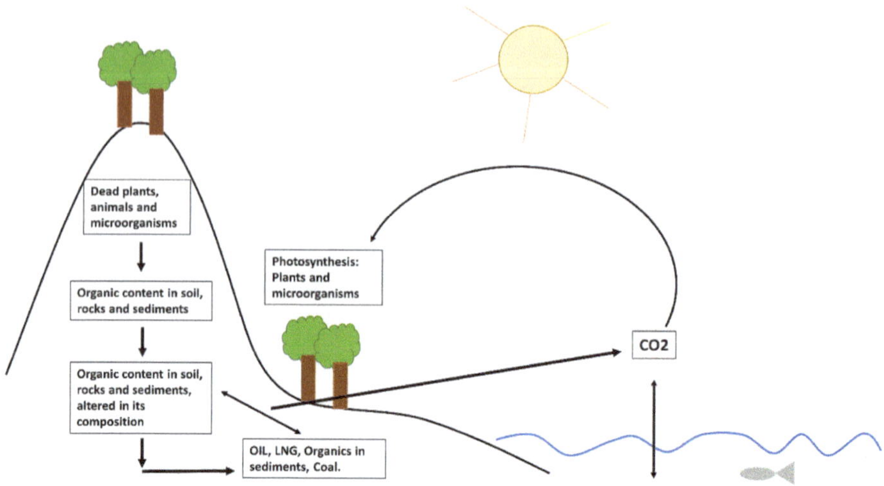


**Fig. 11.3** Carbon cycles and the production of fossil fuels

The usage of lignin and residues from industrial processes have been considered to be much more positive for biofuels production (Liguori and Faraco 2016).

The environmental issues play a major role in the production of biofuels. Considering forest harvesting, the rate of harvesting should not exceed the annual growth in the long term. On the other hand, water resources are also very important to preserve, and the local ecosystems should not be compromised. Furthermore, the production concept should be socially acceptable, meaning that the social acceptability should also be considered here. A commonly accepted criterion is that the utilization and production of the raw materials should not compete with food production, as stated previously. And this is one reason why the first-generation biofuel production should not be significantly increased.

Since the production of biofuels involves the transformation of organic molecules (mainly the ones containing carbon), it is beneficial to remember that carbon is present in many forms in nature and it can be subject to cycles. We remember from the first part of this book that the carbon cycle, and especially the equilibrium between $CO_2$ in the atmosphere, the oceans, and the carbonate rocks, plays a very important role since a shift in this equilibrium can end up with disastrous consequences. In Fig. 11.3, a very simple schematics is depicted with respect to the carbon cycles and the production of fossil fuels, while for a more detailed information on the water–$CO_2$ equilibria, the author invites the reader to consult the introductory section of this book. From these equilibria, it could be even argued that oil and coal could be considered as "renewable," as a matter of fact it takes "only" millions of years for the organic matter to deposit under the ground and then, by a series of geological events, starting with the diagenesis, reach the proper conditions to form oil. This is called catagenesis, and this process is not fast either.

Even rocks are formed by enormous quantities of skeletal shells; in fact, the limestone minerals can be divided into four groups: non-skeletal grains, skeletal grains, micrite, and cement, while, on the other hand, the diagenesis of carbonates involves the processes of cementation, microbial, recrystallization, micritization, neomorphism, dissolution, compaction, and dolomitization (De Blasio 2010). For this reason, limestone can be under the form of calcite, and it can be also in the form of dolomite rocks, $CaMg(CO_3)_2$.

To give some examples of what kind of organisms are forming limestone rocks, we can have radiolarians, diatoms, sponges, and, also in this case, algae. Among the geochemical processes, we have that these processes have importance since the degree of compaction and crystallization of minerals affects the limestone dissolution itself (De Blasio et al. 2018).

Coming back to our catagenesis and the formation of crude oil, if we take into account angiosperms plants living in the Paleogene and Neogene periods (this means from 65 million years to around 3 million years ago), then it would take all this time for this organic deposits to form kaurane (Kashirtsev et al. 2012).

Living cells, or enzymes, will become more common in the future for the purpose of biofuels production. Just to give one example, biofuels can be produced by the fermentation of soybeans (Loman and Ju 2016). Therefore, the fermentation process will become more and more important in the coming years, and that is why the biochemical engineering concepts should be carefully understood by future engineers working in the field of biorefineries. From this need comes one of the goals of this manuscript: the understanding of biological processes in power production and biochemical reaction engineering.

There is a series of companies around the world where biomass is converted and biofuels are produced or converted. Finnish forest and oil companies have developed novel concepts to refine biomass into bioproducts and biofuels. Some of these concepts are Neste's biodiesel and St1's bio-ethanol production concepts. The raw materials that have been mainly utilized in biorefineries include maize, soybean, sugarcane, i.e., the crops which are cultivated on arable land (Cherubini 2010).

Biorefineries can also produce chemicals, biomaterials, and biofuels. In addition, in biorefineries there is the possibility to produce power, heat, or steam as a side product. Considering the actual economical situation and the fact that resources are depleting, the marked for bio-based fuels and chemical will increase in the coming years.

Today, there is more and more regulations on environmental issues and incentives to substitute oil with biomass-based fuels and alternative fuels. Therefore, process integration has become the main drivers for companies to develop sustainable production processes. Especially in the case of biofuels and power, companies are encouraged to deliver eco-friendly solutions by governmental financial support. Furthermore, there is more acceptance from the public toward companies that are utilizing bio-based feedstock. Companies can get advantage when investing in renewable bio-based products and chemicals. One example is the collaboration between the companies Lego and Ikea who have declared the will to replace products with bio-based alternatives (Neste Corporation 2016).

Biodiesel is produced in large quantities by diverse companies, and one example is the Finnish company Neste. The feedstock in this case is vegetable oil or waste oil. Concerning the utilization of materials containing high amount of lignin, enzymes are used in large quantities during the fermentation process for the production of bioethanol (Kubicek and Kubicek 2016). Techniques like the CRISPR–cas9 gene editing system developed by the University of California (Schwartz et al. 2016) together with large amounts of gene sequencing data will greatly improve the efficient production of biofuels (Kubicek and Kubicek 2016). This will be done also because of the always better molecular biology technologies.

Genetics will play a sure role in the development of microorganisms, for instance algae, with the purpose of improving their growth rate, or even their capability of transforming the harvested energy. Even if it will be difficult to improve the almost perfect way, the photosynthesis operates. In fact, artificial photosynthesis will also continue to be researched in the future along with other light-harvesting machines.

As it was provided in the first part of this manuscript, the amount of energy coming from the sun is indeed enormous and we should improve the ways this energy is stored. The sun is driving many processes including the carbon cycles reported previously. New ways of utilizing $CO_2$ will be developed, for instance, by means of methanation, since $CO_2$ can be processed with hydrogen to give methane. Catalysis, of course, will be central in the production of biofuels, and new technologies will be implemented to produce structures where catalysts are allocated.

## References

Cherubini, F. (2010). The biorefinery concept: Using biomass instead of oil for producing energy and chemicals. *Energy Conversion and Management, 51,* 1412–1421. https://doi.org/10.1016/j.enconman.2010.01.015.

De Blasio, C. (2010). *Reactive dissolution of sedimentary rocks in flue gas desulfurization*. Painosalama OY: Modeling and Experimental Investigation.

De Blasio, C., Carletti, C., Salonen, J., & Björklund-Sänkiaho, M. (2018). Ultrasonic power to enhance limestone dissolution in the wet flue gas desulfurization process. Modeling and results from stepwise titration experiments. *ChemEngineering, 2,* 53. https://doi.org/10.3390/chemengineering2040053.

De Blasio, C., De Gisi, S., Molino, A., Simonetti, M., Santarelli, M., & Björklund-Sänkiaho, M. (2019). Concerning operational aspects in supercritical water gasification of kraft black liquor. *Renewable Energy, 130,* 891–901. https://doi.org/10.1016/j.renene.2018.07.004.

De Blasio, C., Lucca, G., Özdenkci, K., Mulas, M., Lundqvist, K., Koskinen, J., et al. (2015). A study on supercritical water gasification of black liquor conducted in stainless steel and nickel-chromium-molybdenum reactors. *Journal of Chemical Technology and Biotechnology, 91,* 2664–2678. https://doi.org/10.1002/jctb.4871.

Kashirtsev, V. A., Gaiduk, V. V., Chalaya, O. N., & Zueva, I. N. (2012). Geochemistry of biomarkers and catagenesis of organic matter of Cretaceous and Cenozoic deposits in the Indigirka-Zyryanka basin (northeastern Yakutia). *Russian Geology and Geophysics, 53,* 787–797. https://doi.org/10.1016/j.rgg.2012.06.006.

Kato, H., Kobayashi, Y., & Miyashiro, H. (2018). Differential voltage curve analysis of a lithium-ion battery during discharge. *Journal of Power Sources, 398,* 49–54. https://doi.org/10.1016/j.jpowsour.2018.07.043.

Khan, S., Siddique, R., Sajjad, W., Nabi, G., Hayat, K. M., Duan, P., et al. (2017). Biodiesel production from algae to overcome the energy crisis. *Hayati Journal of Bioscience, 24,* 163–167. https://doi.org/10.1016/j.hjb.2017.10.003.

Klass, D. L. (1998). *Biomass for renewable energy, fuels, and chemicals*. Academic Press.

Kubicek, C. P., & Kubicek, E. M. (2016). Enzymatic deconstruction of plant biomass by fungal enzymes. *Current Opinion in Chemical Biology Energy Mechanistic Biology, 35,* 51–57. https://doi.org/10.1016/j.cbpa.2016.08.028.

Liguori, R., & Faraco, V. (2016). Biological processes for advancing lignocellulosic waste biorefinery by advocating circular economy. *Bioresource Technology Waste Biorefinery—Advocating Circular Economy, 215,* 13–20. https://doi.org/10.1016/j.biortech.2016.04.054.

Loman, A. A., & Ju, L. -K. (2016). Soybean carbohydrate as fermentation feedstock for production of biofuels and value-added chemicals. *Process Biochemistry, 51,* 1046–1057. https://doi.org/10.1016/j.procbio.2016.04.011.

Neste Corporation. (2016). Neste ja IKEA Sweden kumppaneiksi uusiutuvan biomuovin tuotannossa.

Özdenkçi, K., De Blasio, C., Muddassar, H. R., Melin, K., Oinas, P., Koskinen, J., et al. (2017). A novel biorefinery integration concept for lignocellulosic biomass. *Energy Conversion and Management, 149,* 974–987. https://doi.org/10.1016/j.enconman.2017.04.034.

Palmeros Parada, M., Osseweijer, P., & Posada Duque, J. A. (2017). Sustainable biorefineries, an analysis of practices for incorporating sustainability in biorefinery design. *Industrial Crops Products Challenges in Building a Sustainable Biobased Economy, 106,* 105–123. https://doi.org/10.1016/j.indcrop.2016.08.052.

Pandey, A., Hofer, R., Larroche, C., Taherzadeh, M., & Nampoothiri, M. (2015). *Industrial biorefineries and white biotechnology.* Elsevier.

Schwartz, C. M., Hussain, M. S., Blenner, M., & Wheeldon, I. (2016). Synthetic RNA polymerase III promoters facilitate high-efficiency CRISPR-Cas9-mediated genome editing in Yarrowia lipolytica. *ACS Synthetic Biology, 5,* 356–359. https://doi.org/10.1021/acssynbio.5b00162.

Singh, R., Kemausuor, F., & Wooldridge, M. (2018). Locational analysis of cellulosic ethanol production and distribution infrastructure for the transportation sector in Ghana. *Renewable and Sustainable Energy Reviews, 98,* 393–406. https://doi.org/10.1016/j.rser.2018.09.017.

Statista. (2018a). *Forecast: Global undernourished population 1990–2030, by region* | Statistic [WWW Document]. Statista. https://www.statista.com/statistics/678742/underunourished-population-worldwide-by-region/. Accessed December 31, 2018.

Statista. (2018b). *World—Total population 2007–2017* | Statistic [WWW Document]. Statista. https://www.statista.com/statistics/805044/total-population-worldwide/. Accessed December 31, 2018.

UNICEF, World Health Organization, & The World Bank. (2018). *Levels and trends in child malnutrition.* UNICEF / WHO / World Bank Group Joint Child Malnutrition Estimates Key findings of the 2018 edition.

Uzun, H., Yıldız, Z., Goldfarb, J. L., & Ceylan, S. (2017). Improved prediction of higher heating value of biomass using an artificial neural network model based on proximate analysis. *Bioresource Technology, 234,* 122–130. https://doi.org/10.1016/j.biortech.2017.03.015.

Valmet. (2014). *Valmet technical paper series: Bio-refining.*

# Chapter 12
# Integrated Biorefinery Concepts

*Of course, when nonsense comes out of a computer people have a lot of respect for it, and that is exactly the problem.*
Gianni Astarita, Chem. Eng. Sci. 52, 24, 4681–4698.

## 12.1 An Introduction to Biorefineries and to the Sustainable Utilization of Bioresources

Different solutions have been proposed as biorefinery concepts and therefore we have different technologies available with different products.

One simple classification related to biorefineries divides them in "single product" processes and "integrated process" (Svensson et al. 2015). In an integrated biorefinery, it is possible to get both the bio-based products, for instance pulp or biochemical, and energy products such as electricity, heat, or biofuels (Pandey et al. 2015). It is reported that process integration can have a significant influence to the profitability of the biorefinery and a suitable product diversification can help companies to manage periods of low demand and crisis (Pandey et al. 2015). The integration of processes allows for the reduction of transportation and energy consumption, and in the case there is a surplus of energy produced, this can be sold or utilized for other purposes. In Finland, the pulp and paper industry has a leading role on the economy of the country; in addition, the principal by-product coming from this process, black liquor, is considered as the most important fuel and also the most abundantly used. To give one example, 11% of the total heat and power produced in the nation is derived by this fuel (Helynen et al. 2002). As a matter of fact, black liquor is used in recovery boilers where this fuel is combusted and the chemicals contained in it are recovered by suitable reduction–oxidation processes.

A "multi-feed, multi-purpose" biorefinery concept which involves the processing of black liquor has been proposed by Özdenkci and collaborators (2017). Process integration of the biorefinery concept and improvements of the costs with reduction of the power used can be started from the supply. The supply-chain network structure



C. De Blasio, *Fundamentals of Biofuels Engineering and Technology*,
Green Energy and Technology, https://doi.org/10.1007/978-3-030-11599-9_12

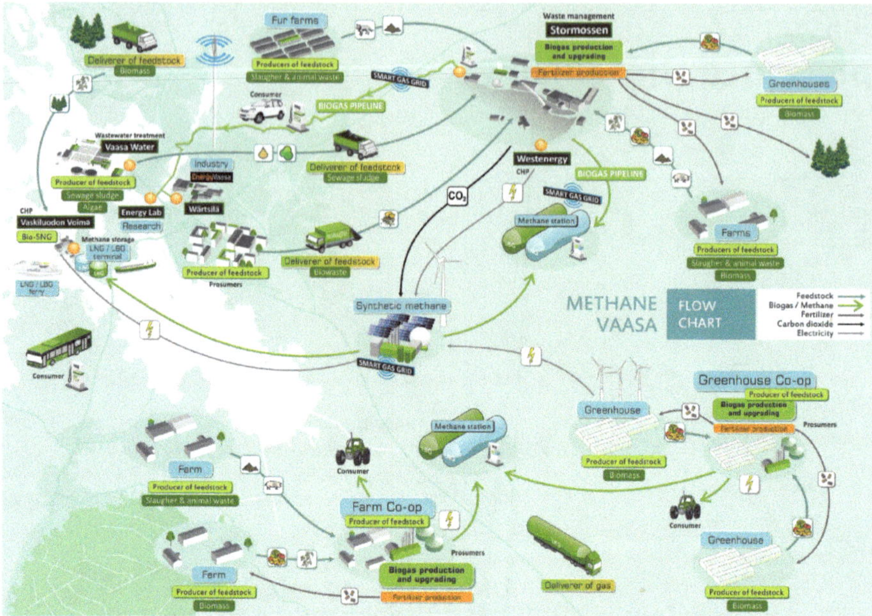


**Fig. 12.1** Integration concept of regional biorefinery. The Vaasa region, Finland. Figure made by Lena Karlsson. Reprinted by permission of Vaasa Water—Public Utility Company

and even the position of the biorefinery have to be planned carefully with respect to the resources and the companies present on the territory. To give one example of suitable conditions for the concept of process integration of biorefinery, the Vaasa region in Finland shown good potential since there is a discreet number of feedstock producers, natural resources, and biogas facilities as shown in Fig. 12.1.

Sectorial integration network can be taken into account together with new technologies; this will include pretreatment of the feedstock where it is available and regional distribution of converted biomass from various sectors. This is followed by the upgrading of the produced biofuels and the separation of the pollutants which will be recovered further. A sectoral integration concept example aimed at a supply-chain network is shown in the Fig. 12.2.

This could reflect the structures of the supply chain; in fact, there are mainly three types of supply-chain networks: the centralized, distributed, and also a hybrid of them known as distributed–centralized. This is described by the following Fig. 12.3.

On the other hand, we can have different approaches concerning the way we utilize biomass. For instance, we can partly decompose the biomass and obtain side streams and products which will be separated end/or upgraded. In alternative, we could proceed to the complete destruction of the biomass and obtain syngas which could, in turn, give further products or it could be utilized in an engine. The third option would be to fractionate the biomass feedstock to give, for instance, cellulose, lignin, and hemicellulose. These compounds could be treated to have chemicals,

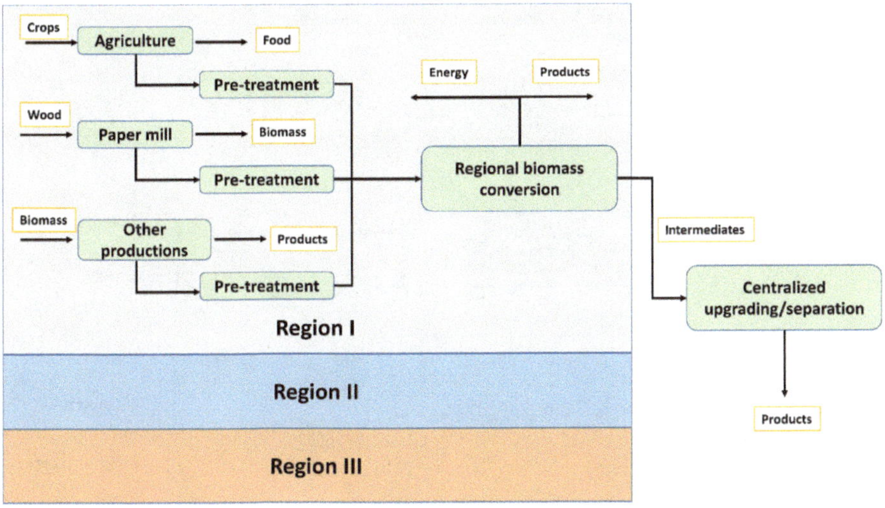


**Fig. 12.2** Sectoral integration concept aimed at the supply-chain network

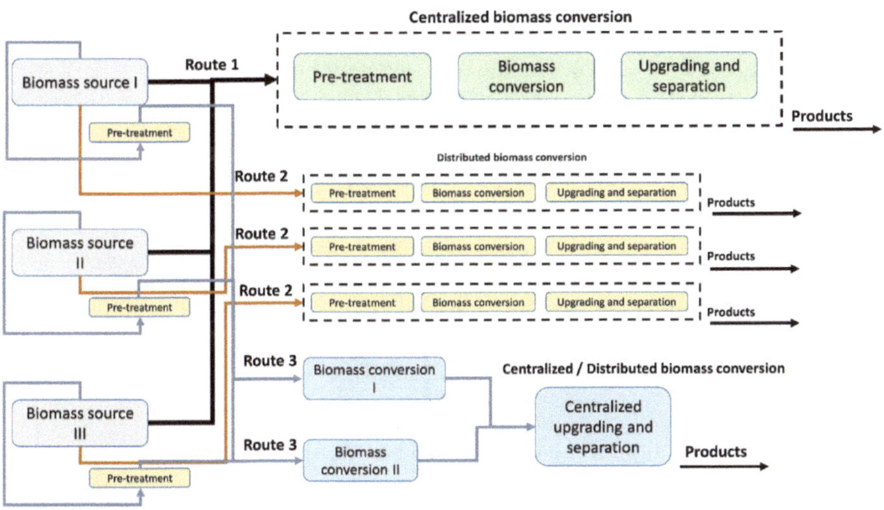


**Fig. 12.3** Supply-chain structures. The centralized biomass conversion (Route 1), the distributed biomass conversion (Route 2), and the centralized–distributed biomass conversion

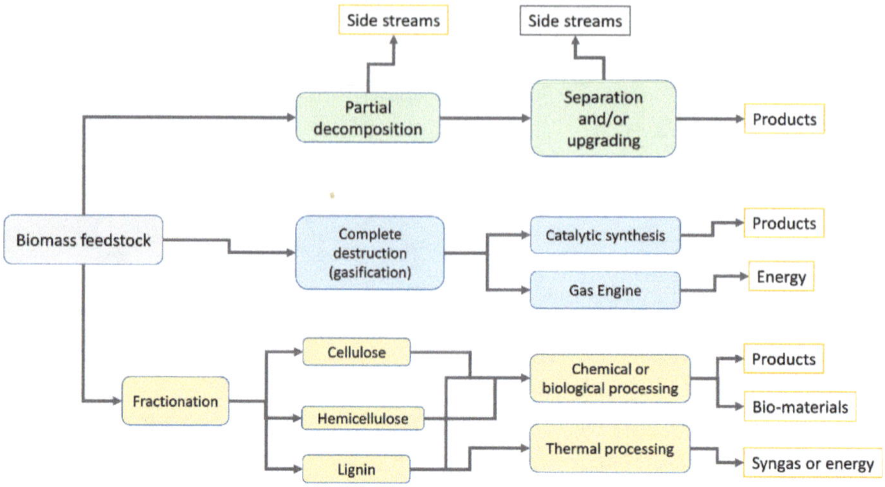


**Fig. 12.4** Different approaches in biomass processing

fibers, or they could be processed with a thermal treatment to give syngas or in addition, sent to a chemical processing unit to obtain biomaterials (Fig. 12.4).

Among the biological processing of biomass, also in this case, we could distinguish diverse kinds of approach (Ishola et al. 2015, 2013; Kim et al. 2016; Li et al. 2016; Zheng et al. 2009):

- saccharification and fermentation. Done at the same time (SSF);
- saccharification and co-fermentation performed simultaneously;
- consolidated bioprocessing (CBP);
- saccharification, followed by filtration and fermentation (SSFF). The steps are done simultaneously;
- genetically engineered yeast (Kwak et al. 2018);
- separate hydrolysis and fermentation (SHF);

while the fractionation of the biomass can be done by means of diverse methods like:

- alkaline treatment (Kim et al. 2016);
- organosolv treatment with alcohol or acid (Choi et al. 2019)

Biomass can be processed by thermal treatment and this could include:

- Pyrolysis. This includes fast pyrolysis (580–980 °C, 10–200 K/s, residence time of 10–200 s) and slow pyrolysis (280–680 °C, 0.1–10 K/s, residence time of 450–550 s) (Balat et al. 2009a);
- thermal gasification (Balat et al. 2009b);
- hydrothermal processes (Özdenkçi et al. 2017)

  • aqueous phase reforming;

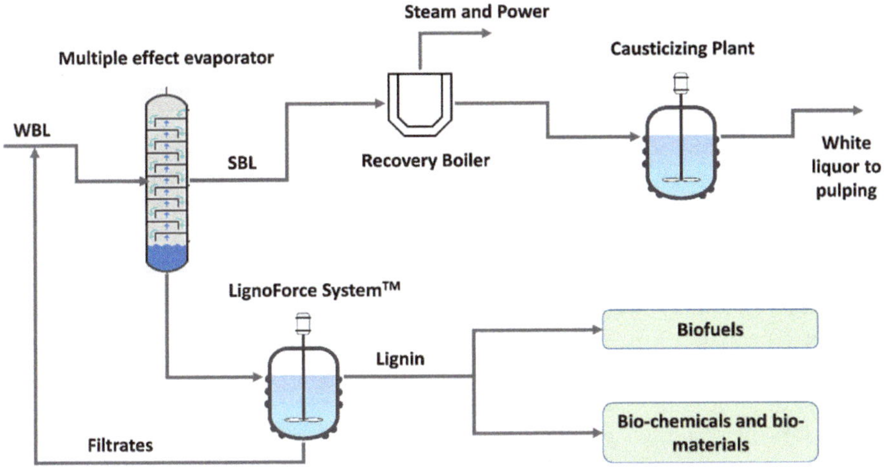


**Fig. 12.5** LignoForce™ System, WBL weak black liquor, SBL strong black liquor

- carbonization;
- hydrothermal liquefaction;
- oxidation and gasification (SCWO and SCWG);

– partial wet oxidation (130–270 °C, 30 min residence time, partial oxygen pressure of 0.5 MPa) (Muddassar et al. 2015, 2014; Özdenkci et al. 2014; Sipponen et al. 2016).

Regarding the PWO of biomass, this is a quite interesting process where the partial oxidation of the biomass fraction takes place. It is often suggested that this treatment could be coupled with other processes since this is an exothermic step where heat could be produced to run side processes which could treat the biomass further. The PWO gives also the possibility to transform difficult feedstock like lignin into carboxylic acids, for instance, and these products will be more easily processed by other methods like SCWG or HTL.

In industry, the LignoForce system is a novel process developed by FPinnovations in Canada and this process is starting with the PWO of the black liquor fraction to oxidize it. The sulfur is also oxidized by this method and, as a result, this is improving the lignin recovery. When the black liquor is acidified, there is the formation of sulfur compounds which could be volatile. By this method, the sulfur compounds are transformed into non-volatile species, and the lignin is separated into two forms, acid form and sodium form. The latter one has been discovered to be particularly suitable as a substitute of phenol–formaldehyde. In addition, lignin could be utilized instead of polyols and polyurethane and in thermoplastics. Lignin could also be used as dispersant and to substitute flocculants and textile industry (Schreiber et al. 2014).

By means of the LignoForce system, lignin can be precipitated by reducing the pH of the partial wet oxidized black liquor. This can be achieved by pumping $CO_2$

or adding sulfuric acid to the solution. The pH value should be reduced to 9–10 from an original value of 13–14 (Kouisni et al. 2016). The process should be followed by a coagulation of the lignin molecules, a filtration process and a washing process which is using sulfuric acid and water.

Compounds like dimethyl sulfide and disulfide, methyl mercaptans, and hydrogen sulfide are totally reduced forms of sulfur (TRS) which are very dangerous, and therefore, in case the acidification of the lignin, slurry is done in presence of pressurized oxygen, this compounds would be oxidized to non-volatile species. The schematics of the process is shown in Fig. 12.5.

In HTL of biomass, we have temperatures between 280 and 380 °C and pressures between 70 and 300 bar. At these conditions, the self-dissociation of water (see the description of hydrothermal gasification explained later in this part of the manuscript) increases significantly and this will mean that there will be more concentration of $H^+$ and $OH^-$ ions allowing for subcritical water to catalyze acid–base reactions. It is referred in literature that this process produces bio-oil with less amount of oxygen (Tekin et al. 2014) and this means that we would need less hydrogen to upgrade it to liquid fuels. In addition, homogeneous and heterogeneous catalysts can be applied also in this case.

Alkali salts are used as catalysts and they are reported to accelerate the water gas shift reaction; on the other hand, potassium salts minimize the formation of char and improve the oil yield (Christensen et al. 2014). As heterogeneous catalysts, we recognize zirconium oxides and nickel alloys.

HTL can lead to a yield of 30–40% of oil (dry basis on the feedstock) if we consider lingo-cellulosic feedstock. The energy recovery is usually around 80% (Toor et al. 2011). However, there is also literature which is not that encouraging with respect to this process. In fact, it is referred that if we take into account a techno-economic analysis of the process, this will be not really competitive with respect to other processes like the ones of petroleum industry (Zhu et al. 2014).

HTL, SCWG, and PWO could be integrated to have a regional system flexible to different kinds of biomass feedstock. An example of an integrated system which takes black liquor as incoming feedstock is shown in Fig. 12.6. The system foresees the recovery of chemicals and energy with the production of syngas and biofuels (Fig. 12.6).

The process conditions for the system proposed in Fig. 12.6 are written in Table 12.1. The conditions suggested by Özdenkci and the author of this book are according to the suggestions given in literature and on the experience gained in research laboratories with particular reference to the case of PWO and SCWG of biomass.

Another example is given here in regard to the utilization of waste organic streams. Waste organic streams could feed algae which, in turn could be used for HTL. The anaerobic fermentation could give volatile fatty acids, VFA, which could be processed by esterification and hydrogenation to give ethanol, butanol, and propanol. The VFAs could be further processed by yeast fermentation and give bio-oil (Fig. 12.7).

According to the cited reference, integration could include also the introduction of plastics into the HTL. However, this option is not taken into account here, considering

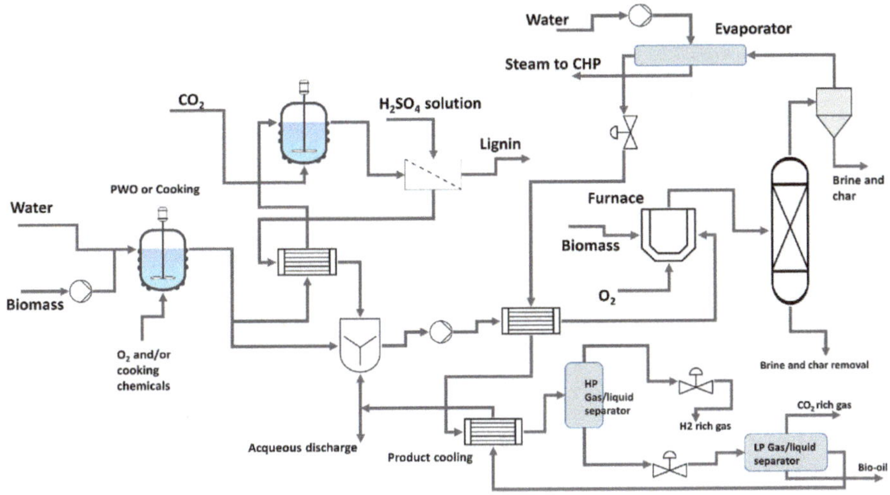


**Fig. 12.6** An example of process for regional biomass conversion. Adapted from Özdenkçi et al. (2017)

**Table 12.1** Reaction conditions for the process of regional biomass conversion

| Unit | Conditions | Comments |
|---|---|---|
| Acidification | *T*: 80 °C<br>*P*: 1 Atm | Lignin precipitation at pH 9–10 according to literature data |
| PWO | *T*: 170–240 °C<br>$P_{O2}$: 0.5–1 MPa<br>*t*: 10–20 min | Heating of the biomass feedstock due to the exothermic conditions |
| The conversion reactor | *T*: 250–350 °C<br>*P*: 4–22 MPa<br>*t*: 1–12 min | HTL |
| The conversion reactor | *T*: 600–700 °C<br>*P*: 25 MPa or above<br>*t*: 1–5 min | SCWG |
| HP G/L separator | *T*: 80–250 °C<br>*P*: 25 MPa or above | $H_2$-rich gas separation, in case of SCWG |
| LP G/L separator | *T*: 80 °C or less<br>*P*: 1 Atm | $CO_2$-rich gas separation, in case of SCWG<br>or<br>Separation of aqueous valuable compounds and bio-oil if we have HTL |

Adapted from Özdenkçi et al. (2017)

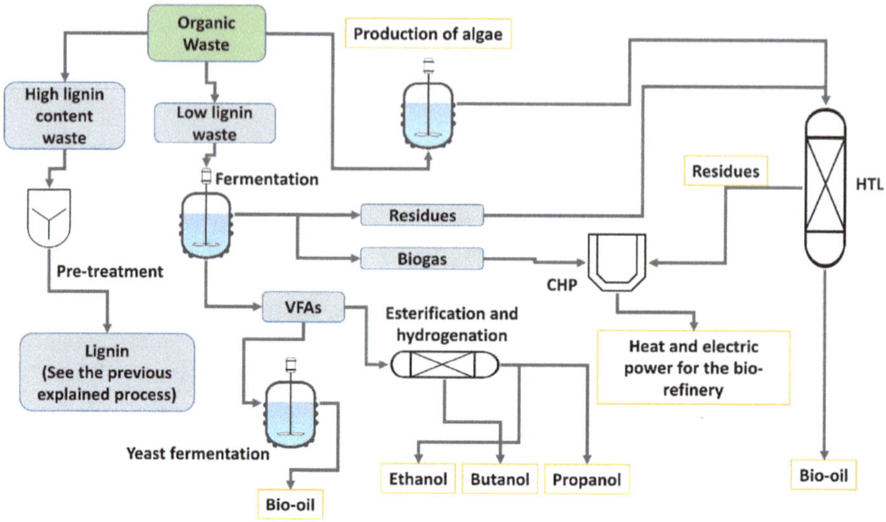


**Fig. 12.7** Integration of fermentation, hydrogenation, HTL, and CHP plant. Adapted from Coma et al. (2017)

that HTL of plastics is not effectively working when, for instance, polyethylene and polypropylene are tested alone, as referred by the same researchers (Coma et al. 2017). Additionally, plastics are a well-known problematic feedstock because it can lead to toxic and corrosive streams and therefore, some other methods could be applied instead like recycling.

In the framework of thermochemical-based biorefineries, biomass gasification plays an important role considering the flexibility of the products quality that can be obtained. The main product of gasification is a gas-phase mixture of hydrogen, carbon monoxide, carbon dioxide, methane, and traces of higher hydrocarbons. This gas-phase product is commonly named as syngas. Depending on the gasification medium, nitrogen may cover a relevant fraction of the gas mixture. For instance, nitrogen content can reach values higher than 60% of the total gas volume when air is used as gasification medium (Pérez et al. 2012). In addition to air, other gasification mediums are steam, pure oxygen, carbon dioxide, or their mixes. Especially in the case of power production, the most common gasification mediums are air or mixtures of air and steam because of economic reasons. Indeed, while pure oxygen reduced the diluting effect of nitrogen, its costs refrains from its utilization. The addition of steam has a low impact on the operating costs of gasifiers if a proper heat recovery system is integrated in the gasification process. Furthermore, it allows increasing the syngas yield and its heating values. An important advantage of using steam is the possibility to produce a hydrogen reach syngas (Prestipino et al. 2017), reducing the efforts for downstream upgrading. Using steam alone as gasification medium leads to allothermal gasification, with additional costs for external heat supply. Hence, pure steam gasification is used only in rare cases for the production of products of high added value (Chiodo et al. 2017; Ning et al. 2018).

Concerning the kinds of feedstock which could be utilized for gasification purposes, wood accounts for the largest source of biomass. However, several studies are under development for enhancing the use of alternative feedstock. Municipal, industrial, and agricultural wastes and residues are promising alternatives that include biorefineries in the circular economy systems (Gądek et al. 2016; Galvagno et al. 2017). Gasification of residues can be also applied and integrated in biochemical biorefineries processes. Indeed, the lignin-rich residues of second generation bioethanol production are good candidates for being used as feedstock for energy and chemicals recovery by thermochemical gasification (Cerone et al. 2017).

The syngas produced from biomass gasification can be used by different kind of downstream processes. Indeed, the gas composition can be also tuned in order to fit the requirements for producing hydrogen, combined cooling heating and power, and liquid biofuels (Chiodini et al. 2017; Kraussler et al. 2018; Palomba et al. 2017).

Different power production units, such as high-temperature fuel cells, internal combustion engines, and gas turbines, can be fed by syngas for delocalized production of energy at different scales (Al-attab and Zainal 2018; Galvagno et al. 2016; Prestipino et al. 2016).

Renewable gas can be produced by utilizing microorganisms growing at anaerobic conditions (Croce et al. 2016). Additionally, biogas, which contains large amounts of carbon dioxide and methane, can be produced by municipal waste. One example of this kind of production is the company Stormossen situated in the Vaasa region in Finland. An additional method is to convert fermentable sugars into products which can be bioethanol or organic acids. This is also done by fermentation.

Furthermore, biofuels like biodiesel can be produced by processing waste vegetable oils or waste in general. The main reactions are esterification and transesterification or for instance, hydrogenation reactions. Later on in this manuscript, these concepts will be covered.

## 12.2 Industrial Examples

### *12.2.1 Biodiesel Production from Bio Oils by Hydrogenation*

As it will be discussed later on in this manuscript, the commonly produced fatty acid methyl ester (FAME) biodiesel is prepared from vegetable oils by transesterification reaction. However, it is reported in literature that the properties of FAME are not optimal for the transport industry and this is because of the high oxygen content and its high cloud point (Ramos et al. 2014).

The hydrogenation of bio-oils to produce biocrude oil and biodiesel will give a product similar to the common diesel and this represents an alternative solution adopted by companies like Neste and UPM (Figs. 12.8 and 12.9).

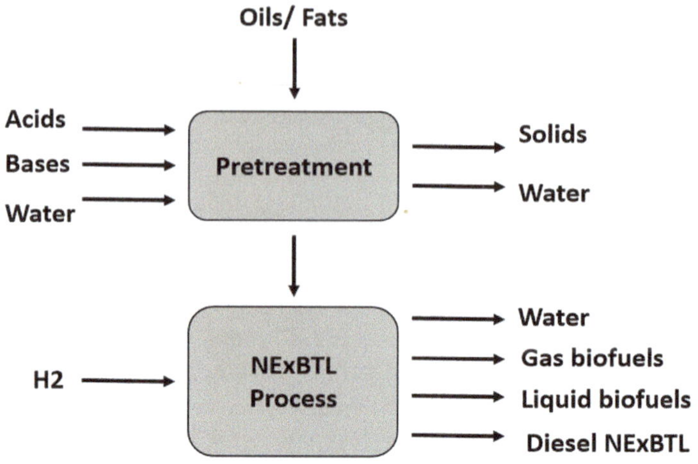


**Fig. 12.8** NEXBTL process used for biodiesel production

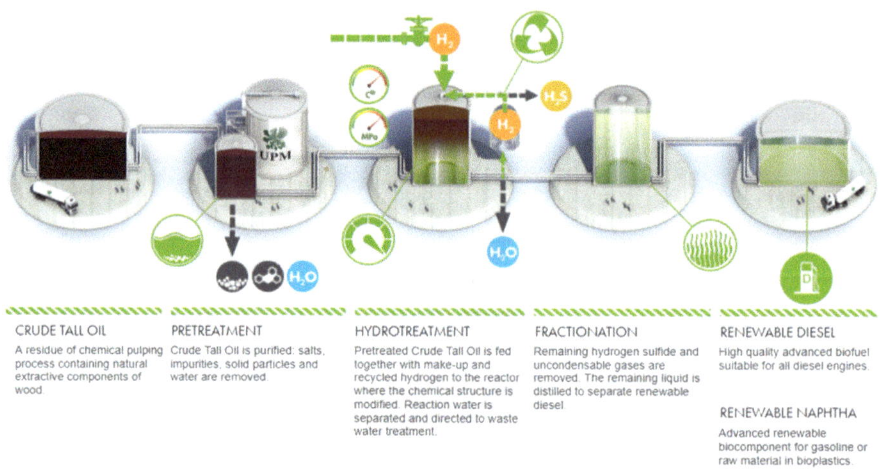


**Fig. 12.9** Production of renewable diesel from tall oil. Permission to use the figure granted by UPM

Neste's next-generation biomass-to-liquid (NExBTL) technology is considered to be a promising technology (Neste Oil 2018). The biodiesel is produced from oils and fats which are processed to remove any of the impurities. Hydrogenation is followed to convert the oils and fats into hydrocarbons and alkenes are obtained together with alkanes; this offers the possibility of refining them to biodiesel or if possible to other products. This process offers also the possibility of producing biogasoline and biopropane. Neste has also investigated the possibility of producing renewable chemicals like bioplastics, always by using the NEXBTL technology (Bezergianni 2013).

Neste is one of the biggest producers of biodiesel; as a matter of fact, the production of biodiesel by Neste is around two million tons per year and the production plants are located in Porvoo (Finland), Rotterdam, and Singapore. The biodiesel is produced from diverse feedstock kinds among which there are vegetable oils and also animal fat from the waste food industry. The vegetable oils are mainly rapeseed, soya, and palm oil (Engman et al. 2016).

UPM, another Finnish company, utilizes similar hydrogenation technology for the production of biodiesel. In this case, the biodiesel is called BioVerno (UPM-Kymmene 2012). The BioVerno diesel is produced in Lappeenranta, Finland with an annual production capacity of 120 mL. For this process, tall oil is utilized as raw material (Vásquez et al. 2017). Tall oil is considered as residual stream in the pulp and paper industry. This feedstock is firstly purified and then hydrogenated. Gases like $H_2S$ and other non-condensable gases are removed during the process. The biodiesel is then separated from the other fractions by distillation. This particular biodiesel is also suitable for driving in winter time since it has the characteristics of the normal diesel (Vauhkonen 2015).

### 12.2.2 Metsä Group's Äänekoski Plant

One example of next generation biorefinery concept is given by Metsä Group. As written on the company Web pages, the biorefinery is built in Äänekoski in Finland. The overall investment has been 1.2 Billion euro, it produces 1.3 million ton per year and it utilizes 6.5 M $m^3$ of wood annually (Engström 2017). The bio-electricity production is 1.8 TWh/year (Nousiainen 2018). The additional goal of this company is to utilize side streams to produce alternative compounds as it can be observed from Fig. 12.10.

Figure 12.10 shows the additional products which could be obtained from the biorefinery and some of these products could even be produced with partners companies. In addition, the company claims to produce a surplus of electricity which can be either sold or utilized in other processes. The claimed aim of the Metsä Group is to achieve a situation in which no waste is produced.

In the Äänekoski plant, also biogas as biofuel will be utilized, the development of the biogas process for Äänekoski plant is done by Envor Protech (Envor Protech OY 2016). The technology involves the fermentation of biomass with high lignin content. Large content of lignin is always problematic when digestion or any other treatment process is required; this is because lignin is a very stable compound formed by aromatic components. We remember from the organic chemistry that an aromatic ring presents more stability since there are resonant forms of the same component.

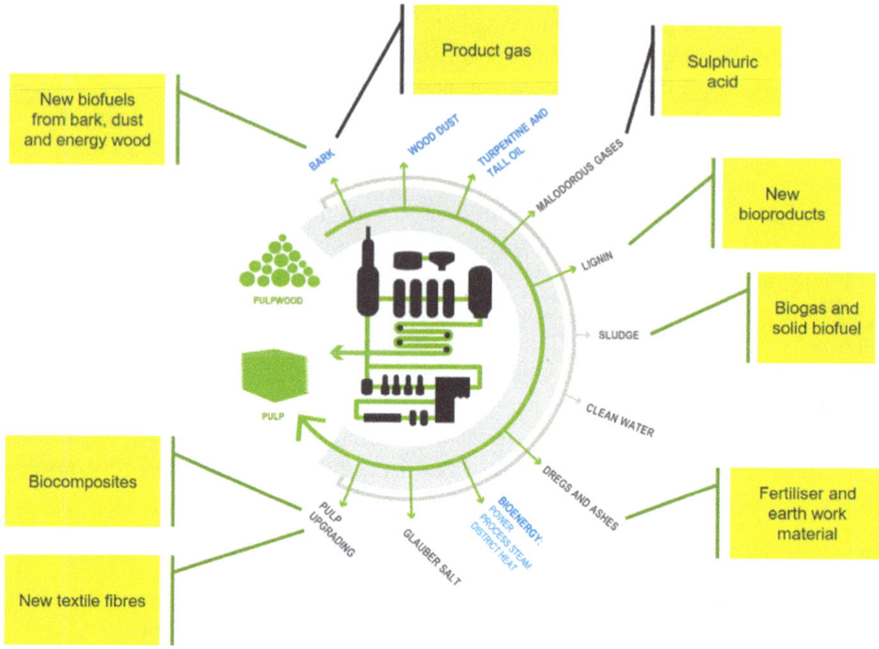


**Fig. 12.10** A next-generation wood biorefinery proposed by Metsä Group. Used with permission (von Weymarn 2016)

## 12.3 Production Concepts for Bioethanol Production

Bioethanol is produced by a fermentation process in which sugars like glucose is converted into ethanol and carbon dioxide. The process conditions are anaerobic. Bioethanol can be mixed to the normal gasoline which is then blended. To give one example of ethanol utilization and mixing in gas fuels, in Finland, there are two possible blended options available: the 5% blend and the 10% blend, while in other European countries, the so-called E10 is not available everywhere (ePURE 2018).

Most of the produced bioethanol is considered as a first-generation biofuel because the raw material for the process is starch or glucose coming mainly from the arable land. To give an example of company producing ethanol in Finland, the Finnish company St1 has developed also new concepts for producing bioethanol. The technology called Ethanolix® for instance is based on the bakeries and breweries waste products and the side products can be utilized for feeding animals (Nair et al. 2017).

St1 has four Etanolix plants in Finland and one in Sweden. The production capacity is, however, quite small since usually this technology is positioned near companies producing food or for instance beer (Fig. 12.11). The production is accounted to be between one to seven million liters per year and for instance, the plant sited in Gothenburg gives five million liters of high-quality ethanol (St1 Nordic Oy 2018, p. 1).

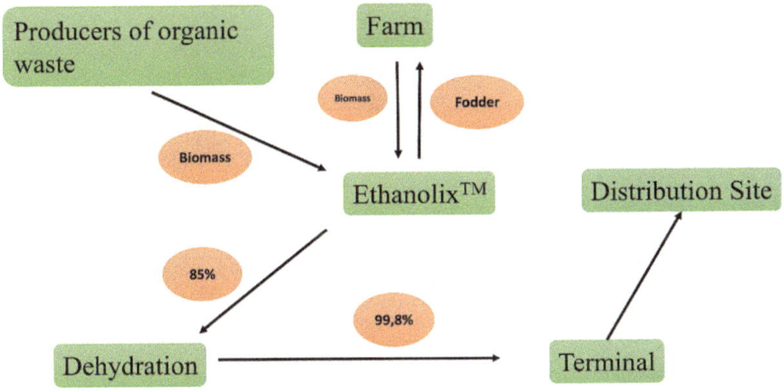


**Fig. 12.11** Ethanolix process and distribution flow

It is possible to see from the figure above that the facilities producing waste are near the ethanol plant. The by-products from the process are utilized as animal feeding and fertilizers; the water is removed in the dehydration section. Finally, the bioethanol is mixed to the gasoline and goes to the gasoline stations.

Another concept developed by St1 for bioethanol production is the Bionolix. The raw materials for the process in this case are municipal and industrial biowaste and also meat products. St1 has one Bionolix plant in Finland. As a side product, it produces electricity, district heat, and fertilizers.

The Cellunolix is an additional technology proposed by the same company (St1 Nordic Oy 2010). The idea behind this process is to produce bioethanol from biomass which contains lignin. As the reader can notice, there is more and more motivation to develop technologies which are able to process lignin for biofuels production. This is because lingo-cellulosic biomass is largely available in some regions of the world like Finland.

In this concept, the lingo-cellulosic material is firstly treated with enzymes in order to release fermentable sugars. Then yeast is utilized to ferment the feedstock in order to produce ethanol. Distillation and water removal are following this step. Using lingo-cellulosic material in bioethanol production is a new technology. Therefore, St1 does not yet have any operating Cellunolix plants. However, the first industrial plant is currently under construction in Kajaani, Finland. The plant will have a capacity of 10,000 $m^3$ per year. The raw material for the process is sawdust and it will come from the local sawmills.

## References

Al-attab, K. A., & Zainal, Z. A. (2018). Micro gas turbine running on naturally aspirated syngas: An experimental investigation. *Renewable Energy, 119,* 210–216. https://doi.org/10.1016/j.renene.2017.12.008.

Balat, M., Balat, M., Kırtay, E., & Balat, H. (2009a). Main routes for the thermo-conversion of biomass into fuels and chemicals. Part 1: Pyrolysis systems. *Energy Conversion and Management, 50,* 3147–3157. https://doi.org/10.1016/j.enconman.2009.08.014.

Balat, M., Balat, M., Kırtay, E., & Balat, H. (2009b). Main routes for the thermo-conversion of biomass into fuels and chemicals. Part 2: Gasification systems. *Energy Conversion and Management, 50,* 3158–3168. https://doi.org/10.1016/j.enconman.2009.08.013.

Bezergianni, S. (2013). *Catalytic hydroprocessing of liquid biomass for biofuels production.* InTECH.

Cerone, N., Zimbardi, F., Contuzzi, L., Prestipino, M., Carnevale, M. O., & Valerio, V. (2017). Air-steam and oxy-steam gasification of hydrolytic residues from biorefinery. *Fuel Processing Technology, 167,* 451–461. https://doi.org/10.1016/j.fuproc.2017.07.027.

Chiodini, A., Bua, L., Carnelli, L., Zwart, R., Vreugdenhil, B., & Vocciante, M. (2017). Enhancements in Biomass-to-Liquid processes: Gasification aiming at high hydrogen/carbon monoxide ratios for direct Fischer-Tropsch synthesis applications. *Biomass and Bioenergy, 106,* 104–114. https://doi.org/10.1016/j.biombioe.2017.08.022.

Chiodo, V., Urbani, F., Zafarana, G., Prestipino, M., Galvagno, A., & Maisano, S. (2017). Syngas production by catalytic steam gasification of citrus residues. *International Journal of Hydrogen Energy*; Advanced Materials for Fuel Cells and Electrolyzers: The E-MRS 2016 Fall Meeting Symposium Q (Vol. 42, pp. 28048–28055), September 19–22, 2016, Warsaw, Poland. https://doi.org/10.1016/j.ijhydene.2017.08.085.

Choi, J.-H., Jang, S.-K., Kim, J.-H., Park, S.-Y., Kim, J.-C., Jeong, H., et al. (2019). Simultaneous production of glucose, furfural, and ethanol organosolv lignin for total utilization of high recalcitrant biomass by organosolv pretreatment. *Renewable Energy, 130,* 952–960. https://doi.org/10.1016/j.renene.2018.05.052.

Christensen, P. S., Peng, G., Vogel, F., & Iversen, B. B. (2014). Hydrothermal Liquefaction of the microalgae phaeodactylum tricornutum: Impact of reaction conditions on product and elemental distribution. *Energy and Fuels, 28,* 5792–5803. https://doi.org/10.1021/ef5012808.

Coma, M., Martinez-Hernandez, E., Abeln, F., Raikova, S., Donnelly, J., …. Chuck, C. (2017). Organic waste as a sustainable feedstock for platform chemicals. *Faraday Discuss, 202*, 175–195. https://doi.org/10.1039/C7FD00070G.

Croce, S., Wei, Q., D'Imporzano, G., Dong, R., & Adani, F. (2016). Anaerobic digestion of straw and corn stover: The effect of biological process optimization and pre-treatment on total bio-methane yield and energy performance. *Biotechnology Advances, 34,* 1289–1304. https://doi.org/10.1016/j.biotechadv.2016.09.004.

Engman, A., Hartikka, T., Honkanen, M., Kiiski, U., Kuronen, M., Lehto, K., et al. (2016). *Neste renewable diesel handbook.* Espoo: Neste Corp.

Engström, J. (2017). Äänekoski a Mill for the 21th Century (No. 36), Spectrum. Andritz.

Envor Protech OY. (2016). *Envor Protech Oy is developing biogas technology to forest industry* [WWW Document]. http://www.envorprotech.fi/en/?newspage=2. Accessed July 14,18.

ePURE. (2018). *Fuel blends | ePURE—European renewable ethanol* [WWW Document]. https://epure.org/about-ethanol/fuel-market/fuel-blends/. Accessed July 14, 18.

Gądek, W., Mlonka-Mędrala, M., Prestipino, M., Evangelopoulos, P., Kalisz, S., & Yang, W. (2016). Gasification and pyrolysis of different biomasses in lab scale system: A comparative study. *E3S Web Conference, 10,* 00024. https://doi.org/10.1051/e3sconf/20161000024.

Galvagno, A., Prestipino, M., Chiodo, V., Maisano, S., Brusca, S., & Lanzafame, R. (2017). Energy performance of CHP system integrated with citrus peel air-steam gasification: A comparative study. In *Energy Procedia, ATI 2017—72nd Conference of the Italian Thermal Machines Engineering Association* (Vol. 126, pp. 485–492). https://doi.org/10.1016/j.egypro.2017.08.233.

Galvagno, A., Prestipino, M., Zafarana, G., & Chiodo, V. (2016). Analysis of an integrated agro-waste gasification and 120 kW SOFC CHP system: modeling and experimental investigation. In *Energy Procedia, ATI 2016—71st Conference of the Italian Thermal Machines Engineering Association* (Vol. 101, pp. 528–535). https://doi.org/10.1016/j.egypro.2016.11.067.

Helynen, S., Sipila, K., Peltola, E., & Holttinen, H. (2002). *Renewable energy sources in Finland by 2030*. Helsinki: Eduskunnan kanslia.

Ishola, M. M., Brandberg, T., & Taherzadeh, M. J. (2015). Simultaneous glucose and xylose utilization for improved ethanol production from lignocellulosic biomass through SSFF with encapsulated yeast. *Biomass and Bioenergy, 77,* 192–199. https://doi.org/10.1016/j.biombioe.2015.03.021.

Ishola, M. M., Jahandideh, A., Haidarian, B., Brandberg, T., & Taherzadeh, M. J. (2013). Simultaneous saccharification, filtration and fermentation (SSFF): A novel method for bioethanol production from lignocellulosic biomass. *Bioresource Technology, 133,* 68–73. https://doi.org/10.1016/j.biortech.2013.01.130.

Kim, J. S., Lee, Y. Y., & Kim, T. H. (2016). A review on alkaline pretreatment technology for bioconversion of lignocellulosic biomass. *Bioresource Technology, 199,* 42–48. https://doi.org/10.1016/j.biortech.2015.08.085.

Kouisni, L., Gagné, A., Maki, K., Holt-Hindle, P., & Paleologou, M. (2016). Ligno force system for the recovery of lignin from black liquor: Feedstock options, odor profile, and product characterization. *ACS Sustainable Chemistry and Engineering, 4,* 5152–5159. https://doi.org/10.1021/acssuschemeng.6b00907.

Kraussler, M., Binder, M., Schindler, P., & Hofbauer, H. (2018). Hydrogen production within a polygeneration concept based on dual fluidized bed biomass steam gasification. *Biomass and Bioenergy, 111,* 320–329. https://doi.org/10.1016/j.biombioe.2016.12.008.

Kwak, S., Jo, J. H., Yun, E. J., Jin, Y. -S., & Seo, J. -H. (2018). Production of biofuels and chemicals from xylose using native and engineered yeast strains. *Biotechnology Advances*. https://doi.org/10.1016/j.biotechadv.2018.12.003.

Li, M.-F., Yang, S., & Sun, R.-C. (2016). Recent advances in alcohol and organic acid fractionation of lignocellulosic biomass. *Bioresource Technology, 200,* 971–980. https://doi.org/10.1016/j.biortech.2015.10.004.

Muddassar, H. R., Melin, K., de Villalba Kokkonen, D., Riera, G. V., Golam, S., & Koskinen, J. (2015). Green chemicals from pulp production black liquor by partial wet oxidation. *Waste Management and Research Journal of the International Solid Wastes Public Cleaning Association ISWA, 33,* 1015–1021. https://doi.org/10.1177/0734242X15602807.

Muddassar, H. R., Melin, K., & Koskinen, J. (2014). Production of carboxylic acids from alkaline pretreatment byproduct of softwood. *Cellulose Chemistry and Technology, 48,* 835–842.

Nair, R. B., Lennartsson, P. R., & Taherzadeh, M. J. (2017). Bioethanol production from agricultural and municipal wastes. In: J. W. –C. Wong, R. D. Tyagi, A. Pandey (Eds.), *Current developments in biotechnology and bioengineering* (pp. 157–190). Elsevier. https://doi.org/10.1016/B978-0-444-63664-5.00008-3.

Neste Oil. (2018). Neste oil's nexbtl diesel.

Ning, S., Jia, S., Ying, H., Sun, Y., Xu, W., & Yin, H. (2018). Hydrogen-rich syngas produced by catalytic steam gasification of corncob char. *Biomass and Bioenergy, 117,* 131–136. https://doi.org/10.1016/j.biombioe.2018.07.003.

Nousiainen, I. (2018). *Metsä Group's bioproduct mill has produced more than 1 million tonnes of pulp, reaching its full production capacity* [WWW Document]. https://www.metsagroup.com/en/media/all-news/Pages/News.aspx?EncryptedId=96853DAD830688F5&Title=MetsaGroupsbioproductmillhasproducedmorethan1milliontonnesofpulp,reachingitsfullproductioncapacity. Accessed January 6, 19.

Özdenkçi, K., De Blasio, C., Muddassar, H. R., Melin, K., Oinas, P., Koskinen, J., et al. (2017). A novel biorefinery integration concept for lignocellulosic biomass. *Energy Conversion and Management, 149,* 974–987. https://doi.org/10.1016/j.enconman.2017.04.034.

Özdenkci, K., Koskinen, J., & Sarwar, G. (2014). Recovery of sodium organic salts from partially wet oxidized black liquor. *Cellulose Chemistry and Technology, 48,* 825–833.

Palomba, V., Prestipino, M., & Galvagno, A. (2017). Tri-generation for industrial applications: Development of a simulation model for a gasification-SOFC based system. *International Journal of Hydrogen Energy*; Advanced Materials for Fuel Cells and Electrolyzers: The E-MRS 2016 Fall Meeting Symposium Q (Vol. 42, pp. 27866–27883), September 19–22, 2016, Warsaw, Poland. https://doi.org/10.1016/j.ijhydene.2017.06.206.

Pandey, A., Hofer, R., Larroche, C., Taherzadeh, M., & Nampoothiri, M. (2015). *Industrial biorefineries and white biotechnology*. Elsevier.

Pérez, J. F., Melgar, A., & Benjumea, P. N. (2012). Effect of operating and design parameters on the gasification/combustion process of waste biomass in fixed bed downdraft reactors: An experimental study. *Fuel, 96,* 487–496. https://doi.org/10.1016/j.fuel.2012.01.064.

Prestipino, M., Chiodo, V., Maisano, S., Zafarana, G., Urbani, F., & Galvagno, A. (2017). Hydrogen rich syngas production by air-steam gasification of citrus peel residues from citrus juice manufacturing: Experimental and simulation activities. *International Journal of Hydrogen Energy, 42,* 26816–26827. https://doi.org/10.1016/j.ijhydene.2017.05.173.

Prestipino, M., Palomba, V., Vasta, S., Freni, A., & Galvagno, A. (2016). A Simulation tool to evaluate the feasibility of a gasification-I.C.E. System to produce heat and power for industrial applications. In *Energy Procedia, ATI 2016—71st Conference of the Italian Thermal Machines Engineering Association* (Vol. 101, pp. 1256–1263). https://doi.org/10.1016/j.egypro.2016.11.141.

Ramos, L. P., Cordeiro, C. S., Cesar-Oliveira, M. A. F., Wypych, F., & Nakagaki, S. (2014). Applications of heterogeneous catalysts in the production of biodiesel by esterification and transesterification, Chap. 16. In *Bioenergy research: Advances and applications* (pp. 255–276). Elsevier, Amsterdam.

Schreiber, M., Vivekanandhan, S., Cooke, P., Mohanty, A. K., & Misra, M. (2014). Electrospun green fibres from lignin and chitosan: a novel polycomplexation process for the production of lignin-based fibres. *Journal Materials Science, 49,* 7949–7958. https://doi.org/10.1007/s10853-014-8481-z.

Sipponen, M. H., Özdenkci, K., Muddassar, H. R., Melin, K., Golam, S., & Oinas, P. (2016). Hydrothermal liquefaction of softwood: Selective chemical production under oxidative conditions. *ACS Sustainable Chemistry and Engineering, 4,* 3978–3984. https://doi.org/10.1021/acssuschemeng.6b00846.

St1 Nordic Oy. (2018). *St1 built a waste-based Etanolix® ethanol production plant in Gothenburg - St1* [WWW Document]. English. https://www.st1.eu/st1-built-a-waste-based-etanolix-ethanol-production-plant-in-gothenburg. Accessed July 14, 18.

St1 Nordic Oy. (2010). st1 produces bioethanol from straw and cellulose. *Focus on Catalysis*, 4–5. https://doi.org/10.1016/S1351-4180(10)70207-5.

Svensson, E., Eriksson, K., & Wik, T. (2015). Reasons to apply operability analysis in the design of integrated biorefineries. *Biofuels, Bioproducts and Biorefining, 9,* 147–157. https://doi.org/10.1002/bbb.1530.

Tekin, K., Karagöz, S., & Bektaş, S. (2014). A review of hydrothermal biomass processing. *Renewable and Sustainable Energy Reviews, 40,* 673–687.

Toor, S. S., Rosendahl, L., & Rudolf, A. (2011). Hydrothermal liquefaction of biomass: A review of subcritical water technologies. *Energy, 36,* 2328–2342. https://doi.org/10.1016/j.energy.2011.03.013.

UPM-Kymmene. (2012). UPM-Kymmene annual report 2011: UPM will build the world's first biorefinery producing wood-based biodiesel. *Focus on Catalyst, 6*. https://doi.org/10.1016/S1351-4180(12)70177-0.

Vásquez, M. C., Silva, E. E., & Castillo, E. F. (2017). Hydrotreatment of vegetable oils: A review of the technologies and its developments for jet biofuel production. *Biomass and Bioenergy, 105,* 197–206. https://doi.org/10.1016/j.biombioe.2017.07.008.

Vauhkonen, V. (2015). *Renewable diesel from tall oil.* UPM The Biofore Company.

von Weymarn, N. (2016). *Metsä group's bioproduct mill: A next generation wood biorefinery in Äänekoski*. Metsä Fibre: Finland.

Zheng, Y., Pan, Z., & Zhang, R. (2009). Overview of biomass pretreatment for cellulosic ethanol production. *International Journal of Agricultural and Biological Engineering, 2*, 51–68. https://doi.org/10.25165/ijabe.v2i3.168.

Zhu, Y., Biddy, M. J., Jones, S. B., Elliott, D. C., & Schmidt, A. J. (2014). Techno-economic analysis of liquid fuel production from woody biomass via hydrothermal liquefaction (HTL) and upgrading. *Applied Energy, 129,* 384–394. https://doi.org/10.1016/j.apenergy.2014.03.053.

# Chapter 13
# Electronegativity and Microbial Catalysis

*Deus sive Natura.*
*God, in other words, the Nature.*
Benedict de Spinoza, 1677, *Of human slavery, that is, of the forces of affections*. In Part 4 of his book: *Ethica Ordine Geometrico Demonstrata.*

## 13.1 Production of Biofuels and Classification of the Related Microorganisms Needed

We have seen previously some examples of different technologies utilized for biofuels productions, and it was possible to see that many of the processes involved are dealing with fermentation, or more generally biomass feedstock is processed by microorganisms to get biofuels. In some of the processes mentioned, the fermentation process is carried out with the aim of enzymes which determine the product yield and rate (Zhao et al. 2018). Enzymes represent a large expense in the process, and therefore, it is of outmost importance to optimize the process in order to have less cost and less energy consumption.

A common fuel produced by these methods is hydrogen. Here, hydrogen is considered first because this particular fuel has the simplest molecular composition and very high enthalpy value, for instance, its enthalpy of formation at the standard conditions is reported as 221.67 kJ/mol (Wagman et al. 1982).

The types of microorganisms utilized for hydrogen production can be classified on the basis of the substrate (food) they are going to oxidize (remembering that a substrate is said to be oxidized when electrons are actually taken from it):

- chemoorganotrophs (the ones that oxidize organic compounds);
- chemolithotrophs (the ones that oxidize non-organic compounds).

The two principal steps in the formation of organic matter are indicated usually as:

- Catabolism (production of energy and ATP);
- Anabolism (use of energy and production of organic matter).



C. De Blasio, *Fundamentals of Biofuels Engineering and Technology*,
Green Energy and Technology, https://doi.org/10.1007/978-3-030-11599-9_13

Catabolism is the process of production of energy and ATP molecules from a substrate; this process is common to humans and animals in general which do not produce organic matter (except when they have to grow to the adult size). On the other hand, anabolism is more familiar to plants and organisms in general, as a matter of fact, they produce organic matter.

Continuing with the main microorganisms' classification, the main biological processes for the conversion of organic substrates are classified commonly as follows:

- aerobic respiration (the electron acceptor in this case is oxygen, and on the other, the product is $CO_2$);
- anaerobic respiration (forms reduced compounds like $H_2$ and $CH_4$ or ethanol) (Tian et al. 2017);
- fermentation (it is commonly referred to have no electron transport chain, and in this case, there is an endogenous electron acceptor which normally is an organic compound (Wong et al. 2016). One good example of fermentation carried on by microorganisms is the one done by cyanobacteria (Stal and Moezelaar 1997);

Each of these biological processes has a relative environment where they can take place. This environment can be:

- aerobic;
- anoxic (provides electron acceptors different than $O_2$);
- anaerobic (absence of $O_2$ and inorganic electron acceptors).

In the process of anaerobic respiration, the electron transport system is similar to the one in aerobic respiration. The difference is that oxygen is not the terminal electron acceptor.

Hydrogen can be produced by a number of microorganisms. The most common microorganisms utilized for the production of hydrogen are reported here with their classification which is done on the basis of the environmental temperature:

- *Caldanaerobacter subterraneus*, thermophilic (Sokolova et al. 2001);
- *Carboxydocella thermautotrophica*, thermophilic (Sokolova et al. 2002);
- *Carboxydothermus hydrogenoformans*, thermophilic (Gerhardt et al. 1991);
- *Citrobacter*, mesophilic (Oh et al. 2003);
- *Rhodopseudomonas palustris*, mesophilic (Jung et al. 1999);
- *Rhodospirillum rubrum*, mesophilic (Zürrer and Bachofen 1979);
- *Rubrivivax gelatinosus*, mesophilic (Uffen 1976);
- *Thermincola carboxydiphila*, thermophilic (Sokolova et al. 2005);
- *Thermococcus strain*, thermophilic (Sokolova et al. 2004b);
- *Thermolithobacter carboxydivorans*, thermophilic (Sokolova et al. 2007);
- *Thermosinus carboxydivorans*, thermophilic (Sokolova et al. 2004a).

Particular attention has to be put on a microorganism called *Pyrococcus furiosus*, discovered in 1986 by Karl Stetter in Italy (Keller et al. 2013). The organism is the subject of recent research because it can be modified genetically to produce hydrogen from $CO_2$ and water. The organism is able to survive in superheated waters or volcano mud. Therefore, it is classified as a thermophile. Furthermore, the *P. furiosus* has been also modified to produce industrial chemicals from $CO_2$ and hydrogen (Keller et al. 2013).

Microorganisms can also produce other kinds of fuels other than hydrogen, for instance, there is a numerous number of microorganisms able to give ethanol. The most common of these organisms are given here with the related scientific reference:

- *Acetobacterium woodii*, mesophilic (BALCH et al. 1977);
- *Butyribacterium methylotrophicum*, mesophilic (Zeikus et al. 1980);
- *Clostridium autoethanogenum*, mesophilic (Abrini et al. 1994);
- *Clostridium glycolicum*, mesophilic (Küsel et al. 2001);
- *Clostridium ljungdahlii*, mesophilic (Barik et al. 1988);
- *Clostridium carboxidivorans*, mesophilic (Liou et al. 2005);
- *Moorella* HUC22-1, thermophilic (Sakai et al. 2004).

## 13.2 Pathways in Biological Processing of Biomass Feedstock

In case we consider glucose as the main nutrient, ethanol is produced by both fermentation and respiration method of digestion. This pattern is described by the Embden–Meyerhof pathway (EMP) (Golder 1979), also called glycolysis, which is represented by the following simplified chemical reaction:

$$C_6H_{12}O_6\left(+2ADP + 2NAD^+\right) \rightarrow 2CH_3COCOOH + 2ATP + 2NADH \quad (13.1)$$

In the reaction above, $2CH_3COCOOH$ refers to the pyruvic acid, and this is the most important part: the production of ATP, the energy carrier. This pathway can take place in all environments. Good representations of this pathway are given in the literature (Madigan et al. 2010).

With reference to Fig. 13.1, the intermediate compounds are: (A) glucose-6-P; (B) fructose-6-P; (C) fructose-1,6-P; (D) dihydroxyacetone-P; (E) glyceraldehyde-3-P; (F) 1,3-bisphosphoglycerate; (G) 3-P-glycerate; (H) 2-P-glycerate; and (I) Phosphoenolpyruvate.

The main enzymes in the same figure are: (1) hexokinase; (2) isomerase; (3) phosphofructokinase; (4) aldolase; (5) triosephosphate isomerase; (6) glyceraldehyde-3-P dehydrogenase; (7) phosphoglycerokinase; (8) phosphoglyceromutase; (9) enolase; (10) pyruvate kinase; (11) lactate dehydrogenase; (12) pyruvate decarboxylase; and (13) alcohol dehydrogenase.

In the fermentation of yeast, glucose gives finally two molecules of ethanol and two $CO_2$ with a release of energy of −239 kJ/(mol of glucose). While for lactic acid bacteria, the glucose gives two molecules of lactate and 196 kJ/mole of glucose of energy. Yeast fermentation is of interest not only for ethanol production, and therefore, for producing alcoholic products, but also these kinds of processes have

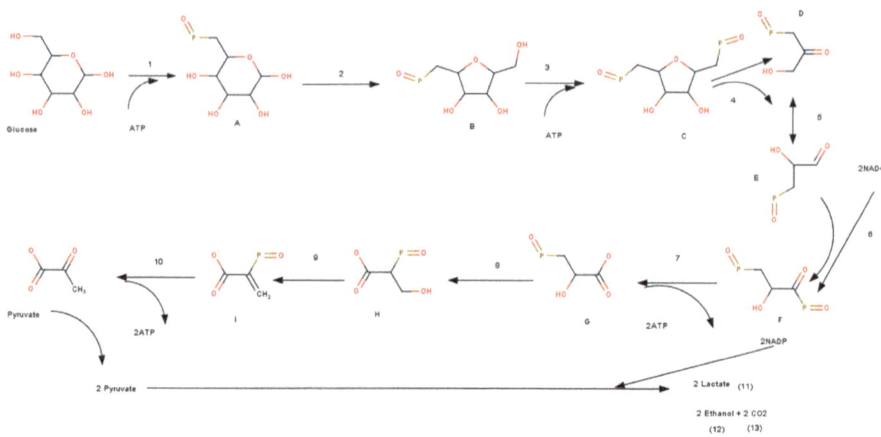


**Fig. 13.1** Embden–Meyerhof pathway (EMP). Adapted from Madigan et al. (2010)

produced a growing interest toward the biowaste utilization and processing for power and fuel generation (Ntaikou et al. 2018).

In Fig. 13.1, the first stage concerns reactions which do not involve oxidation–reduction, while the second stage is the one with this kind of reactions.

The degradation of organic compounds by microorganisms is given in a very detailed way by the Kyoto Encyclopedia of Genes and Genomes. The database KEGG (Kanehisa Laboratories 2018) is available on the web and contains a large number of biological reactions with the related enzymes.

## 13.3 Main Molecules Used as Energy Carriers in Biological Processes

In the first part of this manuscript, the Z-scheme of photosynthesis was described to give an understanding of the importance of the electrochemical potential generation. The substrate (in that case water) is oxidized, and electrons are taken from it. The electrons are then transported through an electron transport chain. This is the needed current which gives power to the molecular functioning. As an introductory concept here, it is beneficial to remember that a substance which has less reduction potential than a reference substance (has a more negative value) is reduced less easily than it is the reference substance. This is quite obvious since this can be imagined as a substance which has "enough" negative electric charges and therefore does not accept easily more electrons.

On the other hand, a substance whose reduction potential is positive (is greater than that of a reference substance) is more easily reduced than the reference. Hydrogen is commonly used as a reference for these purposes. Here, the main molecules utilized

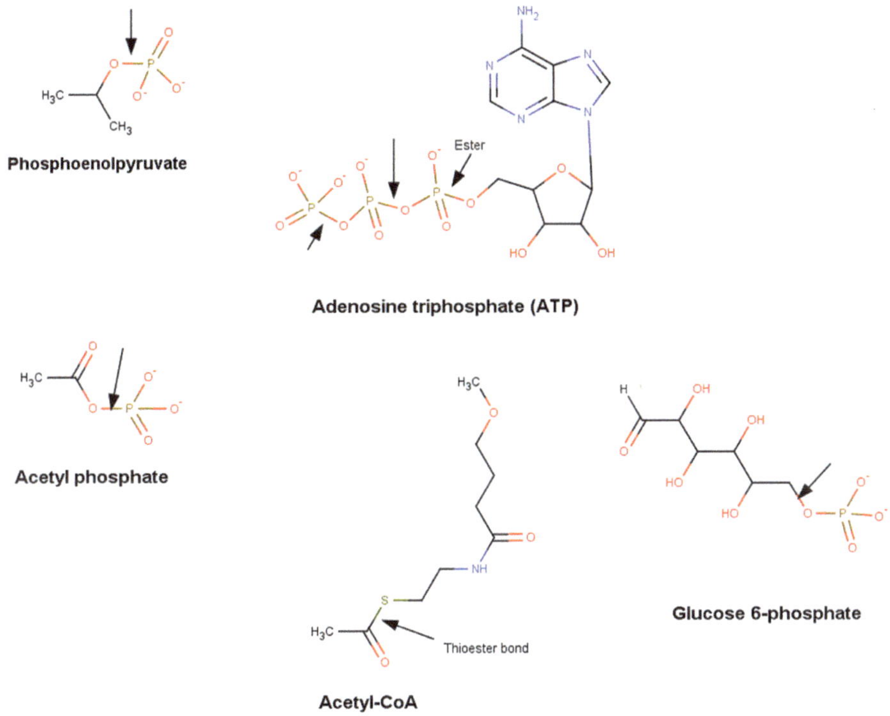


**Fig. 13.2** Most important energetic molecules in living organisms. Adapted from Madigan et al. (2010)

as energy carriers are described in Fig. 13.2. The Gibbs energy values at standard conditions, $\Delta G^0\left(\frac{\text{kJ}}{\text{mol}}\right)$, for the main molecules utilized as energy carriers are as follows (Madigan et al. 2010): phosphoenolpyruvate: −51.6; 1,3-bisphosphoglycerate: −52.0; acetyl phosphate: −44.8; ATP: −31.8; ADT: −31.8; acetyl-CoA: −35.7; AMP: −14.2; and glucose 6-phosphate: −13.8. Some of these molecules are shown also in Fig. 13.2.

The energy shown, $\Delta G^0$, is the energy released when the bonds indicated by the arrows in Fig. 13.2 are hydrolyzed.

On the other hand, an example of electron transport system for purple photosynthetic and green sulfur bacteria is shown in Fig. 13.3.

In reference to this figure, the P870 and P840 are referring to the reaction center, and their excited forms (see first part of this book). We have different types of quinones, Q, cytochromes, Cyt, chlorophyll molecules, BChl, pheophytin, BPh, quinones, A, ferredoxin, FD, plastoquinone, QP, ferredoxin-NADP$^+$ reductase (FNR), clusters, Fe-S, and menaquinone, MQ.

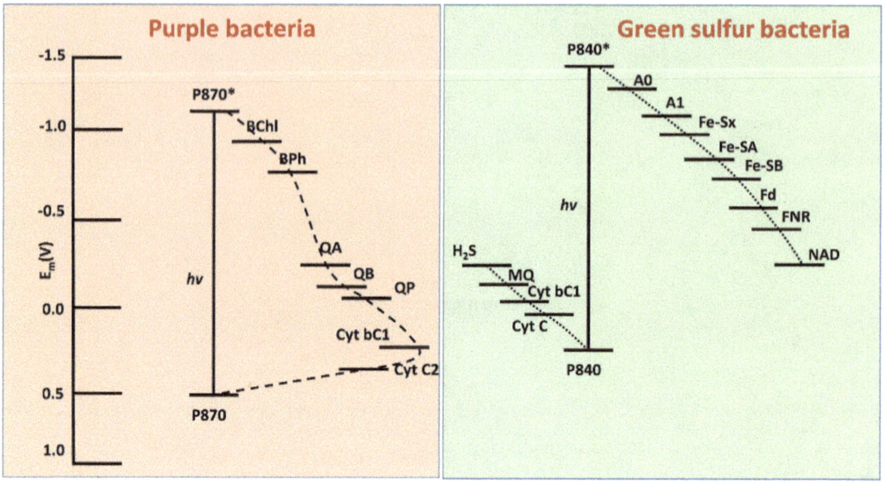


**Fig. 13.3** Example of electron transport system for purple photosynthetic and green sulfur bacteria. Adapted from Rasmussen and Minteer (2014)

## 13.4 Oxidation and Reduction

We can have configurations, for the electron transport chains, where there is a reaction of reduction and NADH is produced. The NADH contains energy because, as we have seen, when some determinate bonds are broken, this will release energy. Here, it is also important to notice that the electrochemical potential falls from values at which is more negative (electronegative) to values where it is at more positive values, this was also mentioned in the first part of this book as "downhill" potential in the Z-scheme of photosynthesis. As a matter of fact, this is in agreement with the theory and the convention adopted: In case we consider electrons possessing a negative charge, then, the more electrons are present in a determinate region, and the more negative will be their electrochemical potential.

The electrons are subtracted from a substrate and sent to the electron transport chain, and finally, we need electron acceptors. This is obvious, since in very simple words, the electrons have to end up somewhere. As examples here, the nitrate and sulfate acceptance of electrons in non-oxygenic processes (Camacho 2009) is described by the following reactions:

$$5e^- + NO_3^- + 6H^+ \rightarrow \frac{1}{2}N_2 + 3H_2O \quad (13.2)$$

$$8e^- + SO_4^{2-} + 10H^+ \rightarrow H_2S + 4H_2O \quad (13.3)$$

Sulfur-based compounds are not normally utilized by the living organisms to produce organic matter, but instead these types of molecules are often utilized for their ability to release energy and to give or accept electrons. Sulfur is usually associated with very strong pollutants which are very famous within the power conversion of problematic fuels like coal. As a matter of fact, coal can have very high sulfur content and it is well-known that when combusted, $SO_2$ can give the formation of acidic rains. Now, when wet flue gas desulfurization is applied to reduce the $SO_2$ emissions, the key step is the limestone dissolution and the formation of calcium ions which, in turn, react with the sulfate ions to form gypsum (De Blasio et al. 2016). Indeed, the sulfite oxidation which takes in processes like flue gas desulfurization could be coupled with the oxidative degradation of organic compounds, and this idea was already reported in 1987 (Lee and Rochelle 1987).

This was mentioned to show that sulfate ions are indeed present in other very different processes. Coming back to our biological processes, the electrochemical potentials for the equations written previously should be positive for reactions written in this way if phenomena are spontaneous.

The oxidation process produces ATP molecules (our fuel) and, on the basis of the particular enzyme (the catalyst) used to perform this operation, a certain amount of moles or number of NADH or $FADH_2$ and ATP will be given. If we consider palmitic oil, for instance, from this oil the palmitate is formed and transformed to palmitoyl-CoA which is then completely oxidized to the final products. The table which follows shows the number of fuel molecules produced by complete oxidation of palmitoyl-CoA to $CO_2$ and $H_2O$ (Kamel and Halperin 2017, p. 5) (Table 13.1).

**Table 13.1** Moles of ATP produced on the basis of the particular enzyme which catalyzes the oxidation reactions

| Enzyme | NADH $FADH_2$ (number) | ATP (number) |
|---|---|---|
| A-CoA | 7 $FADH_2$ | 14 |
| β-H-CoA | 7 NADH | 21 |
| Iso-D | 8 NADH | 24 |
| α-K-D | 8 NADH | 24 |
| S-CoA synthetase | | 8 |
| S-D | 8 $FADH_2$ | 16 |
| M-D | 8 NADH | 24 |

A-CoA is the acyl-CoA dehydrogenase, β-H-CoA is the β-hydroxyacyl-CoA dehydrogenase, Iso-D is the isocitrate dehydrogenase, S-CoA is the succinyl-CoA synthetase, S-D is the succinate dehydrogenase, α-K-D is the α-ketoglutarate dehydrogenase, and M-D is the malate dehydrogenase

Adapted from Nelson and Cox (2012)

## 13.5 An Introduction to Modeling of Digestion Processes

An organic substrate can be digested by microorganisms which will grow in number depending on the availability of the "food" and the environmental conditions adopted.

Other than analytical models (which will be described more in details later in this manuscript); in the literature also, empirical models can be found for the digestion of a substrate. To give one example, a model for biomass (cells) formation in yeasts can be described by Eq. (13.4) (van Dijken and Scheffers 1986):

$$737C_6H_{12}O_6 + 680NH_3 + 6H_2SO_4 + 931NADPH + 1349NAD^+ \rightarrow 1000C_4H_{7.32}O_{2.24}N_{0.68}S_{0.006}(100\ \text{g cells}) + 424CO_2 + 931NADP^+ + 1349NADH + 1358H_2O + 418H^+ \quad (13.4)$$

Figure 13.4 shows the main steps of the pathway for the anaerobic digestion performed by bacteria, from the hydrolysis of complex organic compounds, a series of shorter chain molecules, are obtained. The fermentation is achieved after the hydrolysis process and it is followed by acetogenesis where volatile acids and fatty acids are converted to acetic acid $H_2$ and $CO_2$, while the methanogenesis allows for the release of methane. The two processes can be also carried out at the same time by enriched microbiological cultures. As a matter of fact, the hydrogenotrophic methanogens are indeed able to carry on the reaction:

$$4H_2 + CO_2 \rightarrow CH_4 + 2H_2O; \quad (13.5)$$

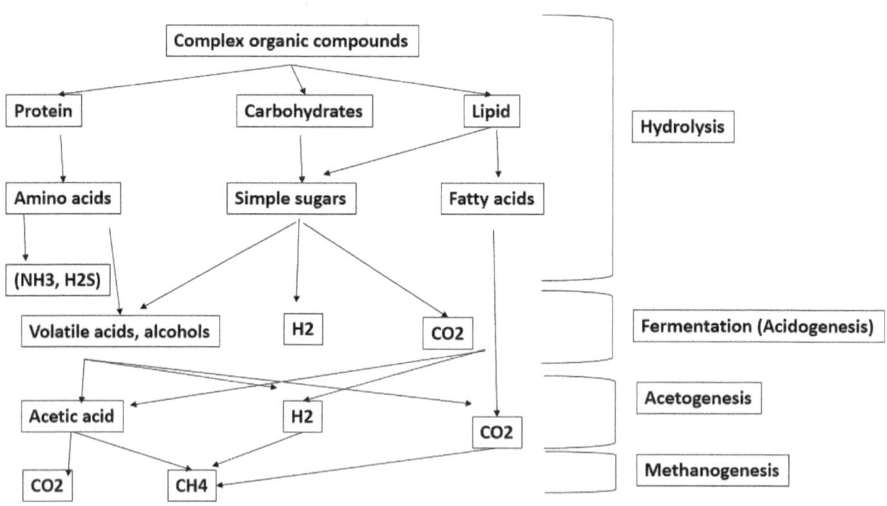


**Fig. 13.4** Main steps of the pathway for the anaerobic digestion performed by bacteria

with a Gibbs free energy of −131 kJ/mol at standard conditions (Ma et al. 2019). While the homoacetogens are able to finalize the reaction which leads to the formation of acetate:

$$4H_2 + 2CO_2 \rightarrow CH_3COO^- + 2H_2O + H^+ \tag{13.6}$$

with a Gibbs free energy of −95 kJ/mol at standard conditions. By coupling these two kinds of microbes, we can achieve the carbon reduction and energy conservation at the same time (Siriwongrungson et al. 2007).

To carry on these processes, since reduction is involved, we need to provide electrons to the converting species and this is a fundamental requirement. As reducing species, zero-valent metals are utilized, for instance, zero-valent iron (Kondratenko et al. 2013).

The anaerobic digestion process is commonly very unstable (Lyberatos and Skiadas 1999), and for these reasons, good mathematical models are then required in order to have a suitable control of the process. As we will see later in this manuscript, the mathematical models for the substrate depletion and product formation can be different on the basis of which parameter is the limiting step during the process. For instance, the product yield can be inhibited by the same substrate or the product, like in the case of ethanol production from *Saccharomyces cerevisiae*. Table 13.2 demonstrates provided mathematical models in which it is assumed that the methanogen organisms are inhibited by the substrate concentration. All the mathematical models provided in this table follow a Monod kinetics which will be analyzed later in this manuscript.

More specifically and for the purposes of this book, it is also possible to individuate and classify the microorganisms' species on the basis of the fuel produced. The microbial modeling of biofuel production is done as follows:

- fermentative bacteria and archaea (which produce $H_2$ and $CH_4$);
- fungi and algae.

We can have also a classification of biological processes related to the biofuels production as shown in Fig. 13.5, for instance, alcoholic fermentation:

$$C_6H_{12}O_6 \rightarrow 2C_2H_5OH + 2CO_2 \tag{13.7}$$

As preliminary information, it will be mentioned here that the main parameters used for biomass growth are commonly referred to as:

- $\nu$, number specific growth rate [1/h]
- $\mu$, biomass specific growth rate [1/h]

and this will be analyzed further during the following chapters.

On the other hand, the most known fermenting organism is the *S. cerevisiae* also represented by some biologist by the picturesque symbol given in Fig. 13.6.

Also in this case, alcoholic fermentation is studied and simulated to be integrated with biorefineries. In the literature, alcoholic fermentation is sometimes coupled with

**Table 13.2** Some references where mathematical models are given for different processes involved in anaerobic digestion and where there is substrate inhibition

| Organism type and (substrate) | Process | Model proposed by | Inhibited by |
|---|---|---|---|
| Acetoclastic methanogens (unionized VFA as acetate) | Methanogenesis | Graef and Andrews (1974) | Unionized VFA or external inhibitor |
| Acid formers (glucose)<br>Producing methane (unionized VTA as acetate) | Hydrolysis of insoluble organics, acidogenesis and methanogenesis | Hill and Barth (1977) | Unionized VTA<br>Unionized VTA and unionized NH3 |
| Acid formers (soluble organics)<br>Producing methane (acetate) | Acidogenesis<br>Methanogenesis | Kleinstreuer and Poweigha (1982) | Unionized acetate, toxic substances for both processes |
| Acid formers (glucose)<br>Producing methane (acetate) | Acidogenesis<br>Methanogenesis | Moletta et al. (1986) | Unionized acetate<br>Unionized acetate |
| Acid formers (soluble organics)<br>Producing methane (unionized VFAs) | Acidogenesis<br>Methanogenesis | First order<br>Andrews (1969) | Total VFA<br>Unionized VFAs |

*VFA* indicated the volatile fatty acids
Adapted from Lyberatos and Skiadas (1999)

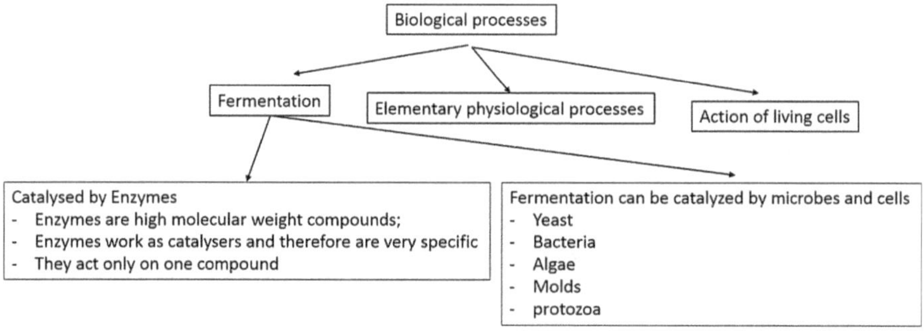


**Fig. 13.5** Classification of biological processes related to the biofuels production. Adapted from Levenspiel (1999)

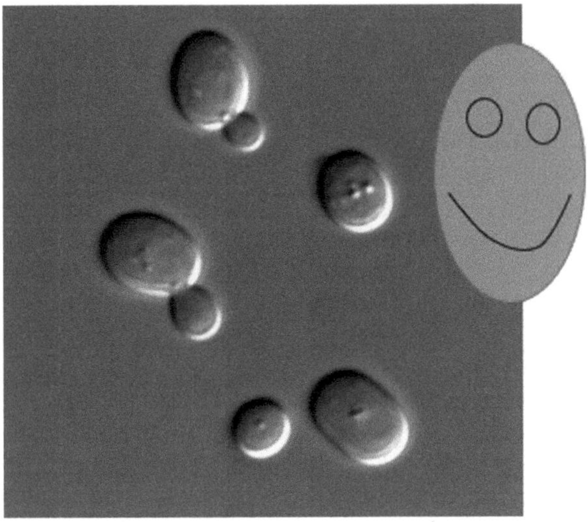

**Fig. 13.6** Commonly used symbol for the *Saccharomyces cerevisiae*

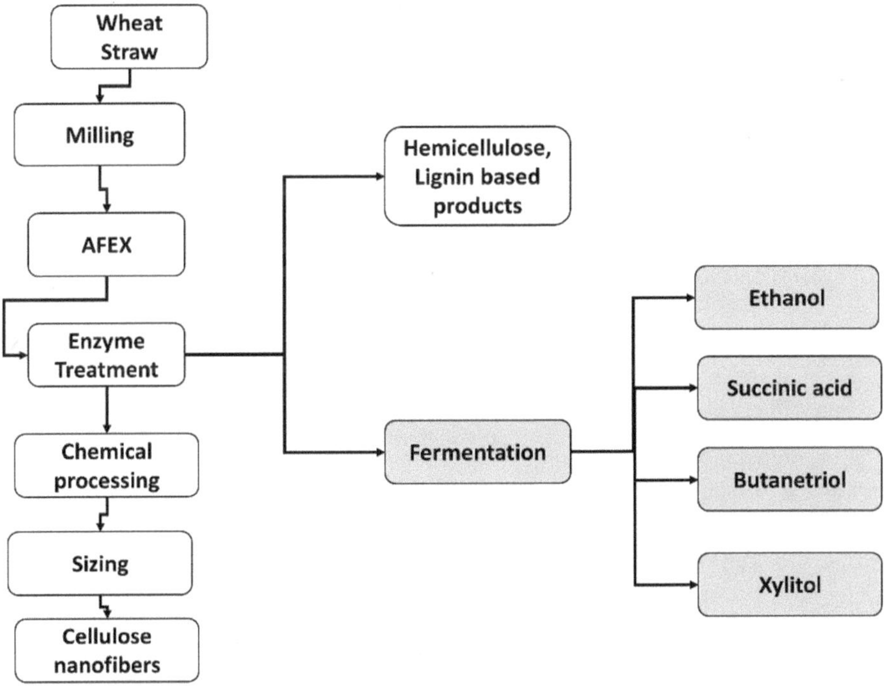


**Fig. 13.7** Example of ethanol production plant integrated with a cellulose nanofibers production plant. AFEX, ammonia fiber expansion. Adapted from Leistritz et al. (2006)

hydro-processing, pyrolysis, gasification, electrolysis, and even synthesis of transport fuels and chemicals. An example of process integration for a cellulose nanofibers production plant with a bio-ethanol plant is demonstrated in Fig. 13.7.

To better understand how biofuels production is taking place, for instance, the microbial fermentation to produce ethanol or biogas, the author will focus the discussion here on the basics of the bioreactions engineering approach. This kind of knowledge is necessary when we need a design of the reactors to be used; furthermore, this is the expertise the energy engineer should have in order perform the above-mentioned design; and with "design," it is meant here the optimization of the working features of the reactors. Therefore, the discussion will start here with the explanation of the main difference between the reactors configurations and then more in details the kinetics of microbial growth and fuel yield will be treated.

## References

Abrini, J., Naveau, H., & Nyns, E.-J. (1994). *Clostridium autoethanogenum*, sp. nov., an anaerobic bacterium that produces ethanol from carbon monoxide. *Archives of Microbiology, 161,* 345–351. https://doi.org/10.1007/BF00303591.

Andrews, J. F. (1969). Dynamic model of the anaerobic digestion model. *Journal of the Sanitary Engineering Division American Society Civil Engineers SA, 1,* 95–116.

Balch, W. E., Schoberth, S., Tanner, R. S., & Wolfe, R. S. (1977). Acetobacterium, a new genus of hydrogen-oxidizing, carbon dioxide-reducing, anaerobic bacteria. *International Journal of Systematic and Evolutionary Microbiology, 27,* 355–361. https://doi.org/10.1099/00207713-27-4-355.

Barik, S., Prieto, S., Harrison, S. B., Clausen, E. C., & Gaddy, J. L. (1988). Biological production of alcohols from coal through indirect liquefaction. *Applied Biochemistry and Biotechnology, 18,* 363–378. https://doi.org/10.1007/BF02930840.

Camacho, A. (2009). Sulfur bacteria. In G. E. Likens (Ed.), *Encyclopedia of inland waters* (pp 261–278). Oxford: Academic Press. https://doi.org/10.1016/B978-012370626-3.00128-9.

De Blasio, C., Carletti, C., Lundell, A., Visuri, V.-V., Kokkonen, T., Westerlund, T., et al. (2016). Employing a step-wise titration method under semi-slow reaction regime for evaluating the reactivity of limestone and dolomite in acidic environment. *Minerals Engineering, 86,* 43–58. https://doi.org/10.1016/j.mineng.2015.11.011.

Gerhardt, M., Svetlichny, V., Sokolova, T. G., Zavarzin, G. A., & Ringpfeil, M. (1991). Bacterial CO utilization with $H_2$ production by the strictly anaerobic lithoautotrophic thermophilic bacterium *Carboxydothermus hydrogenus* DSM 6008 isolated from a hot swamp. *FEMS Microbiology Letters, 83,* 267–271.

Golder, R. H. (1979). Patterns in the Embden-Meyerhof pathway of glycolysis. *Biochemical Education, 7,* 7–8. https://doi.org/10.1016/0307-4412(79)90007-4.

Graef, S. P., & Andrews, J. F. (1974). Stability and control of anaerobic digestion. *Journal of Water Pollution Control Federation, 46,* 666–683.

Hill, D. T., & Barth, C. L. (1977). A Dynamic model for simulation of animal waste digestion. *Journal of Water Pollution Control Federation, 49,* 2129–2143.

Jung, G. Y., Jung, H. O., Kim, J. R., Ahn, Y., & Park, S. (1999). Isolation and characterization of *Rhodopseudomonas palustris* P4 which utilizes CO with the production of $H_2$. *Biotechnology Letters, 21,* 525–529. https://doi.org/10.1023/A:1005560630351.

Kamel, K. S., Halperin, M. L. (2017). Ketoacidosis, Chap. 5. In K. S. Kamel & M. L. Halperin (Eds.), *Fluid, electrolyte and acid-base physiology* (5th ed., pp. 99–139). Philadelphia: Elsevier. https://doi.org/10.1016/B978-0-323-35515-5.00005-1.

Kanehisa Laboratories. (2018). *KEGG: Kyoto encyclopedia of genes and genomes* [WWW Document]. https://www.kegg.jp/. Accessed July 15, 18.

Keller, M. W., Schut, G. J., Lipscomb, G. L., Menon, A. L., Iwuchukwu, I. J., Leuko, T. T., et al. (2013). Exploiting microbial hyperthermophilicity to produce an industrial chemical, using hydrogen and carbon dioxide. *Proceedings of National Academy of Sciences, 110,* 5840–5845. https://doi.org/10.1073/pnas.1222607110.

Kleinstreuer, C., & Poweigha, T. (1982). Dynamic simulator for anaerobic digestion processes. *Biotechnology and Bioengineering, 24,* 1941–1951. https://doi.org/10.1002/bit.260240903.

Kondratenko, E. V., Mul, G., Baltrusaitis, J., Larrazábal, G. O., & Pérez-Ramírez, J. (2013). Status and perspectives of $CO_2$ conversion into fuels and chemicals by catalytic, photocatalytic and electrocatalytic processes. *Energy & Environmental Science, 6,* 3112–3135. https://doi.org/10.1039/C3EE41272E.

Küsel, K., Karnholz, A., Trinkwalter, T., Devereux, R., Acker, G., & Drake, H. L. (2001). Physiological ecology of *Clostridium glycolicum* RD-1, an Aerotolerant acetogen isolated from sea grass roots. *Applied and Environment Microbiology, 67,* 4734–4741. https://doi.org/10.1128/AEM.67.10.4734-4741.2001.

Lee, Y. J., & Rochelle, G. T. (1987). Oxidative degradation of organic acid conjugated with sulfite oxidation in flue gas desulfurization: Products, kinetics, and mechanism. *Environmental Science and Technology, 21,* 266–272. https://doi.org/10.1021/es00157a007.

Leistritz, F. L., Senechal, D. M., Stowers, M. D., McDonald, W. F., Saffron, C. M., & Hodur, N.M. (2006). *Preliminary feasibility analysis for an integrated biomaterials and ethanol biorefinery using wheat straw feedstock (No. 23500).* Agribusiness & Applied Economics Report. North Dakota State University, Department of Agribusiness and Applied Economics.

Levenspiel, O. (1999). *Chemical reaction engineering*. New York, US: Wiley.

Liou, J. S.-C., Balkwill, D. L., Drake, G. R., & Tanner, R. S. (2005). *Clostridium carboxidivorans* sp. nov., a solvent-producing clostridium isolated from an agricultural settling lagoon, and reclassification of the acetogen *Clostridium scatologenes* strain SL1 as *Clostridium drakei* sp. nov. *International Journal of Systematic and Evolutionary Microbiology, 55,* 2085–2091. https://doi.org/10.1099/ijs.0.63482-0.

Lyberatos, G., & Skiadas, I. V. (1999). Modeling of anaerobic digestion—A review. *Global NEST International Journal, 1,* 63–76.

Ma, L., Zhou, L., Ruan, M.-Y., Gu, J.-D., & Mu, B.-Z. (2019). Simultaneous methanogenesis and acetogenesis from the greenhouse carbon dioxide by an enrichment culture supplemented with zero-valent iron. *Renewable Energy, 132,* 861–870. https://doi.org/10.1016/j.renene.2018.08.059.

Madigan, M. T., Martinko, J. M., Stahl, D. A., & Clark, D. P. (2010). Brock biology of microorganisms (13th ed.,). Benjamin Cummings.

Moletta, R., Verrier, D., & Albagnac, G. (1986). Dynamic modelling of anaerobic digestion. *Water Research, 20,* 427–434. https://doi.org/10.1016/0043-1354(86)90189-2.

Nelson, D. L., & Cox, M. M. (2012). *Lehninger principles of biochemistry* (6th ed.). W.H: Freeman.

Ntaikou, I., Menis, N., Alexandropoulou, M., Antonopoulou, G., & Lyberatos, G. (2018). Valorization of kitchen biowaste for ethanol production via simultaneous saccharification and fermentation using co-cultures of the yeasts *Saccharomyces cerevisiae* and *Pichia stipitis*. *Bioresource Technology, 263,* 75–83. https://doi.org/10.1016/j.biortech.2018.04.109.

Oh, Y.-K., Seol, E.-H., Kim, J. R., & Park, S. (2003). Fermentative biohydrogen production by a new chemoheterotrophic bacterium *Citrobacter* sp. Y19. *International Journal of Hydrogen Energy, 28,* 1353–1359. https://doi.org/10.1016/S0360-3199(03)00024-7.

Rasmussen, M., & Minteer, S. D. (2014). Photobioelectrochemistry: Solar Energy conversion and biofuel production with photosynthetic catalysts. *Journal of the Electrochemical Society, 161,* H647–H655. https://doi.org/10.1149/2.0651410jes.

Sakai, S., Nakashimada, Y., Yoshimoto, H., Watanabe, S., Okada, H., & Nishio, N. (2004). Ethanol production from H2 and $CO_2$ by a newly isolated thermophilic bacterium, *Moorella* sp. HUC22-1. *Biotechnology Letters, 26,* 1607–1612. https://doi.org/10.1023/B:BILE.0000045661.03366.f2.

Siriwongrungson, V., Zeng, R. J., & Angelidaki, I. (2007). Homoacetogenesis as the alternative pathway for $H_2$ sink during thermophilic anaerobic degradation of butyrate under suppressed methanogenesis. *Water Research, 41,* 4204–4210. https://doi.org/10.1016/j.watres.2007.05.037.

Sokolova, T. G., González, J. M., Kostrikina, N. A., Chernyh, N. A., Slepova, T. V., Bonch-Osmolovskaya, E. A., et al. (2004a). *Thermosinus carboxydivorans* gen. nov., sp. nov., a new anaerobic, thermophilic, carbon-monoxide-oxidizing, hydrogenogenic bacterium from a hot pool of Yellowstone National Park. *International Journal of Systematic and Evolutionary Microbiology, 54,* 2353–2359. https://doi.org/10.1099/ijs.0.63186-0.

Sokolova, T. G., Jeanthon, C., Kostrikina, N. A., Chernyh, N. A., Lebedinsky, A. V., Stackebrandt, E., et al. (2004b). The first evidence of anaerobic CO oxidation coupled with $H_2$ production by a hyperthermophilic archaeon isolated from a deep-sea hydrothermal vent Extrem. *Life Extreme Cond., 8,* 317–323. https://doi.org/10.1007/s00792-004-0389-0.

Sokolova, T. G., González, J. M., Kostrikina, N. A., Chernyh, N. A., Tourova, T. P., Kato, C., et al. (2001). *Carboxydobrachium pacificum* gen. nov., sp. nov., a new anaerobic, thermophilic, CO-utilizing marine bacterium from Okinawa Trough. *International Journal of Systematic and Evolutionary Microbiology, 51,* 141–149. https://doi.org/10.1099/00207713-51-1-141.

Sokolova, T., Hanel, J., Onyenwoke, R. U., Reysenbach, A.-L., Banta, A., Geyer, R., et al. (2007). Novel chemolithotrophic, thermophilic, anaerobic bacteria *Thermolithobacter ferrireducens* gen. nov., sp. nov. and *Thermolithobacter carboxydivorans* sp. nov Extrem. *Life Extreme Condition, 11,* 145–157. https://doi.org/10.1007/s00792-006-0022-5.

Sokolova, T. G., Kostrikina, N. A., Chernyh, N. A., Kolganova, T. V., Tourova, T. P., & Bonch-Osmolovskaya, E. A. (2005). *Thermincola carboxydiphila* gen. nov., sp. nov., a novel anaerobic, carboxydotrophic, hydrogenogenic bacterium from a hot spring of the Lake Baikal area. *International Journal of Systematic and Evolutionary Microbiology, 55,* 2069–2073. https://doi.org/10.1099/ijs.0.63299-0.

Sokolova, T. G., Kostrikina, N. A., Chernyh, N. A., Tourova, T. P., Kolganova, T. V., & Bonch-Osmolovskaya, E. A. (2002). *Carboxydocella thermautotrophica* gen. nov., sp. nov., a novel anaerobic, CO-utilizing thermophile from a Kamchatkan hot spring. *International Journal of Systematic and Evolutionary Microbiology, 52,* 1961–1967. https://doi.org/10.1099/00207713-52-6-1961.

Stal, L. J., & Moezelaar, R. (1997). Fermentation in cyanobacteria. *FEMS Microbiology Reviews, 21,* 179–211. https://doi.org/10.1016/S0168-6445(97)00056-9.

Tian, T., Qiao, S., Yu, C., Tian, Y., Yang, Y., & Zhou, J. (2017). Distinct and diverse anaerobic respiration of methanogenic community in response to $MnO_2$ nanoparticles in anaerobic digester sludge. *Water Research, 123,* 206–215. https://doi.org/10.1016/j.watres.2017.06.066.

Uffen, R. L. (1976). Anaerobic growth of a Rhodopseudomonas species in the dark with carbon monoxide as sole carbon and energy substrate. *Proceedings of the National Academy of Sciences of the United States of America, 73,* 3298–3302.

van Dijken, J. P., & Scheffers, W. A. (1986). Redox balances in the metabolism of sugars by yeasts. *FEMS Microbiology Letters, 32,* 199–224. https://doi.org/10.1111/j.1574-6968.1986.tb01194.x.

Wagman, D. D., Evans, W. H., Parker, V. B., Schumm, R. H., Halow, I., Bailey, S. M., et al. (1982). The NBS tables of chemical thermodynamic properties. Selected values for inorganic and C1 and C2 organic substances. *Journal of Physical and Chemical Reference Data.*

Wong, M. T., Cheng, D., Wang, R., & Hsing, I.-M. (2016). Modifying the endogenous electron fluxes of *Rhodobacter sphaeroides* 2.4.1 for improved electricity generation. *Enyzme and Microbial Technology, 86,* 45–51. https://doi.org/10.1016/j.enzmictec.2016.01.009.

Zeikus, J. G., Lynd, L. H., Thompson, T. E., Krzycki, J. A., Weimer, P. J., & Hegge, P. W. (1980). Isolation and characterization of a new, methylotrophic, acidogenic anaerobe, the marburg strain. *Current Microbiology, 3,* 381–386. https://doi.org/10.1007/BF02601907.

Zhao, T., Tashiro, Y., Zheng, J., Sakai, K., & Sonomoto, K. (2018). Semi-hydrolysis with low enzyme loading leads to highly effective butanol fermentation. *Bioresource Technology, 264,* 335–342. https://doi.org/10.1016/j.biortech.2018.05.056.

Zürrer, H., & Bachofen, R. (1979). Hydrogen production by the photosynthetic bacterium *Rhodospirillum rubrum. Applied and Environment Microbiology, 37,* 789–793.

# Chapter 14
# Main Reactors Configurations

*Who is aiming high, eventually, will more differentiate himself. That is when addressing at the great book of Nature, which is the own object of the philosophy; this is the way to rise our view. In the same book, although all that is read, it is as done by an Omnipotent Architect, and therefore the most proportionated.*
Galileo Galilei, 1632, Dialogue Concerning the Two Chief World Systems.

## 14.1 Introduction to Reactor Design and the Different Configurations Used

This section focuses on what are the main reactors' configurations and the modeling of the residence time needed and therefore the size for the reactor necessary to complete a wanted conversion of substrate to give the target products. It is good to remember here that the volume of the reactor is represented by the volume of the actual mass of reagents, products, and catalysts in the reactor.

To use a simple metaphor, the working principles of the chemical reaction engineering could be applied also to normal daily life activities like cooking: you need a suitable pot (the reactor) and the size of the pot is chosen on the basis of how much food we would like to prepare, some good ingredients, and salts are also welcome. You need oil or water and some time is needed before you can serve a nice dish.

The concepts of depletion of a substrate and production of a product can be applied to any chemical engineering process. Among the general view of the chemical engineering processes, the biochemical reaction engineering represents our particular case here. The three main types of reactors used in chemical engineering are illustrated in Fig. 14.1.

Starting from the left of Fig. 14.1, in the batch reactor, there is no steady-state conditions: the reagents and the following products are all processed in the reactor and after a determinate time, the products are then obtained from it. The batch reactor is also stirred, meaning that the concentrations of reagents and products are yes changing with time, but in all areas of the reactor, the concentrations are the same.



C. De Blasio, *Fundamentals of Biofuels Engineering and Technology*,
Green Energy and Technology, https://doi.org/10.1007/978-3-030-11599-9_14

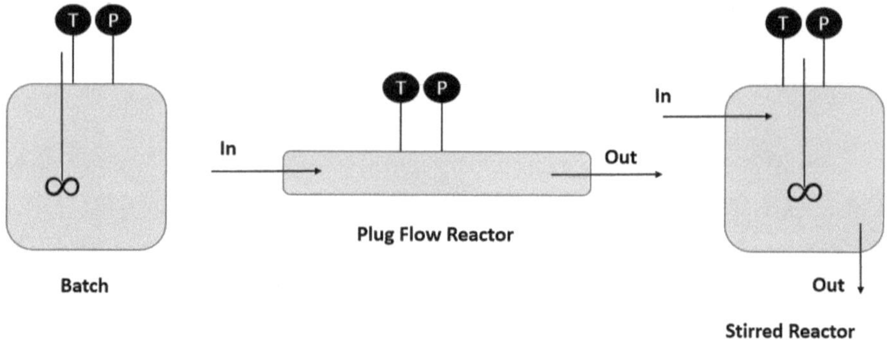


**Fig. 14.1** Three main reactors configuration for chemical engineering processes

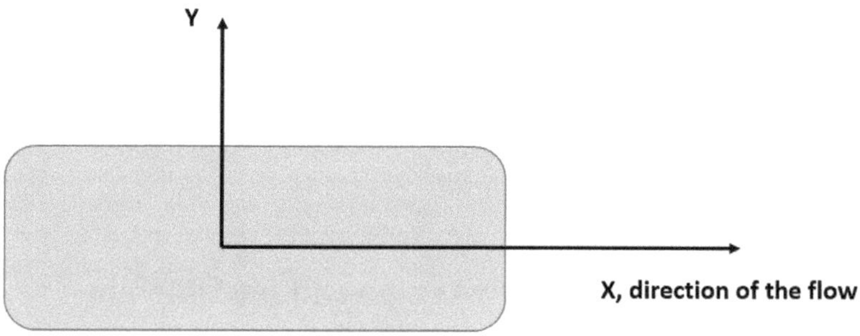


**Fig. 14.2** Example of plain axis in PFR reactor

Therefore, the concentrations of reagents and products have a time dependency, but not a space dependency.

The plug flow reactor (Fig. 14.1—center), PFR, and the Stirred Tank Reactor (Fig. 14.1—right), STR, are steady-state reactors; this means that reagents and products are entering and exiting the reactors with the same concentrations at all time (when the steady state is reached). Nevertheless in the case of plug flow reactors (also called tubular reactors), the concentrations of reagents and products are varying along the axes in the direction of the flow. While for the stirred tank reactor, there is no dependency on the position into the reactor.

It has to be acknowledged that in the case of PFR, it is usually assumed that the concentrations of the reagents and products are not varying along the direction which is perpendicular to the flow direction (the one also perpendicular to the walls of the tubular reactor). This is more clear considering Fig. 14.2 where the $x$-axes are the one in the direction of the flow and the $y$-axes are the one perpendicular to it.

In this manuscript, we focus on how microorganisms are utilized in the production of biofuels from a reactor design point of view. With reactor design, it is meant here that we want to know the volume of the reactor (and its type) necessary to have a particular mass flow and desired concentrations of the products.

The treatment of this theory is started here with the representation of microbial and enzymatic fermentation. An organic feed is inserted in our reactor to give products. An enzyme is used as a catalyst:

$$(\text{organic feed}, A) \longrightarrow \text{enzime } E(\text{product chemical}, R); \tag{14.1}$$

$$(\text{organic feed}, A) \longrightarrow \text{microbe } C(\text{product chemical}, R + \text{more cells}, C). \tag{14.2}$$

From Eqs. (14.1) and (14.2), we see that from an organic feed $A$, for instance sugar, with the help of an enzyme, we produce a fuel $R$. In the first case, the enzyme takes part to the formation of the fuel but the overall amount of it remains unchanged; therefore, the enzyme is very well assumed to be a catalyst in this case. While in the second case, we have that microorganisms' cells are reproducing to form the fuel product and also more cells.

For systems which are not with a constant density (the volume is varying), for instance, when the number of moles in the system is changing by time (notice that we are talking about moles here and not mass), we have by definition (Coker and Kayode 2001):

$$\varepsilon_A = \frac{V_{X_A=1} - V_{X_A=0}}{V_{X_A=0}}; \tag{14.3}$$

where $\varepsilon_A$ is the fractional change in volume between no conversion and complete conversion of the reactant $A$. In case, the volume is linearly a function of the conversion we have:

$$V = V_0(1 + \varepsilon_A X_A); \tag{14.4}$$

however, this information will be given by experiments and in theory could also have another function. Considering Eqs. (14.3) and (14.4), we obtain:

$$\frac{C_A}{C_{A0}} = \frac{1 - X_A}{1 + \varepsilon_A X_A}. \tag{14.5}$$

The demonstration of Eq. (14.4) is coming from the definitions of $C_A$ and $C_{A0}$, with Eq. (14.5) and it is as follows:

Remember that the conversion of an element $A$ at some time $t$ is defined as: $X_A = 1 - \frac{n_A}{n_{A0}}$ where $n_A$ and $n_{A0}$ are the moles of $A$ at time $t$ and the moles of $A$ at time zero. This becomes $X_A = 1 - \frac{C_A}{C_{A0}}$ when the volume (of reaction) remains constant.

As a matter of fact:

$$\frac{C_A}{C_{A0}} = \frac{\frac{n_A}{v}}{\frac{n_{A0}}{v_0}}; \tag{14.6}$$

and from Eq. (14.4)

$$v_0 = \frac{v}{1 + \varepsilon_A x_A}; \tag{14.7}$$

Therefore, we obtain:

$$\frac{n_A}{v} \cdot \frac{v_0}{n_{A0}} = \frac{n_A}{v} \cdot \frac{v}{(n + \varepsilon_A x_A) n_{A0}}; \tag{14.8}$$

which gives

$$\frac{C_A}{C_{A0}} = \frac{(1 - x_A)}{1 - \varepsilon_A x_A}. \tag{14.9}$$

From Eq. (14.9), we have

$$X_A = \frac{C_{A_0} - C_A}{C_{A_0} + \varepsilon_A C_A}; \tag{14.10}$$

and therefore by differentiating $X_A$ with respect to $C_A$, the following relations are then obtained:

$$\mathrm{d}x_A = \frac{-C_{A0}(1 - \varepsilon_A)}{(C_{A0} + \varepsilon_A C_A)^2} \mathrm{d}C_A; \tag{14.11}$$

$$\mathrm{d}C_A = -C_{A0} \cdot \frac{(1 - \varepsilon_A)}{(1 + \varepsilon_A X_A)^2} \mathrm{d}x_A. \tag{14.12}$$

A little premise should be done here to explain in a very simple way, what is the reaction rate order. In the following treatment of this theory, it will be referred at the reaction rate as equal to:

$$\text{Disappearance} = (-r_A)(\text{Reactor Volume}) = (\text{moles of } A \text{ reacting})/\text{time}. \tag{14.13}$$

This will be repeated later on when the derivation of the residence time will be performed for different configurations. Now, it is important to notice that the reaction rate is a function of the concentration of the reagents and in other words, it is directly proportional to the concentration of the reagents, or products, or both. Let us assume now that the moles of reagent $A$ reacting over time is directly proportional to the concentration of $A$. That will be:

$$(-r_A) = kC_A^1; \tag{14.14}$$

Where the exponent 1 refers to the reaction order. In this way, for reactions of zero order, we have:

$$(-r_A) = k; \tag{14.15}$$

And so on. Notice also that the concentration of the substance is actually equal to the number of moles on the volume of the reactor (which is the volume of substance into the reactor, not the physical volume of the container) and this volume can change significantly depending on the number of moles present in the system.

As an example of utilization of the fractional change in volume in our calculations, one case is given here. The example is reported here with the complete solution for the residence time calculation.

Consider that the reaction $A \rightarrow 3R$ is taking place. Concerning the feed: 50% is formed by the $A$ component and 50% is given by an inert compound. However, let us assume that the 50% of the moles is inert, from the reaction we have that the fractional change in volume is:

$$\varepsilon_A = \frac{4-2}{2} = 1. \tag{14.16}$$

Suppose that we want a conversion of 80% of the component $A$, therefore, for a reactor working in the plug flow mode, PFR, the residence time is expressed as:

$$\tau = C_{A_0} \int_0^{x_{Af}} \frac{\mathrm{d}x_A}{-\Gamma_A} = C_{A_0} \int_0^{x_{Af}} \frac{\mathrm{d}X_A}{KC_{A_0}^{1/2}\left(\frac{1-x_A}{1+\varepsilon_A x_A}\right)^{0.5}} = \frac{C_{A_0}^{1/2}}{K} \cdot \int_0^{0.8} \left(\frac{1+x_A}{1-X_A}\right)^{1/2} \mathrm{d}X_A. \tag{14.17}$$

The complete analytical solution of the integral is given here:

Let us do the transformation:

$$(1-x_A)^{1/2} = t. \tag{14.18}$$

This will give:

$$1 - x_A = t^2 \rightarrow x_A = 1 - t^2 \rightarrow \mathrm{d}x_A = -2t\ \mathrm{d}t; \tag{14.19}$$

we substitute the expressions obtained in Eq. (14.17) and now the integral on the right hand side of Eq. (14.17) looks as follows:

$$\int_0^{0.8} \frac{(1+1-t^2)^{0.5}}{t} \cdot (-2t)\,\mathrm{d}t = \int -2 \cdot \sqrt{2-t^2}\mathrm{d}t; \tag{14.20}$$

integrating by parts, which means, $\int_a^b \left[f'(x)g(x)\right]\mathrm{d}x = \left[f(x)g(x)\right]|_a^b - \int_a^b \left[f(x)g'(x)\right]\mathrm{d}x$

This gives:

$$-2\left(t\cdot\sqrt{2-t^2}+\int\frac{t^2}{\sqrt{2-t^2}}\mathrm{d}t\right)=-2\int\sqrt{2-t^2}\mathrm{d}t. \tag{14.21}$$

Notice that the derivative of $\frac{\mathrm{dArcsen}x}{\mathrm{d}x}$ is:

$$\frac{\mathrm{dArcsen}x}{\mathrm{d}x}=\frac{1}{\sqrt{1-x^2}}; \tag{14.22}$$

therefore:

$$\int\sqrt{2-t^2}\mathrm{d}t=t\cdot\sqrt{2-t^2}+\int\frac{t^2}{\sqrt{2-t^2}}\mathrm{d}t\rightarrow-\int\frac{2-E^2-2}{\sqrt{2-t^2}}; \tag{14.23}$$

and

$$\int\sqrt{2-t^2}\mathrm{d}t=t\cdot\sqrt{2-t^2}-\int\sqrt{2-t^2}+2\int\frac{1}{\sqrt{2-t^2}}\mathrm{d}t; \tag{14.24}$$

from which we have:

$$2\int\sqrt{2-t^2}\mathrm{d}t=t\cdot\sqrt{2-t^2}+2\int\frac{1}{\sqrt{2-t^2}}\mathrm{d}t; \tag{14.25}$$

this will give:

$$\int\sqrt{2-t^2}\mathrm{d}t=\left[\frac{1}{2}t\cdot\sqrt{2-t^2}+\mathrm{Arcsin}\left[\frac{t}{\sqrt{2}}\right]\right]\Bigg|_0^{0.8}. \tag{14.26}$$

Considering again that we have done some transformations in Eqs. (14.18)–(14.19), the final formula for the residence time at these conditions will be:

$$\tau=\frac{C_{A0}^{1/2}}{K}\left[\frac{1}{2}(1-x_A)^2\cdot\sqrt{2-(1-x_A)}+\mathrm{Arcsin}\left[\frac{(1-x_A)^{1/2}}{\sqrt{2}}\right]\right]\Bigg|_0^{0.8}. \tag{14.27}$$

## 14.2 Performance of a Batch Stirred Tank Reactor

In the batch stirred tank reactor, BSTR, as previously mentioned, there is no inlet and outlet during the reaction process. The feedstock is inserted together with the reagents in the reactor and then only after some time has passed, the products are

**Fig. 14.3** Illustration of the batch stirred tank reactor

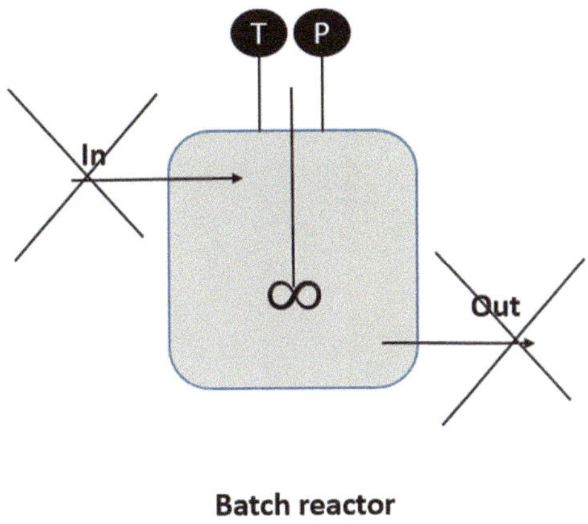


then recovered from the reactor. The illustration of this kind of reactors is given in Fig. 14.3.

In this case, the general balance of moles for our system is:

$$\text{In} - \text{Out} - (\text{or} +)\text{disappearance} = \text{accumulation}. \tag{14.28}$$

Notice that in Eq. (14.28), the term "disappearance" can have a minus or plus sign depending if the number of moles is increasing or decreasing. I would like to stress here that this is valid in case we are talking about number of moles and **not** mass. As it is well known from the thermodynamics principles, the mass remains unchanged during the process. However, the number of moles of the system can indeed change. For instance, if we consider a simple reaction like $C+O_2 = CO_2$ we notice that from two moles (one of carbon and one of $CO_2$), we obtain one mole. This is particularly important when the volume of the system is taken into account: we remember that in gas phase, for instance, the volume of the gas is directly proportional to the number of moles, i.e., if we have half moles, the volume occupied will be much decreased.

Let us look at all the terms given in the balance:

$$\text{Input of } A\left(\frac{\text{moles}}{\text{time}}\right) = F_{A0}; \tag{14.29}$$

$$\text{Out}(A) = F_{A0}(1 - X_a); \tag{14.30}$$

where $X_a$ is the mole fraction of the $A$ component. The disappearance will be:

$$\text{Disappearance} = (-r_A)V = \frac{\text{mol } A \text{ reacting}}{\text{time}} \text{Volume}_{\text{reactor}}. \tag{14.31}$$

In the case of batch mode of operation, we do not have inlet and outlet of material; therefore, in this case, the disappearance is equal to the accumulation of moles:

$$(-r_A)V = \frac{dn_A}{dt};\tag{14.32}$$

and:

$$\frac{dC_A}{dt} = -r_A;\tag{14.33}$$

Therefore,

$$dt = \frac{dC_A}{-r_A}.\tag{14.34}$$

At this point, it is good to remember the definition of conversion:

$$x_A = \frac{(C_{A0} - C_A)}{C_{A0}}.\tag{14.35}$$

For this reason, the expression for the residence time in this case is:

$$\tau = \int_0^{t_{\text{final}}} t = C_{A0}\int_{x_{A0}}^{X_{\text{final}}} \frac{dx_A}{-r_A}.\tag{14.36}$$

It has to be noticed that, in order to solve the integral, the expression for the term $(-r_A)$ is needed and $(-r_A)$ is a function of the kinetic expression for what it was called before "disappearance." Later, this will become more clear reading the following pages.

## 14.3 Performance of a Continuous Stirred Tank Reactor

In a continuous stirred tank reactor, there is inlet of feedstock and outlet of products (Fig. 14.4) and in this case, the inlet and the outlet are proceeding within steady-state conditions.

As done previously, we start from the general mass balance:

$$\text{In} - \text{Out} - \text{disappearance} = \text{accumulation};\tag{14.37}$$

and in the case of a continuous stirred tank reactor, we have:

$$\text{Input of } A \text{ (moles/time)} = F_{A0};\tag{14.38}$$

$$\text{Out }(A) = F_A = F_{A0}(1 - X_A);\tag{14.39}$$

**Fig. 14.4** Simple schematics of a STR

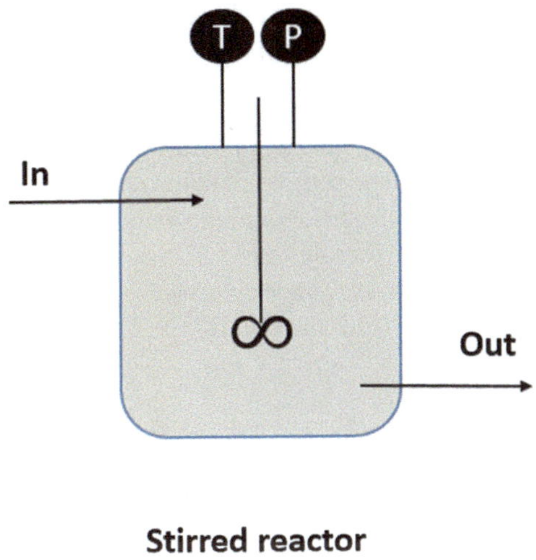


$$\text{Disappearance} = (-r_A)V = \frac{\text{mol } A \text{ reacting}}{\text{time}} \frac{V_{\text{reactor}}}{\text{Volume fluid}}. \tag{14.40}$$

In this case, the term related to the accumulation in the reactor is missing because there are steady-state conditions. This means that at any time, we have the same inlet flow, the same outlet (at the exit), and the concentrations in the reactor are not changing with time. Therefore, the balance will give:

$$F_{A0}X_A = (-r_A)V \rightarrow \frac{V}{F_{A0}} = \frac{X_A}{-r_A} \tag{14.41}$$

And the residence time to have a final conversion $X_A$ will be:

$$\tau = \frac{V}{v_0} = \frac{VC_{A0}}{F_{A0}} = \frac{C_{A0}X_A}{-r_A}; \tag{14.42}$$

where

$V_0$ volumetric inlet
$V$ reactor volume, (notice that, here we intend the volume of the substance into the reactor)

For first-order chemical reaction with no change in volume: $\tau = \frac{C_{A0}-C_A}{kC_A}$ and this is because $-r_A = kC_A$.

## 14.4 First-Order Chemical Reaction in Stirred Tank Reactor with Changing Reaction Volume

In this section, the case of a first-order chemical reaction is taken into account for STR in the case of change in volume over time. The volume of substance into the reactor is directly proportional to the amount of moles within the reactor and therefore, since the amount of moles can vary significantly, the volume of the reactor (the substance into the reactor) can vary also.

We start from the definition of the expansion factor (Davis and Davis 2003):

$$\varepsilon = \frac{V_{xA=1} - V_{xA=0}}{V_{xA=0}}; \tag{14.43}$$

where $x_A = 1$ indicates that the conversion of the component $A$ (sugar for instance) is one, while for $x_A = 0$, we have a conversion of zero and this takes place at the reactor inlet. It will be shown here that for first-order reaction and the case of STR, the residence time is expressed as follows:

$$\tau = \frac{X_A(1 + \varepsilon_A X_A)}{k(1 - X_A)} \quad \text{for any} \quad \varepsilon_A. \tag{14.44}$$

This is demonstrated as follows, the mass balance done previously gives:

$$\tau = \frac{V}{v_0} = \frac{VC_{A0}}{F_{A0}} = \frac{C_{A0}X_A}{-r_A}; \tag{14.45}$$

remember that $X_A$ is the conversion at the exit and also the conversion measured hypothetically in the reactor volume. This is because the concentration at the exit of the reactor is the one which is formed in the reactor after the residence time has passed. In case we assume that the reactor slurry volume has a linear dependence from the initial volume we have:

$$V = V_0(1 + \varepsilon_A X_A). \tag{14.46}$$

Substituting Eq. (14.46) in Eq. (14.45), will give:

$$\frac{C_A}{C_{A0}} = \frac{1 - X_A}{1 + \varepsilon_A X_A}; \tag{14.47}$$

for first-order reaction and the case of continuous stirred reactor, we have previously written:

$$\tau = \frac{V}{v_0} = \frac{C_{A0}X_A}{kC_A}; \tag{14.48}$$

this gives:

$$k\tau = \frac{C_{A0}}{C_A} X_A = \frac{X_A(1 + \varepsilon_A X_A)}{1 - X_A}. \tag{14.49}$$

Now, let us consider the more complicated case in which the reaction is of second order:

$$-r_A = kC_A^2; \tag{14.50}$$

in this case:

$$\tau = \frac{V}{v_0} = \frac{C_{A0} X_A}{-r_A} = \frac{C_{A0} X_A}{kC_A^2}; \tag{14.51}$$

from which

$$k\tau = \frac{C_{A0} X_A}{C_A C_A} = \frac{(1 + \varepsilon_A X_A)}{(1 - X_A)} \frac{X_A}{C_A}. \tag{14.52}$$

Remember that:

$$\frac{C_A}{C_{A0}} = \frac{1 - X_A}{1 + \varepsilon_A X_A}; \tag{14.53}$$

and therefore,

$$k\tau = \frac{(1 + \varepsilon_A X_A)}{(1 - X_A)} \cdot \frac{X_A}{\frac{C_{A0}(1 - X_A)}{(1 + \varepsilon_A X_A)}} = \frac{(1 + \varepsilon_A X_A)^2}{(1 - X_A)^2 C_{A0}}. \tag{14.54}$$

## 14.5 Plug Flow Reactor

As mentioned previously, in the PFR model, we have steady-state conditions for the inlet and the outlet. This means that the concentrations and the mass flow at the inlet and outlet are always the same as regards to time. However, even if it is true that the concentrations are not varying with time for any of the sections of the reactor, the conversion and the concentrations are changing along with the axes in the flow direction (see Fig. 14.2).

Notice here that we have a similitude with the chromatographic columns treated in the first part of this book, in that case also, a residence time in the chromatographic column was associated to an equivalent length of the column itself. It is good to keep in mind that the more will be the reactor length, the more will be the residence time

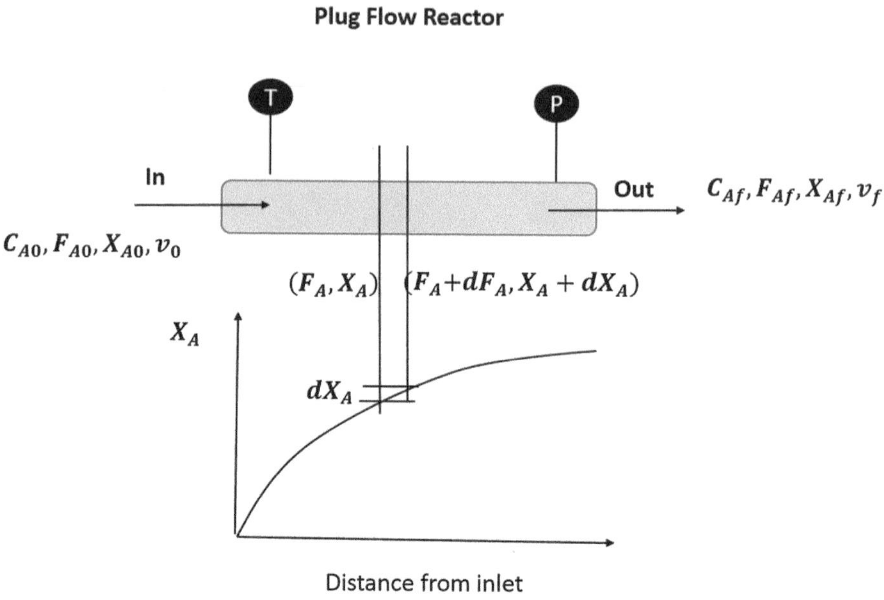


**Fig. 14.5** Illustration of a PFR and its differential change in conversion and concentration with respect to the compound $A$ along the distance from the inlet axes

and the conversion in the reactor; therefore, length and time go hand to hand in this case (Fig. 14.5).

Also in this case, we do the general mass balance but this time for a differential section in the tubular reactor. The concentration of each component is varying with the length of the reactor: considering two sections along with the flow direction, $x_1$ and $x_2$, if we consider one fixed reaction section, the concentrations will not vary with time, nevertheless if we focus on the next section then the concentrations will be different with respect to the previous section.

As it was done before:

In – Out + Generation = 0 (the accumulation is 0 in steady state) and:

$$F_A = (F_A + \mathrm{d}F_A) + (-r_A)\mathrm{d}V. \tag{14.55}$$

Also in this case, there is no accumulation in the differential volume, which means that at any fixed section of the PFR, the concentrations are the same at any time.

Considering:

$$\mathrm{d}F_A = \mathrm{d}[F_{A0}(1 - X_A)] = -F_{A0}\mathrm{d}X_A; \tag{14.56}$$

we obtain:

$$F_{A0}\mathrm{d}X_A = (-r_A)\mathrm{d}V; \tag{14.57}$$

therefore:

$$\tau = \frac{V}{v_0} = \frac{VC_{A0}}{F_{A0}} = C_{A0}\int_0^{X_{Af}} \frac{\mathrm{d}X_A}{-r_A}. \tag{14.58}$$

This is valid for constant density systems. For first-order irreversible homogeneous reactions:

$$-r_A = kC_A; \tag{14.59}$$

this gives:

$$\tau = C_{A0}\cdot\int_0^{X_A} \frac{\mathrm{d}X_A}{kC_A} = \int_0^{X_A} \frac{C_{A0}\mathrm{d}X_A}{kC_A}; \tag{14.60}$$

considering Eq. (14.53), we have:

$$\frac{C_A}{C_{A0}} = \frac{1-X_A}{1+\varepsilon_A X_A}; \tag{14.61}$$

and this will help us by giving:

$$k\tau = \int_0^{X_A} \frac{(1+\varepsilon_A X_A)}{(1-X_A)}\mathrm{d}X_A. \tag{14.62}$$

Solving the integral:

$$k\tau = \int_0^{X_A} \frac{1}{1-X_A}\mathrm{d}X_A + \int_0^{X_A} \frac{\varepsilon_A X_A}{(1-X_A)}\mathrm{d}X_A = \int_0^{X_A} \frac{1}{1-X_A}\mathrm{d}X_A$$
$$-\varepsilon_A \int_0^{X_A} \frac{-X_A}{(1-X_A)}\mathrm{d}X_A \tag{14.63}$$

Which gives:

$$k\tau = \int_0^{X_A} \frac{1}{1-X_A}\mathrm{d}X_A - \varepsilon_A \int_0^{X_A} \frac{1-X_A-1}{(1-X_A)}\mathrm{d}X_A = \int_0^{X_A} \frac{1}{1-X_A}\mathrm{d}X_A$$

$$- \varepsilon_A \int_0^{X_A} dX_A + \varepsilon_A \int_0^{X_A} \frac{1}{1 - X_A} dX_A \tag{14.64}$$

Equation (14.65) is derived:

$$k\tau = -\ln(1 - X_A) - \varepsilon_A X_A - \varepsilon_A \ln(1 - X_A) = -(1 + \varepsilon_A)\ln(1 - X_A) - \varepsilon_A X_A. \tag{14.65}$$

## 14.6 The Case of Reversible Reaction

Consider the following case:

$$A \underset{\to k_1}{\overset{\leftarrow k_2}{\rightleftharpoons}} rR; \tag{14.66}$$

where $r$ is the stoichiometric coefficient for the product $R$, $k_1$ the forward chemical reaction constant, and $k_2$ the backward chemical reaction constant. Then for first-order chemical reactions we have:

$$-r_A = k_1 C_A - k_2 C_R. \tag{14.67}$$

For convenience, it is possible to define the following variable $M$:

$$\frac{C_{R0}}{C_{A0}} = M; \tag{14.68}$$

consider also that $x_A$ is the conversion at equilibrium and as usual remember the relation:

$$\frac{C_A}{C_{A0}} = \frac{1 - X_A}{1 + \varepsilon_A X_A}. \tag{14.69}$$

Keeping this in mind:

$$r(C_{A0} - C_A) = (C_R - C_{R0}) \to C_R = r[C_{A0} - C_A] + C_{R0}; \tag{14.70}$$

the following result is obtained

$$\frac{C_R}{C_{A0}} = r\left[1 - \frac{C_A}{C_{A_0}}\right] + \frac{C_{R0}}{C_{A0}} \to \frac{C_R}{C_{A0}} = r\left[1 - \frac{(1 - x_A)}{(1 + \varepsilon_A x_A)}\right] + M; \tag{14.71}$$

and then:

$$rC_A = rC_{A0} - C_R + C_{R0} \rightarrow C_A = C_{A0} - \frac{1}{r}C_R + \frac{1}{r}C_{R0}. \quad (14.72)$$

We substitute the expression for $C_R$ obtained, Eq. (14.71), and we have:

$$C_A = C_{A0} - \frac{1}{r}\left[r \cdot C_{A0}\left(1 - \frac{(1 - X_A)}{(1 + \varepsilon_A X_A)}\right) + C_{A0} \cdot M\right] + \frac{1}{r} \cdot C_{R0}; \quad (14.73)$$

at the equilibrium:

$$k_1 C_A = k_2 C_r \rightarrow \frac{k_1}{k_2} = K = \frac{C_R}{C_A}; \quad (14.74)$$

where $K$ is the equilibrium constant. After this, Eq. (14.75) is obtained:

$$k_2 = k_1 \cdot \frac{C_A}{C_R} = k_1 \cdot \frac{\left[C_{A0} \cdot \frac{1}{r}\left[rC_{A0}\left(1 - \frac{(1-x_A)}{(1+\varepsilon_A X_A)} + C_{A_0} \cdot M\right)\right] + \frac{1}{r}C_{R_0}\right]}{\left[C_{R0} \cdot r\left(1 - \frac{(1-x_A)}{(1+\varepsilon_A x_A)}\right) + C_{A0} \cdot M\right]}. \quad (14.75)$$

In these equations previously written, the $x_A$ refers to the fraction at equilibrium. At this point, we have that $k_2 = f(k_1)$, where $f$ means "function of". For first-order forward and backward chemical reactions we have:

$$\frac{dC_A}{dt} = k_1 C_A - k_2 C_R \rightarrow k_2 = k_1 \cdot f(X_A). \quad (14.76)$$

Remember, as usual, the relation:

$$\frac{C_A}{C_{A0}} = \frac{1 - X_A}{1 + \varepsilon_A X_A}; \quad (14.77)$$

Therefore, integrating the resulting expression, it will be possible to obtain $k_1$ as a function of the fractional change in volume, the fraction of the feedstock, etc.

$$k_1 = f(\varepsilon_A X_A, \ldots); \quad (14.78)$$

in the same way, from the relation:

$$\frac{k_1}{k_2} = \frac{C_R}{C_A}; \quad (14.79)$$

we have that $k_2$ is a function of the other variables and parameters:

$$k_2 = f(\varepsilon_A X_A, \ldots); \quad (14.80)$$

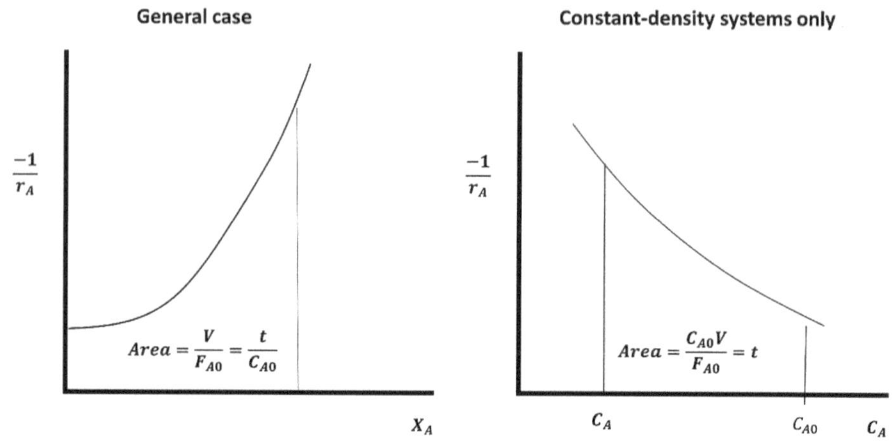


**Fig. 14.6** Inverse of the reaction rate as a function of the conversion (left) and concentration (right) for the component $A$

and it will be possible to evaluate $\tau$.

## 14.7 Performance Equations for Plug Flow Reactors

For PFR, the mass balances can be illustrated by plotting the conversion against the reaction rate. In Fig. 14.6, it has to be noticed that when we consider the concentrations (plot on the right), the volume of the reactor (with reactor volume it is meant what is inside the reactor) is considered to remain constant, in the drawing on the right, the concentration of the feedstock should be read from the right to the left ($C_{A0}$–$C_A$) while in the drawing on the left, the conversion goes from zero at the inlet and goes to $X_A$ at a generic time $t$.

Before proceeding further, it is suggested to remember that: $x_A = 1 - \frac{n_A}{n_{A0}}$ this can be considered as: $x_A = 1 - \frac{C_A}{C_{A0}}$ when the volume (of reaction) remains constant. The residence time for PFR is expressed as follows:

$$\tau = \frac{V}{v_0} = \frac{VC_{A0}}{F_{A0}} = C_{A0}\int_0^{X_{Af}} \frac{\mathrm{d}X_A}{-r_A}. \tag{14.81}$$

For zero-order homogeneous reactions in a plug flow, we have:

$$\tau = \frac{kC_{A0}V}{F_{A0}} = C_{A0}X_A. \tag{14.82}$$

This is because for reactions of order zero, $-r_A = k$. the reaction rate is a constant.

## 14.8 Autocatalitic Reactions

Microorganisms, which are used to produce biofuels like hydrogen for instance, utilize a determinate feedstock to reproduce and to discard a product which is our biofuel of interest. This category of processes can be represented by so-called autocatalytic reactions of this kind:

$$A + C \rightarrow C + C. \tag{14.83}$$

where $C$ is the concentration of microorganisms in the system and $A$ is as usual our feedstock, for instance sugar. The general case of reaction rate for this kind of reactions is described by:

$$-r_A = kC_A^a C_c^c. \tag{14.84}$$

Consider now the general equation for the residence time of $A$ in PFR, Eq. (14.81) and the definition of conversion when there is no volume change in the reactor: $x_A = 1 - \frac{C_A}{C_{A0}}$. Equation (14.35), when the stoichiometric coefficients are both equal to one becomes:

$$-r_A = k \cdot C_{A0}(1 - X_A) \cdot C_{A0}(1 + X_A). \tag{14.85}$$

This would be reasonable since, in the assumptions taken here, when there is total conversion of moles for the nutrient (sugar for instance), the overall number of moles for the $C$ component (the microorganism) will be doubled. Therefore, Eq. (14.81) becomes now:

$$\tau = \frac{V}{v_0} = \frac{VC_{A0}}{F_{A0}} = C_{A0}\int_0^{X_{Af}} \frac{\mathrm{d}X_A}{-r_A} = C_{A0}\int_0^{X_{Af}} \frac{\mathrm{d}X_A}{k \cdot C_{A0}^2 \cdot (1 - X_A)(1 + X_A)}$$

$$= \frac{1}{k \cdot C_{A0}}\int_0^{X_{Af}} \frac{\mathrm{d}X_A}{(1 - X_A)(1 + X_A)} \tag{14.86}$$

At this point, the integral on the right side of Eq. (14.86) will be solved considering:

$$\frac{1}{(1 - X_A)(1 + X_A)} = \frac{1 - X_A + X_A}{(1 - X_A)(1 + X_A)} = \frac{1}{(1 + X_A)} + \frac{2 \cdot X_A}{2(1 - X_A)(1 + X_A)}; \tag{14.87}$$

where
$(1 - X_A)(1 + X_A) = X_A^2 + 1.$
Now Eq. (14.86) will be given by Eq. (14.88):

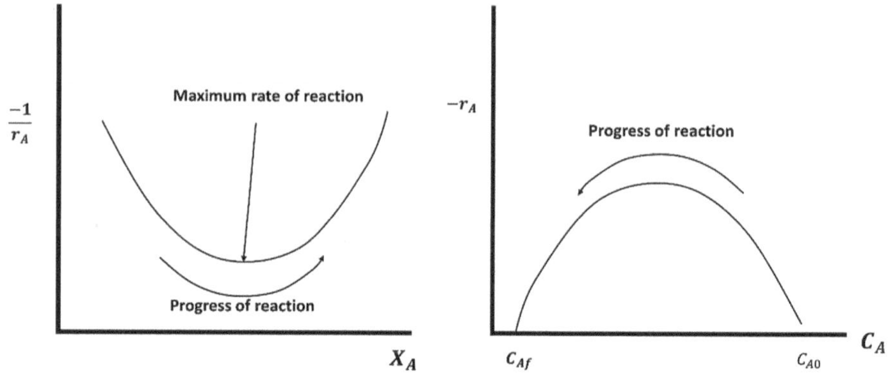


**Fig. 14.7** General trend of the reaction rate as a function of the conversion for autocatalytic reactions

$$\tau = \frac{1}{k \cdot C_{A0}} \left[ \int_0^{X_{Af}} \frac{1}{(1 + X_A)} + \frac{1}{2} \int_0^{X_{Af}} \frac{2X_A}{X_A^2 + 1} \right]; \tag{14.88}$$

which will give:

$$\tau = \frac{1}{k \cdot C_{A0}} \cdot \ln\left[ (1 + X_{Af})(X_{Af}^2 + 1) \right]. \tag{14.89}$$

The general trend of the reaction rate for the sugar depletion is described in Fig. 14.7.

## 14.9 Optimal Choice of Reactors

At this stage, it is good to remember that for stirred tank reactors, the residence time is:

$$\tau = \frac{V}{v_0} = \frac{VC_{A0}}{F_{A0}} = \frac{C_{A0}X_A}{-r_A}; \tag{14.90}$$

while for plug flow reactors, we have:

$$\tau = \frac{V}{v_0} = \frac{VC_{A0}}{F_{A0}} = C_{A0} \int_0^{X_{Af}} \frac{\mathrm{d}X_A}{-r_A}; \tag{14.91}$$

therefore, we see that the bigger is the residence time, the bigger is the reactor volume. For stirred reactors, Eq. (14.92) is valid:

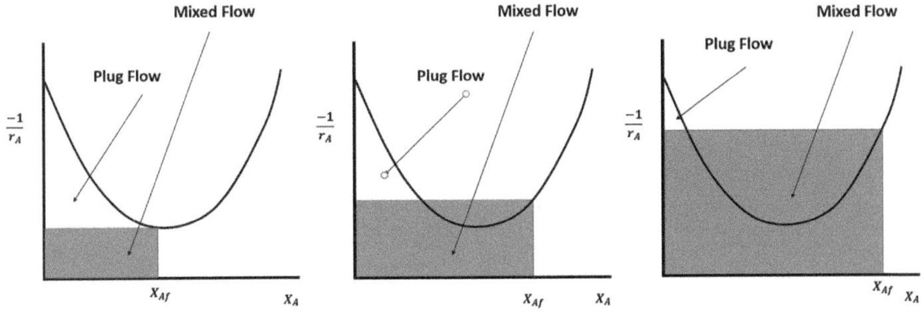


**Fig. 14.8** Comparison between reactor volumes of PFR and STR at a fixed conversion. Optimal choice of the reactors for a given conversion with autocatalytic reactions

$$\frac{V}{F_{A0}} = \frac{X_A}{-r_A}. \tag{14.92}$$

From what was derived previously, it is possible to see very clearly that the ratio $\frac{V}{F_{A0}}$ is proportional to the integral area of the curve described by $\frac{1}{-r_A}$ in the first case, while it is the area of the rectangle given by $\frac{1}{-r_A} \cdot X_A$ in the second case (Fig. 14.8).

Therefore, for auto-catalytic reactions, it is better to use a stirred tank reactor for low conversions and a plug flow reactor for high conversions.

## References

Coker, A. K., Kayode, C. A. (Eds.). (2001). Introduction to reactor design fundamentals for ideal systems (Chap. 5). In *Modeling of chemical kinetics and reactor design* (pp. 260–423). Woburn: Gulf Professional Publishing. https://doi.org/10.1016/B978-088415481-5/50007-3.

Davis, M. E., Davis, R. J. (2003). *Fundamentals of chemical reaction engineering*. McGraw-Hill Higher Education.

# Chapter 15
# Enzyme Kinetics

*Pavia, 12. August. 1895.*
*Very Illustrious Professor!.*
*Sir Eng. Vitali coming back from Turin referred to me that he had the precious occasion of communicate with Your Illustrious Lordship and that my name was mentioned there. Seen such benevolence from you, it makes me bold to prey you to write me a little recommendation letter to be presented to Professor Weber of Zurich, who I would like to visit before the study courses will be opened at that university. This is to overcome the difficulties of attending that program due to my young age. Thanking you in advance and with the highest consideration of Your Illustrious Lordship.*
*Devotedly, Alberto Einstein.*

A letter written by Albert Einstein to Galileo Ferraris (known as the Father of three-phase current) when he was only sixteen. The letter was published by the Italian journal of finance, Il Sole 24 Ore on the 27th of Nov. 2005.

## 15.1 Product Formation in Enzyme Catalyzed Biological Processes

The production of biofuels largely involves the utilization of enzymes and therefore, from this point of view, it would be beneficial to know more on the way they actually work. Even if the scope of this book is not to focus on the biological aspects related to the formation, production, and conservation of such entities, it will be mentioned here that they could be utilized in a variety of processes including the production of biodiesel where they are used as alternative catalysts. As a preliminary example and to underline the importance of the enzymes usage in the production of biofuels, Fig. 15.1 describes the production of free fatty acids, FFA, and fatty acid methyl esters, FAME from castor oil by means of enzymes catalyzed route.

The process of biodiesel production will be described later in this manuscript; however, it is good here to have an idea of the importance of reactor design and process integration in this kind of productions. Enzymes represent also a big part



C. De Blasio, *Fundamentals of Biofuels Engineering and Technology*,
Green Energy and Technology, https://doi.org/10.1007/978-3-030-11599-9_15

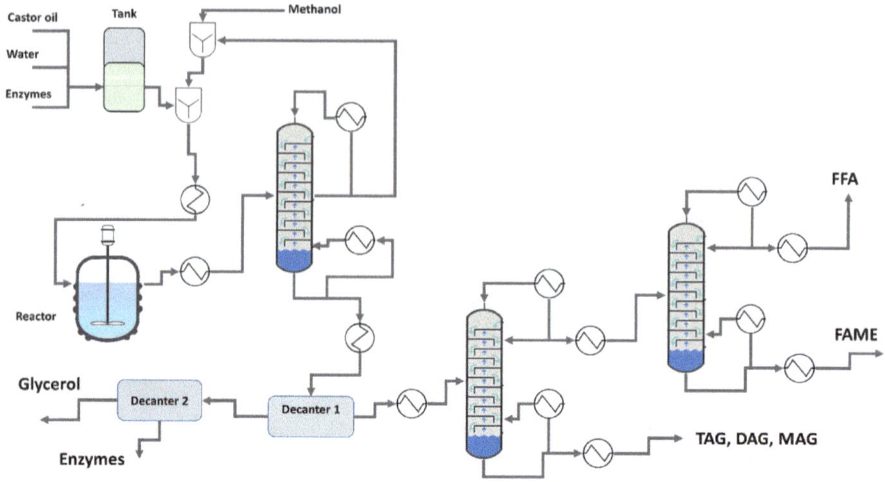


**Fig. 15.1** Transesterification of castor oil driven with enzymes as catalysts. Adapted from Andrade et al. (2019)

of the expense related to the biofuel generation; therefore, it is a big interest for the companies working in this field to have an optimized system where the reactors volumes are kept to a minimum.

Enzymes are indeed catalysts for the reactions in which they are involved. When modeling the reactions yields and rates, their total concentration is assumed to be constant in the sense that their total number will be preserved as it will be shown later. Of course, this is valid in the theoretical treatment of the process and in practice, a loss of enzymes is always taking place.

It is reasonable to consider the concentration of enzymes constant, because the enzymes, like catalysts, are imagined to interact with the substrate, facilitate the reactions involved, and then they liberate the products. After this, they can be used for one additional reaction. The enzyme reacts with the substrate and gives the products $P$, according to:

$$E + S \overset{k_1, k_{-1}}{\longleftrightarrow} C \overset{k_2}{\rightarrow} E + P; \tag{15.1}$$

where

| | |
|---|---|
| $E$ | enzyme concentration |
| $S$ | substrate concentration |
| $K_{1,\,-1,\,2}$ | reaction rate constant [1/time for first order] |
| $P$ | product concentration |
| $C$ | $[E{\cdot}S]$ complex |

In this way, if we consider the last reaction as a first-order chemical reaction:

$$\frac{\mathrm{d}P}{\mathrm{d}t} = k_2 C \tag{15.2}$$

and considering the same assumption we have:

$$\frac{\mathrm{d}C}{\mathrm{d}t} = k_1 E \cdot S - k_{-1} C - k_2 C. \tag{15.3}$$

Assuming that the complex $C$ is soon converted in $P$ and $E$ we obtain:

$$\frac{\mathrm{d}C}{\mathrm{d}t} \sim 0. \tag{15.4}$$

To be noticed that this does not mean that the concentration of $C$ is zero but just that the concentration of the complex remains constant. To be noticed also that the equation above written are simply stating that the variation of $P$ is directly proportional to the concentration of $C$ and that the concentration of $C$ is proportional to $E$, $S$, and $P$ taking into account the amount of this terms that is actually disappearing and what is the element which is produced instead.

## 15.2 Enzime Kinetics. The Michaelis–Menten Model

Within this model, the enzyme is reacting with a substrate giving an enzyme–substrate intermediate. This gives again the free form of the enzyme and the product according to:

$$E + S \overset{k_1, k_2}{\longleftrightarrow} ES \overset{k_3}{\rightarrow} E + P; \tag{15.5}$$

where

$E$ enzyme
$S$ substrate
$ES$ enzyme–substrate

The rate of reaction written for the products is:

$$r = \frac{\mathrm{d}C_p}{\mathrm{d}t} = k_3 C_{ES}. \tag{15.6}$$

Let us assume that the enzyme–substrate is in equilibrium with both: the enzyme and the substrate according to:

$$k_1 C_s C_E = k_2 C_{ES}. \tag{15.7}$$

In addition, assume that the total concentration of enzymes is conserved

$$C_{E0} = C_E + C_{ES}; \tag{15.8}$$

therefore

$$C_{ES} = \frac{C_{E0}C_S}{\frac{k_2}{k_1} + C_s}; \tag{15.9}$$

where

$\frac{k_2}{k_1} = K_m$ is the dissociation constant (equilibrium constant).

Substituting these into the rate of formation of the product, Eq. (15.6), we obtain:

$$\frac{\mathrm{d}C_p}{\mathrm{d}t} = V = k_3\frac{C_{E0}C_S}{K_m + C_S} = \frac{V\max C_S}{K_m + C_S}. \tag{15.10}$$

It is understandable that in the beginning of the reaction, we have the max production of *P* over time, this equals $k_3C_{E0}$. This is because at the start (at time zero), we have that the enzyme is utilizing all the substrate (for instance sugar) at its disposition and therefore, the *ES* complex is at its maximum concentration.

## 15.3 Enzime Kinetics, Briggs–Haldane Approach

Also in this case, the enzyme is reacting with a substrate giving an enzyme–substrate intermediate which in turn gives again the free form of the enzyme and the product according to:

$$E + S \overset{k_1,k_2}{\longleftrightarrow} ES \overset{k_3}{\rightarrow} E + P. \tag{15.11}$$

The rate of formation for the products will be:

$$r = \frac{\mathrm{d}C_p}{\mathrm{d}t} = k_3C_{ES}. \tag{15.12}$$

In this case, we assume that the change of $C_{ES}$ with time is really small compared to the change in the concentrations of the product and substrate, $C_p$ and $C_s$.

$$\frac{\mathrm{d}C_{ES}}{\mathrm{d}t} = k_1C_SC_E - k_2C_{ES} - k_3C_{ES} = 0. \tag{15.13}$$

It is assumed, like previously, that the total concentration of enzymes is conserved

$$C_{E0} = C_E + C_{ES}; \tag{15.14}$$

therefore:

$$C_{ES} = \frac{C_{E0} C_S}{\frac{k_2+k_3}{k_1} + C_S}; \tag{15.15}$$

and we obtain the rate of formation for the products as a function of the concentration of the substrate (sugar) and the initial concentration of enzyme:

$$\frac{dC_p}{dt} = V = k_3 \frac{C_{E0} C_S}{K_m + C_S} = \frac{V\text{max}\, C_S}{K_m + C_S}; \tag{15.16}$$

with

$$K_m = \frac{K_2 + K_3}{K_1}; \tag{15.17}$$

and

$$V\text{max} = k_3 C_{E0}; \tag{15.18}$$

If $k_2 \gg k_3$, then the limiting reaction is the last one and $K_m$ becomes the same as the Michaelis–Menten constant.

Consider a mixed flow reactor where fermentation occurs and let us focus, in this case, on the development of the microorganisms. The production rate of microbes is:

$$r_C = \frac{k C_A C_C}{C_A + C_M}; \tag{15.19}$$

where

$C_A$ concentration of food
$C_C$ concentration of cells
$C_M$ concentration of $A$ when the cells reproduce at ½ of their maximum rate

with the cell growth rate as:

$$r_c = \frac{dC}{dt}; \tag{15.20}$$

a simplified drawing of this case is shown in Fig. 15.2.

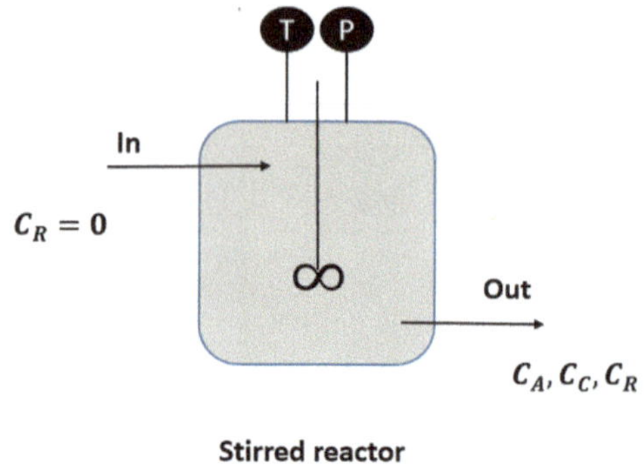


**Fig. 15.2** Simple schematics showing the major components in fermentation

## 15.4 The Monod Model

From here on, we will take into consideration how the reaction (conversion) rate is related to other variables of our system. The variation of the concentration of microorganisms is affected by the same concentration of microbes (c = cells) and by the concentration of sugar. With a good probability the most used model for the cell growth rate is the Monod model. The model is named after Jacques Monod who at the time of his research was working at the Pasteur institute in Paris, France (Monod 1949). In this model, the cell growth rate is expressed as follows:

$$r_C = \frac{kC_A C_C}{C_A + C_M}; \tag{15.21}$$

where

| | |
|---|---|
| $r_c = \frac{dC_c}{dt}$ | cell growth |
| $C_A$ | concentration of food |
| $C_C$ | concentration of cells |
| $C_M$ | concentration of $A$ where the cells reproduce at ½ of their maximum rate. |

In case, the concentration of cells would be proportional to the amount of food (which is actually reasonable), $C_A = k_1 C_c$, the solution for Eq. (15.21) would be obtained as follows:

$$\frac{dC}{dt} = \frac{k \cdot k_1 C_c C_C}{k_1 C_c + C_M}; \tag{15.22}$$

with $a = k \cdot k_1$, and $b = C_M$, $C_c = C$, Eq. (15.22) becomes:

$$\frac{dC}{dt} = \frac{aC^2}{k_1 C + b}. \tag{15.23}$$

Rearranging we have:

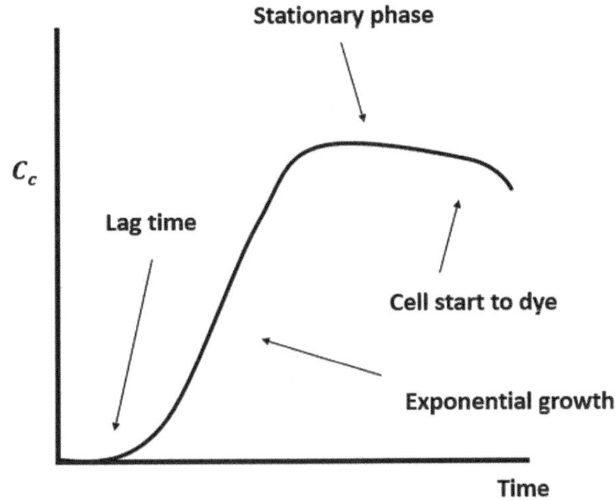


**Fig. 15.3** Main phases in the variation of the cell concentration with time. Adapted from Levenspiel (1999)

$$\left(\frac{k_1 C + B}{C^2}\right)\mathrm{d}C = A \cdot \mathrm{d}t; \tag{15.24}$$

Which will give:

$$\int_{C_0}^{C} \frac{k_1}{C} + B\int_{C_0}^{C} \frac{1}{C^2} = A \cdot \int_{0}^{t} \mathrm{d}t; \tag{15.25}$$

and finally:

$$k_1 \ln \frac{C}{C_0} + B\left(\frac{1}{C_0} - \frac{1}{C}\right) = A \cdot t. \tag{15.26}$$

Figure 15.3 describes in a general way the variation of the cell concentration with time, it is possible to notice three phases, a period in which time is needed to start reproducing (called lag time), an exponential growth, in which the cells utilize effectively the food given, a stationary phase, and a phase where the cells start dying (called decay phase).

While in Fig. 15.4, it is possible to notice that the production rate of product follows different curves on the basis of if we have high food concentration or low food concentration.

We consider:

- $A$ substrate, (also indicated as $S$ in some references)
- $E$ enzyme
- $R$ product (also indicates as $P$ in some references)

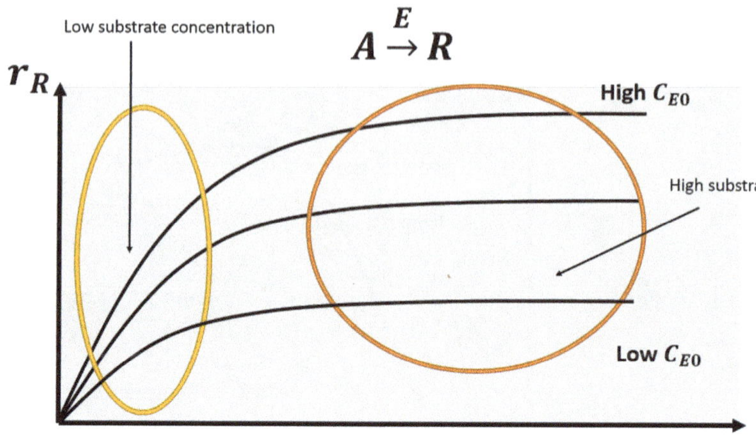


**Fig. 15.4** Production rate of product as a function of the substrate concentration. Adapted from Levenspiel (1999)

At high concentrations of the substrate $A$, we have that the rate is not dependent on the substrate concentration and this is because if the substrate is a large quantity, then it will not matter, it will remain a large quantity anyway. At low concentration of the substrate, the rate of conversion is proportional to the concentration of the substrate and therefore, the order of reaction can be considered to be of the first order.

## 15.5 Main Characteristics of the Briggs–Haldane Equation

The Briggs–Haldane approach (Briggs and Haldane 1925), also known as the quasi-steady-state approach to enzyme kinetics, is also used to model the product yield for biochemical processes; this approach is very much similar to the Monod reasoning or the Michaelis–Menten one (Michaelis and Menten 1913).

To be noticed that in some references, the products are indicated with the letters $R$ or $P$ while $r_{\max}$ has the same meaning of our previous $V_{\max,}$ and sometimes $K_m$ has the same meaning of $C_M$, the substrate is written as $C_s$ sometimes instead of $C_{A.}$

With this in mind, we have:

$$\frac{\mathrm{d}R}{\mathrm{d}t} = V = k_3 \frac{C_{E0} C_A}{C_m + C_A} = \frac{V\max C_A}{C_m + C_A} \tag{15.27}$$

Consider that:

$$r_R = \frac{\mathrm{d}R}{\mathrm{d}t} \tag{15.28}$$

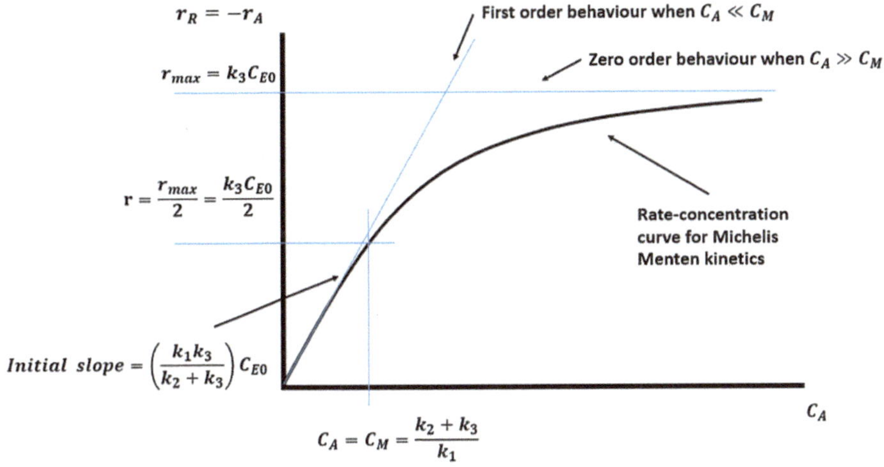


**Fig. 15.5** Feedstock consumption rate function and its main parts

a detailed representation of the function for the feed consumption rate is represented in Fig. 15.5.

In this manuscript, the detailed explanation of the terms shown in figure is derived. At $C_A = 0$, the initial slope is calculated by the following limit:

$$\lim_{C_A \to 0}\left(\frac{\mathrm{d}}{\mathrm{d}C_A}\left[\frac{k_3 C_{E0} C_A}{C_m + C_A}\right]\right) \tag{15.29}$$

while at $C_A = C_{M,}$ we substitute this value in the expression for $\frac{\mathrm{d}C_p}{\mathrm{d}t}$ above to obtain $r$. When $C_A$ tends to infinite we have an indeterminate form of the $-r_A$ rate since we have an infinite value which divides an infinite value. Applying the theorem of Guillaume de l'Hôpital, we readily obtain:

$$-r_A \quad \text{will be} \quad r_{\mathrm{Max}} = k_3 C_{E0}. \tag{15.30}$$

Considering the expression:

$$-\frac{\mathrm{d}A}{\mathrm{d}t} = V = k_3 \frac{E_0 A}{C_m + A} = \frac{V\mathrm{max} \cdot A}{C_m + A}; \tag{15.31}$$

we can solve easily the differential equation between an initial time and a general time $t$, therefore:

$$C_M \ln \frac{C_{A0}}{C_a} + (C_{A0} - C_A) = k_3 C_{E0} t \tag{15.32}$$

Rearranging, we obtain a form which can be plotted. As shown in Fig. 15.6.

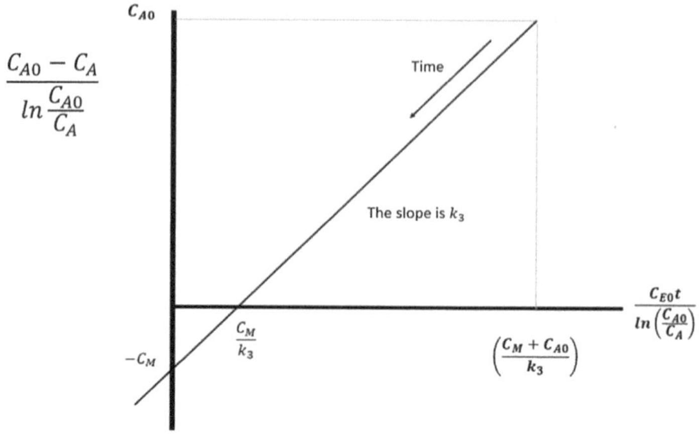


**Fig. 15.6** Linearized function for the concentration of the substrate as a function of time to find the constant parameters

$$\frac{C_{A0}-C_A}{\ln\frac{C_{A0}}{C_A}} = -C_M + k_3 C_{E0}\frac{t}{\ln\frac{C_{A0}}{C_A}} \tag{15.33}$$

Equation (15.33) allows us to have a linear form where it is more easy to find the key constants $C_M$ and $k_3$.

To better understand the terms in the drawing above, when the concentration of the substrate is equal to $C_{A0}$, we are facing again an indeterminate form at the ordinate value, by solving this limit we obtain:

$$\lim_{C_A\to C_{A0}}\left[\frac{C_{A0}-C_A}{\ln\frac{C_{A0}}{C_A}}\right] = C_{A0}. \tag{15.34}$$

After deriving this, it will be possible to evaluate the abscissa from:

$$\frac{C_{A0}-C_A}{\ln\frac{C_{A0}}{C_A}} = C_{A0} = -C_M + k_3 C_{E0}\frac{t}{\ln\frac{C_{A0}}{C_A}}. \tag{15.35}$$

## 15.6 Residence Time in a Mixed Flow Fermentor

From the original form of the balance on the substrate depletion,

$$-\frac{\mathrm{d}A}{\mathrm{d}t} = V = k_3\frac{E_0 A}{C_m + A} = \frac{V\max A}{C_m + A}; \tag{15.36}$$

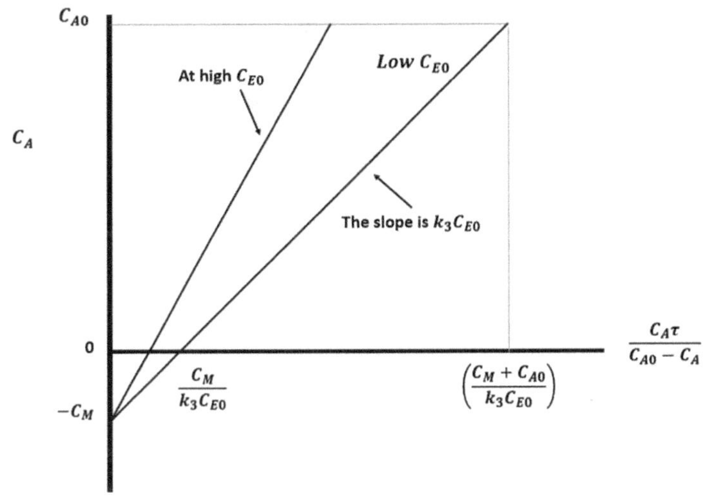


**Fig. 15.7** Linearized form of the substrate concentration as a function of the residence time expression at low and high enzyme concentration

we remember that the residence time in a continuous stirred tank reactor (with no volume change between $t_0$ and $t$) is:

$$\tau = \frac{C_{A0} - C_A}{-r_A} \rightarrow k_3 C_{E0}\tau = \frac{(C_{A0} - C_A)(C_M + C_A)}{C_A}; \tag{15.37}$$

where $C_M$ is the Michaelis–Menten constant. Rearranging we have:

$$C_A = -C_M + k_3\left(\frac{C_{E0}C_A\tau}{C_{A0} - C_A}\right); \tag{15.38}$$

From Fig. 15.7, it is possible to see how $C_M$ can be calculated from linearized forms of the $C_A$ concentration. The slope of the straight line will change according to the initial enzyme concentration since this depends on that concentration. In the same figure, notice that the $C_A$ concentration is plotted against $\frac{C_A\tau}{C_{A0}-C_A}$ and therefore, it is quite simple to derive the terms shown on the abscissa axis: for $C_A = C_{A0}$, we obtain one value of the abscissa, for $C_A = 0$ another value is obtained and the slope of the curve is obtained by simply applying the property of the straight lines.

## References

Andrade, T. A., Martín, M., Errico, M., & Christensen, K. V. (2019). Biodiesel production catalyzed by liquid and immobilized enzymes: Optimization and economic analysis. *Chemical Engineering Research and Design, 141,* 1–14. https://doi.org/10.1016/j.cherd.2018.10.026.

Briggs, G. E., & Haldane, J. B. (1925). A note on the kinetics of enzyme action. *Biochemical Journal, 19,* 338–339.
Levenspiel, O. (1999). *Chemical reaction engineering*. New York, US: Wiley.
Michaelis, L., & Menten, M. (1913). Die kinetik der invertinwirkung. *Biochemistry Zeitung, 49,* 333–369.
Monod, J. (1949). The growth of bacterial cultures. *Annual Review of Microbiology, 3,* 371–394.

# Chapter 16
# Balances on Microbial Fermentation

*And then you see that sometimes, to undermine the false authority of an absurd proposition which is against any reasoning, even a laughter can be a right instrument. Often the laughter also serves to confuse evil persons and make their foolishness shine.*

Umberto Eco, 1980. The Name of the Rose, Fabbri publisher.

## 16.1 Cell Balances

The microbial fermentation is a complicated series of several processes, and each of them could be represented by some form of reactions, and however, in our case the scope of this textbook is not to explain all the biological steps involved but simply model the entire process to have the minimum level of accuracy in order to dimension the reactors used. In a fermentation process, a substrate $A$, for instance sugar, is consumed by microorganisms who will reproduce (until some point) and produce the product $R$ as follows:

$$A \xrightarrow{C} C + R; \tag{16.1}$$

where we recall

$A =$ substrate
$C =$ the living organism
$R =$ product

This can be rewritten also as:

$$A + C \rightarrow C + R; \tag{16.2}$$

and we could consider this as an autocatalytic reaction (in case we do not make a difference between $C$ and $R$). In other words, a simplification for Eq. (16.1) is possible if we consider no difference between the microorganisms and the products



C. De Blasio, *Fundamentals of Biofuels Engineering and Technology*,
Green Energy and Technology, https://doi.org/10.1007/978-3-030-11599-9_16

$$A + R \rightarrow R + R. \tag{16.3}$$

Considering the mass balance on the substrate:

$$-r_A = -\frac{dC_A}{dt} = kC_AC_R; \tag{16.4}$$

the sum of the concentration of $C_A$ and $C_R$ is constant. This is an assumption: It means that is the substrate $A$ is decreasing, the product $R$ is actually increasing, and this is most likely true. Therefore:

$$C_0 = C_A + C_R = C_{A0} + C_{R0} = \text{constant}; \tag{16.5}$$

and

$$-\frac{dC_A}{dt} = kC_A(C_0 - C_A). \tag{16.6}$$

Integrating between a time zero and a general time $t$:

$$\ln\frac{C_R/C_{R0}}{C_A/C_{A0}} = C_0kt = (C_{A0} + C_{R0})kt. \tag{16.7}$$

In terms of fractional conversion $X_A$, the equation obtained can be written as:

$$\ln\frac{M + X_A}{M(1 - X_A)} = C_{A0}(M + 1)kt = (C_{A0} + C_{R0})kt; \tag{16.8}$$

with

$$M = \frac{C_{R0}}{C_{A0}};$$

and

$$x_A = \frac{(C_{A0} - C_A)}{C_{A0}}.$$

Figure 16.1 is a more detailed version of Fig. 5.3 and reports the conversion of the substrate against time for autocatalytic reactions of this kind, it is possible to notice that at the beginning of the process, the conversion remains quite low with time, and this is because the microorganisms utilize in the beginning only a fraction of the feed at their disposal. In other words, it is like they have to "realize" that there is food for them. After this phase, microorganisms utilize the food in a very effective way also because they reproduce at this time. Finally, the consumption rate decreases and this

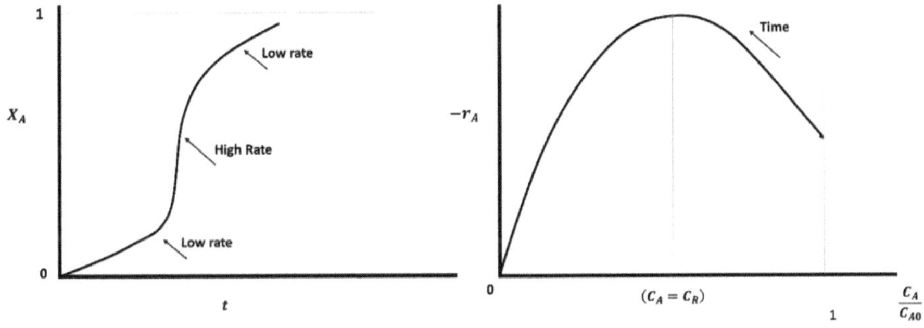


**Fig. 16.1** Conversion of the substrate against time for autocatalytic reactions

could be because of two reasons: One is that the food is decreasing, or on the other hand that the microorganisms are actually dying.

## 16.2 Models for Microbial Growth Used in Power Generation

Concerning the rate of biomass formation, as we have seen previously, for first-order chemical reaction we have that the rate of formation is equal to a reaction rate constant multiplied by a concentration; that is:

$$r_C = k \cdot C; \tag{16.9}$$

where $C$ is the concentration of biomass expressed as g/L.

In case we follow the Monod model, this can be expressed as:

$$k = \frac{k\mathrm{max}\, C_s}{K_s + C_s}; \tag{16.10}$$

with

- $k$ specific growth rate coefficient (check the dimensions always on the basis of the reaction order);
- $k_{\mathrm{max}}$ maximum specific growth rate (the equivalent of $V_{\mathrm{max}}$ cited previously);
- $C_s$ substrate concentration (this refers usually to the substrate that is soluble and it is the equivalent of $A$ mentioned previously);
- $K_s$ half saturation constant.

Inhibition can be provoked also by the same substrate (Edwards 1970), for instance when the concentration is too high. In case we have substrate inhibition, the specific growth rate coefficient can be expressed as:

$$k = \frac{k\text{max}\, C_s}{K_s + C_s + \frac{C_s^2}{K_\text{I}}} \tag{16.11}$$

where $K_\text{I}$ is one inhibition coefficient.

The substrate actually can inhibit the process and slow down the cell growth, and therefore, the inhibition coefficient is taken into account.

Product inhibition is quite common, for instance in the case of ethanol production, the ethanol itself is inhibiting the further production and the concentration of ethanol in these kinds of processes cannot go further than fixed values (Luong 1985). In this case, the reaction rate constant is given by:

$$k = \frac{k\text{max}}{\left(1 + \frac{K_s}{C_s}\right)}\left(1 - \frac{C_p}{C_\text{pcrit}}\right)^n \tag{16.12}$$

Here, $C_p$ is the product concentration, while the $C_\text{pcrit}$ is the concentration of products which will cause inhibition of the growth. The constant $n$ is called inhibition factor.

It is also possible to have inhibition from other compounds, and in this case, the modeling is as follows:

$$k = \frac{k\text{max}\, C_s}{C_s\left(1 + \frac{C_\text{I}}{K_\text{I}}\right) + K_s} \tag{16.13}$$

with $C_\text{I}$ as the concentration of the inhibitor, while $K_\text{I}$ is a constant related to the inhibition. In this textbook, it is preferred not to give explicitly the dimensions of $C_\text{I}$ or $K_\text{I}$ because they could vary depending on the mathematical model adopted. For instance, they could have the dimensions of mg/L, but not necessarily this can be always the case.

In addition to the cases presented above, it is also possible to have inhibition from complementary substrates (Shijie 2016); in this case, the microorganism is eating two kinds of food and the modeling is done as follows:

$$k = k\text{max}\left(\frac{C_{s1}}{K_{s1} + C_{s1}}\right)\left(\frac{C_{s2}}{K_{s2} + C_{s2}}\right) \tag{16.14}$$

where the number 1 is referring to the substrate 1, and the 2 is related to the substrate number 2.

## 16.3 Kinetic Rate Expressions

A simple mass balance on the biomass done in case organisms is consuming two kinds of substrate can be expressed as follows. From the above discussion we have:

$$r_{CB} = kC_B \tag{16.15}$$

now taking into consideration, the following relation

$$k = k_{\max}\left(\frac{C_{s1}}{K_{s1} + C_{s1}}\right)\left(\frac{C_{s2}}{K_{s2} + C_{s2}}\right) \tag{16.16}$$

It is possible then to obtain a complete relationship for the rate of production of biomass.

In case we consider the rate of growth-associated product formation, the following equation should also be taken into account:

$$r_{Cp(g)} = \mathrm{d}C_{p(g)}/\mathrm{d}t = K_{Cp(g)}r_{CB} = K_{Cp(g)}kC_B \tag{16.17}$$

in this case, $C_p$ refers to the concentration of the product, while $(g)$ indicates that the growth-associated product formation is taken into account.

$K_{Cp(g)}$ is the growth-associated product formation ratio (sometimes referred to as mass of product/mass of biomass).

The growth-associated product formation ratio is obtained from the literature. For example, for the fermentation of sugar done by Kluyveromyces Marxianus, the yield coefficient was found to be 0.51 g of ethanol on grams of sugar when the hydraulic residence time, HRT, was 64.4 h (Ozmihci and Kargi 2008). Sometimes in the literature, the yield coefficients for different microorganisms at different conditions are plotted against the dilution rate, *D*, which is defined as the medium flow rate divided by the culture volume.

In the same way, the rate of non-growth-associated product formation (Biazar et al. 2003) can be derived as follows:

$$r_{Cp(n)} = K_{Cp(n)}C_B \tag{16.18}$$

Here, the $(n)$ indicates that the non-growth-associated product formation is taken into account. The $K_{Cp(n)}$ is the non-growth-associated product formation constant.

Notice that the dimensions of $K_{Cp(n)}$ are a bit different from the previous case because there is no $k$ in this case.

The meaning is that this rate is associated with the amount of biomass present, but it is not directly proportional to the variation of the concentration, d*C*/d*t*, or, in other words, the rate of biomass organisms produced.

Taking into account the growth and non-growth-associated product formation, this yields:

$$r_{Cp} = K_{Cp(g)}kC_B + K_{Cp(n)}C_B \tag{16.19}$$

The biomass yield and product yield are defined as: $\gamma_{CB/Cs}$ and $\gamma_{Cp/Cs}$, and they are expressed as a function of the substrate depletion

$$r_{Cs} = \frac{r_{CB}}{\gamma_{CB/Cs}} = \frac{kC_B}{\gamma_{CB/Cs}} \tag{16.20}$$

with

$$\gamma_{Cp/Cs} = r_{Cp}/r_{Cs}$$

Biomass yield and product yield have been calculated for a series of organisms, for instance, $\gamma_{x/s}$ of yeast is 0.084 C-mole-biomass/C-mole-sucrose and $\gamma_{Cp/Cs} = 0.58$ C-mole-ethanol/C-mole-sucrose (Krzystek and Ledakowicz 1998). Here, C-mole means that the comparison is done on the moles of carbon which, in some cases, would give a more accurate assessment of the fermentation performances.

It is also possible to calculate the rate of $O_2$ utilization with suitable models like:

$$r_{Co} = \left(1 - Y_{\frac{CB}{Cs}}\right) r_{Cs} = \frac{1 - Y_{\frac{CB}{Cs}}}{Y_{\frac{CB}{Cs}}} kC_B \tag{16.21}$$

Certainly $\left(1 - Y_{\frac{CB}{Cs}}\right)$ is positive since the rate of substrate consumption has to be larger than the biomass formation. Some strategies of mixed substrate utilization in microorganisms are given in (Harder et al. 1982).

## 16.4 Decay of Organisms

Organisms are growing in different kinds of environment, from seawater to sand, and naturally, they also decay. The decay kinetics can be modeled mathematically, and also in this case, diverse models can be provided (Zhang et al. 2015). During the process of biofuels production, the decay of the microorganisms has been studied in detail by several authors. One common model can be written as:

$$r_{CB(d)} = -bC_B \tag{16.22}$$

in this case, $r_{CB(d)}$ refers to the rate of decay for the biomass concentration. Remember that with biomass concentration, we refer here to the concentration of cells into the system. The term $b$ represents the decay constant.

The overall balance in batch reactors would be then:

$$kC_B - bC_b = \frac{dC_B}{dt} \tag{16.23}$$

## 16.5 Overall Balance in Continuous STR

Previously, we have seen that in the case of continuous STR, we can consider the simplified case of steady state for the substrate and product (that is still valid). However, not necessarily this has to be applied for the microorganisms. In this case, the mass balance would be described as follows:

$$C_{Bi}\frac{\dot{V}}{V} - C_B\frac{\dot{V}}{V} + kC_B - bC_B = \frac{\mathrm{d}C_B}{\mathrm{d}t} \tag{16.24}$$

where $C_{Bi}$ is the concentration of organisms which is inserted into the reactor from the inlet stream.

The term $\frac{\dot{V}}{V}$ is the ratio of the volumetric flow and the reactor volume. With reactor volume, it is always good to stress that we indicate actually the volume of what is inside the reactor and not the physical reactor volume.

If the biomass inlet is zero, meaning that we just add food in the reactor, and if we have steady-state conditions:

$$k = \frac{1}{\tau} + b \tag{16.25}$$

Remember also the definition of hydraulic retention time

$$\tau = \frac{V}{\dot{V}} \tag{16.26}$$

## 16.6 Effluent Concentration of the Substrate

With reference to the following mass balances, the substrate (food) will be indicated with the letter *s*. From the Monod relation:

$$k = \frac{k\mathrm{max}\,C_s}{K_s + C_s} \tag{16.27}$$

And considering Eq. (16.25), we obtain:

$$C_s = \frac{K_s\left(\frac{1}{\tau} + b\right)}{k\mathrm{max} - \left(\frac{1}{\tau} + b\right)} \tag{16.28}$$

And therefore:

$$\tau = \frac{K_s + C_s}{C_s(k\mathrm{max} - b) - K_s b} \tag{16.29}$$

The effluent concentration is, of course, the concentration inside the stirred tank reactor. The balance of substrate (sugar for instance) in continuous steady-state STR is done as follows. Consider:

$$C_{si}\frac{\dot{V}}{V} - C_s\frac{\dot{V}}{V} - \frac{kC_B}{Y_{\frac{CB}{Cs}}} = \frac{dC_s}{dt} \tag{16.30}$$

In this case, we just add food to the reactor, and therefore, the balance given is correct. Remember also the relations previously written:

$$r_{Cs} = \frac{r_{CB}}{\gamma_{CB/Cs}} = \frac{kC_B}{\gamma_{CB/Cs}} \tag{16.31}$$

where

$$\gamma_{Cp/Cs} = r_{Cp}/r_{Cs};$$

and

$$k = \frac{1}{\tau} + b;$$

In consideration of the previously written relations. The following expression is obtained:

$$C_B = \frac{Y_{CB/Cs}(C_{si} - C_s)}{1 + \tau b} \tag{16.32}$$

Concerning the balance of products ($H_2$, ethanol, etc.). In a continuous steady-state STR, we have:

$$C_{pi}\frac{\dot{V}}{V} - C_p\frac{\dot{V}}{V} + k_{Cp(g)}kC_B + k_{Cp(n)}C_B = \frac{dC_p}{dt} \tag{16.33}$$

Since there is no product inserted into the reactor, the first term on the left side of the equation above can be neglected. We remember the relations:

$$r_{Cp(n)} = K_{Cp(n)}C_B; \tag{16.34}$$

$$r_{Cp(g)} = K_{Cp(g)}r_{CB} = K_{Cp(g)}kC_B; \tag{16.35}$$

and the non-growth-associated (NGA) product formation $p$:

$$C_{p(\mathrm{NGA})} = k_{Cp(n)}C_B\tau, \tag{16.36}$$

expressed as g/L, and the growth-associated (GA) $p$ formation will be:

$$C_{p(\mathrm{GA})} = k_{Cp(g)} k C_B \tau \quad (16.37)$$

## 16.7 Production of Ethanol by Digestion

Ethanol is produced normally by digestion of sugars which are derived from different kinds of feedstock. The most common organism utilized for the ethanol production is the *Saccharomyces cerevisiae*. These organisms can process several number of sugars (Rodicio and Heinisch 2009) among which we remember maltotriose, sucrose, galactose, mannose, maltose, glucose, and fructose. The common way of producing ethanol by this organism is the Embden–Meyerhof–Parnas or EMP pathway (Peretó 2014) also known as the glycolytic pathway. A simplified overall reaction is expressed as:

$$C_6H_{12}O_6 + 2\mathrm{Pi} + 2\mathrm{ADP} \rightarrow 2C_2H_5OH + 2CO_2 + 2\mathrm{ATP} + 2H_2O \quad (16.38)$$

From Eq. (16.38), it is possible to calculate the related yield for the overall process. The glycolytic pathway for ethanol production in *S. cerevisiae* (Drapcho et al. 2008) is demonstrated in Fig. 16.2.

The same ethanol is inhibiting the microorganism. This means that the microbe is able to grow until a certain concentration of ethanol.

$$\mu_{Sg} = \mu_{Sog}\left[1 - \left(\frac{C_s}{C_s\mathrm{max}}\right)\right]^n \quad (16.39)$$

$$f_{Sf} = f_{Sof}\left[1 - \left(\frac{C_s}{C^*_{s\mathrm{max}}}\right)\right]^{n^*} \quad (16.40)$$

In the equations presented, the fermentation rates, $f$, and the growth rates, $\mu$, are directly proportional to their value considered when there is no by-product (the alcohol) and also proportional to a term (in parenthesis) which slows down the same rate (Nabais et al. 1988). With the term "max," we indicate here the maximum concentration allowable for the alcohol product.

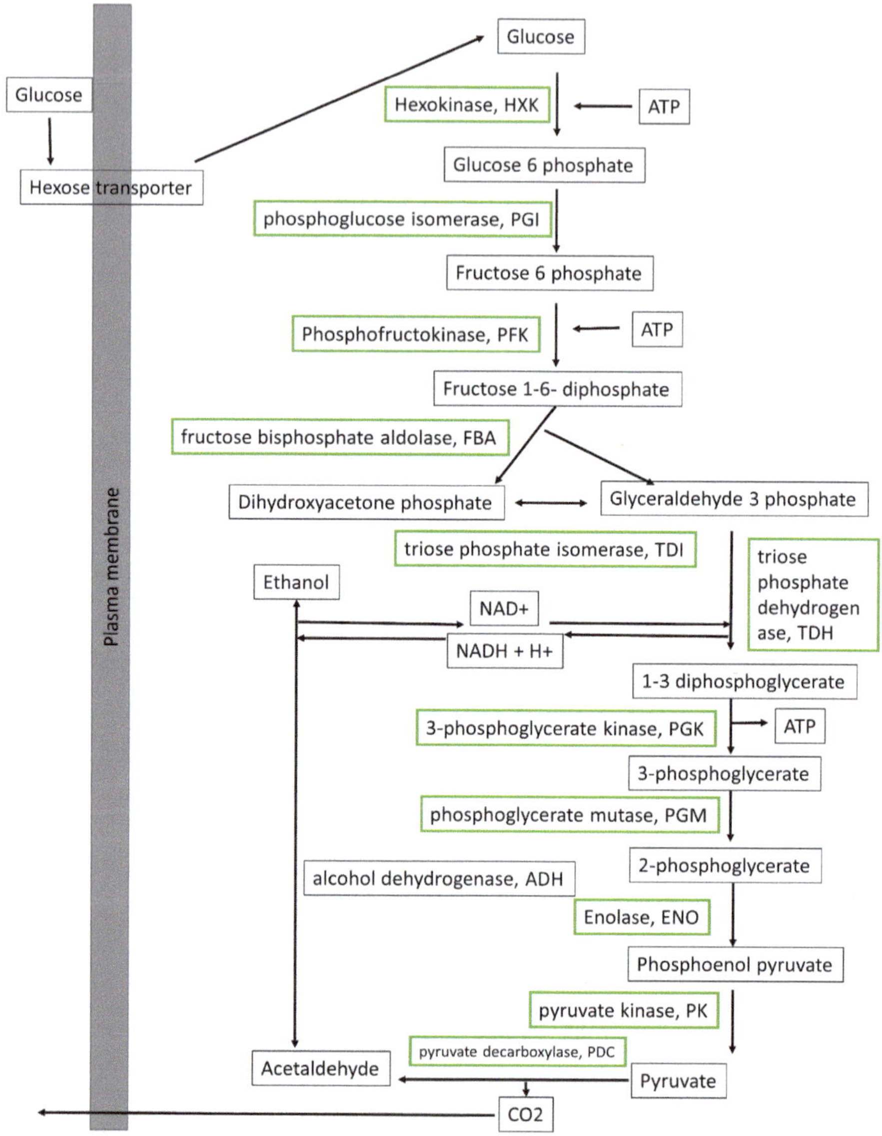


**Fig. 16.2** Glycolytic pathway for ethanol production in *S. cerevisiae*. Adapted from Drapcho et al. (2008)

# References

Biazar, J., Tango, M., Babolian, E., & Islam, R. (2003). Solution of the kinetic modeling of lactic acid fermentation using Adomian decomposition method. *Applied Mathematics and Computation, 144,* 433–439. https://doi.org/10.1016/S0096-3003(02)00418-6.

Drapcho, C. M., Nghim, N. P., & Walker, T. (2008). *Biofuels, bioproducts and biorefining*. McGraw-Hill.

Edwards, V. H. (1970). The influence of high substrate concentrations on microbial kinetics. *Biotechnology and Bioengineering, 12,* 679–712. https://doi.org/10.1002/bit.260120504.

Harder, W., Dijkhuizen, L., & Postgate, J. R. (1982). Strategies of mixed substrate utilization in microorganisms [and discussion]. *Philosophical Transactions of the Royal Society of London. Series B, Biological Sciences, 297,* 459–480.

Krzystek, L., & Ledakowicz, S. (1998). Yield and maintenance coefficients in *S. cerevisiae* cultures. *Journal of Chemical Technology and Biotechnology, 71,* 197–208. https://doi.org/10.1002/(SICI)1097-4660(199803)71:3%3c197:AID-JCTB825%3e3.0.CO;2-H.

Luong, J. H. (1985). Kinetics of ethanol inhibition in alcohol fermentation. *Biotechnology and Bioengineering, 27,* 280–285. https://doi.org/10.1002/bit.260270311.

Nabais, R. C., Sá-Correia, I., Viegas, C. A., & Novais, J. M. (1988). Influence of calcium ion on ethanol tolerance of *Saccharomyces bayanus* and alcoholic fermentation by yeasts. *Applied and Environment Microbiology, 54,* 2439–2446.

Ozmihci, S., & Kargi, F. (2008). Ethanol production from cheese whey powder solution in a packed column bioreactor at different hydraulic residence times. *Biochemical Engineering Journal, 42,* 180–185. https://doi.org/10.1016/j.bej.2008.06.017.

Peretó, J. (2014). Embden-Meyerhof-Parnas pathway. In: Amils R. et al. (Eds.), *Encyclopedia of Astrobiology* (pp. 1–1). Berlin, Heidelberg: Springer. https://doi.org/10.1007/978-3-642-27833-4_503-2.

Rodicio, R., & Heinisch, J. J. (2009). Sugar metabolism by *Saccharomyces* and non-*Saccharomyces* yeasts. In *Biology of microorganisms on grapes, in must and in wine* (pp. 113–134). Berlin, Heidelberg: Springer. https://doi.org/10.1007/978-3-540-85463-0_6.

Shijie, L. (2016). *Bioprocess engineering. Kinetics, sustainability, and reactor design* (2nd ed.). Elsevier.

Zhang, Q., He, X., & Yan, T. (2015). Differential decay of wastewater bacteria and change of microbial communities in beach sand and seawater microcosms. *Environmental Science and Technology, 49*, 8531–8540. https://doi.org/10.1021/acs.est.5b01879.

# Chapter 17
# Processes of Bioethanol Production

*Creativity is strictly a human characteristics and it belongs to the neocortical component. Innovation is a man's advantage to be able to make new technological developments and uses the creativity.*

Rita Levi Montalcini, Nobel laureate in medicine, an interview in 2009 for her first 100 years.

## 17.1 Some Systems Adopted and Product Compositions

One of the most produced biofuels worldwide is bioethanol. The fuel can be produced from biomass feedstock like sugarcane but also from other sources like corn (Kumar and Singh 2019), microorganisms (Onay 2018), yeast (Mohd Azhar et al. 2017), and even marine biomass (Greetham et al. 2018). Some fuel properties are given in Table 17.1.

To give some ideas of the volumes implied in this production: as an example, Brazil is surely among the biggest producers of ethanol in the world. The total cost of ethanol production in Brazil is estimated to be between 0.25 US dollars and 0.30 US dollars per L (including all inputs and factors) (Watanabe 2009). In Brazil, 10% of the total cultivated land (this accounts for 13.8 million acres) is utilized for sugarcane production. From this land, about 7.4 million of these acres are used to produce around 30 billion L of ethanol per year. The liquid effluent (which is called vinasse) is used as a fertilizer and irrigation supply to the cane fields, thereby eliminating costs for wastewater treatment. In 2007 worldwide, 11.6 BG (billion gallons) of ethanol were produced from 2 billion ton of corn (Drapcho et al. 2008a).

When we want to produce bioethanol from sugarcane and sugar beets, the sugar is first isolated by crushing the stalks with specialized rollers to release the juice. The first operation is the cleaning of the sugarcane, and this can be operated by dry cleaning and removal of solid external entities like rocks. These operations are carried by suitable rolls which include pneumatic separation systems. For instance, these operations are done to remove the presence of leaves and tops. Continuing with the operations, the extraction of sugarcane juice is done by roll mills and these can be designed for large-scale operations and small-scale operations.



C. De Blasio, *Fundamentals of Biofuels Engineering and Technology*,
Green Energy and Technology, https://doi.org/10.1007/978-3-030-11599-9_17

**Table 17.1** Some properties of bioethanol derived from different kinds of feedstock

| Feedstock | Density ($g/cm^3$) | Flash point (°C) | pH value (mg/L) | Heating value (kcal/kg) | Viscosity (cSt) |
|---|---|---|---|---|---|
| Bioethanol standard | 0.789 | 12 | 6.5–9.0 | 6380 | 1.53 |
| Mucasea | 0.824 | 12 | 5.8 | 5524 | 2.15 |
| *Maranta arundinacea* Linn. | 0.796 | 14 | 4.6 | 5969 | 3.05 |
| *Solanum lyycopersicum* | 0.817 | 13 | 5.3 | 5696 | 2.84 |
| *Amorphophallus campanulatus* | 0.823 | 19 | 4.7 | 5892 | 2.38 |
| *Amorphophallus variabilis* | 0.858 | 15 | 2.5 | 6786 | 2.76 |
| *Saccharum officinarum* Linn. | 0.801 | 13 | 8.5 | 8756 | 1.64 |
| *Alocasia macrorrhizos* | 0.836 | 16 | 3.8 | 6445 | 1.96 |

Adapted from Sutjahjo (2018)

The pretreatment of the sugarcane bagasse can be done by operations like steam explosion or delignification. The cellulose is also hydrolyzed by using acids, while among the mechanical treatment we can recall here the comminution process which reduces the bagasse size and therefore, increases the surface of these particles. In addition, also the cellulose crystallinity is reduced by this method.

The principle behind the steam explosion process is that steam usually is used to increase the temperature of the feedstock so that the water contained in it becomes hotter. The pressure will be quickly dropped and, because of this, the steam will basically go to the steam phase very rapidly like an explosion. This method is also used in other processes and in the food industry. For instance, in the tomato processing industry when the tomatoes are peeled.

The delignification is operated by chemical treatment, for example, Organosolv process. The method includes ethanol and in some cases sulfuric acid which are used as a pretreatment media (Fig. 17.1).

The juice obtained can be cleaned by physical operations, like filtration or by using a hydrocyclone. The other possibility is to use chemical treatment with the usage of phosphoric acid. This acid helps the removal of impurities during the settlement operation. This processing can include steps like sulfitation, liming, and clarification with the consequent separation of the mud. The sulfitation is adopted to purify the cane juice produced, and this is done by using sulfur dioxide gas. In practice, the juice is heated up at certain temperatures and with a particular pH (for instance, a temperature around 70 °C and a pH of around 5) and it is processed with $SO_2$ and lime

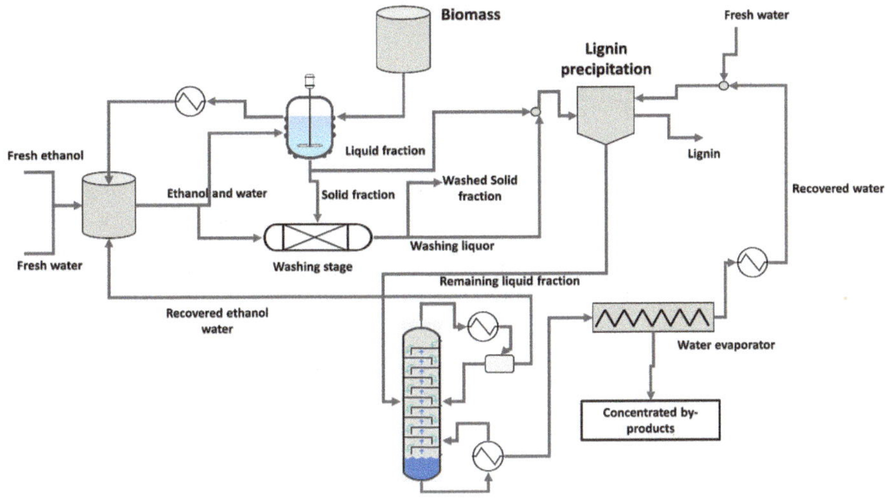


**Fig. 17.1** Schematics for the organosolv process with ethanol. Adapted from Alriols et al. (2010)

until the sugar is then separated from the other compounds. At the end of the process, the clarification separates the mud at certain conditions. For instance, one method could be to heat the juice at a near boiling temperature and operate a juice clarifier with a flash tank. The stirring conditions and the design of the clarification unit are also important to optimize the consumption of the plant and the air entrainment in the flash unit should also be minimized since the air could prevent the settling of the mud.

A more detailed example of the preliminary operations performed to obtain the clarified juice is shown in Fig. 17.2.

In regard to the juice produced after the preliminary treatment of the sugarcane feedstock, the juice is concentrated (this is called evaporation) and it is also sterilized to remove germs and bacteria possibly present into the biomass (Fig. 17.3).

In reference to Fig. 17.3, the syrup stream after the cooking process is sent to the evaporation section. In this section, the crystallization occurs, and this is described in more detail in Fig. 17.4.

A remark should be introduced here on the sterilization process. This is one of the most important processes in the food industry and one of the most energy consuming. Concepts related to condensation and heat transfer are particularly important in this sense. There are many types of sterilization processes; for instance, they can be classified in thermal and not thermal. In addition, the thermal processes can be classified on the basis of if the medium used for the sterilization is directly in contact with the food or it is in non-direct contact with the medium to be sterilized.

Let us be a bit more specific here: if the juice is a ready-made product and should be sterilized, then the procedure is usually to put the product into a suitable container and then proceed on the sterilization of the product by indirect thermal sterilization.

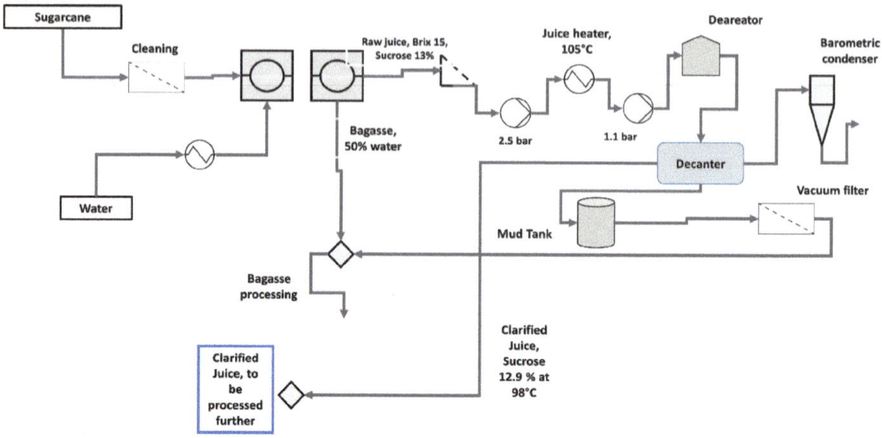


**Fig. 17.2** Pretreatment and derivation of clarified juice in bioethanol production from sugarcane. Adapted from Cortes-Rodríguez et al. (2018)

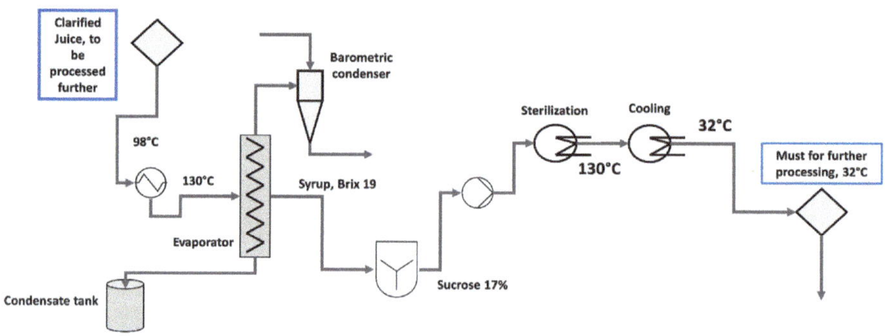


**Fig. 17.3** Juice concentration and sterilization process in bioethanol production from sugarcane. Adapted from Palacios-Bereche et al. (2014)

For example, aluminum cans are commonly used in the food and beverage industry, and this shape is used for a precise reason. The cylindrical shape of the can offers the highest volume/surface area with respect to the possible commercial shapes available. In fact, the sphere is even better but it is not a commercially suitable shape for obvious reasons. In this way, we spare on the investments related to the metal of the container.

Another choice is the aluminum metal; aluminum offers a heat transfer rate between the external medium (the steam) and the internal juice which is much higher than steel. Without going into details with the proof of this; the difference between aluminum and steel is readily evident when we use the different metals for cooking. As an example, if two moka pots of the same volumetric capacity are used, the one made in stainless steel will require more time to make the coffee ready. This has to do also with the different heat capacities of the two metals and even with the internal heat conductivity.

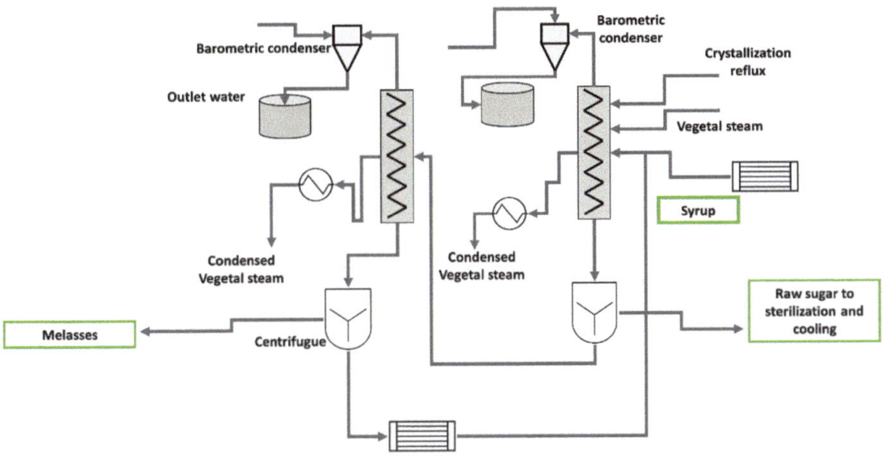


**Fig. 17.4** A scheme of the crystallization process where molasses and raw sugar is produced. Adapted from Velasquez et al. (2013)

When sterilized, the cans are sent to rolls where steam is injected on their surfaces; steam offers the possibility of "injecting" calories into the can at higher temperature compared to water since it can be over-heated. After the sterilization, the containers should be cooled down to ambient temperature. In some cases, this should be done as fast as possible since when particular products are produced, there could be the development of some bacteria which grow at certain high temperatures. This is the case of the tomato industry products like peeled tomato and related.

The cooling is done also on rolls, where droplets of water are used in this case. This is also done with a precise reason: droplets allow for the maximization of the contact surface between the water and the can and the minimization of the water consumption. There is also another reason why droplets are used and this is related to the fact that when droplets are falling into air, they will be cooled even further eventually, and this is because part of the water from the droplet is evaporating and for this process, some energy is needed and lost by the droplet itself which will be brought at even lower temperature.

To have a quantitative determination of this process, the reader is suggested to consult some of the psychrometric charts and the related theory available in literature. The reader is also suggested to look further on gathering the basis for the understanding of the heat transfer concepts which are explained in details elsewhere in literature (Bird et al. 2006).

Coming back to the overall process for the production of bioethanol from sugarcane, after the sterilization and the cooling of the produced must, the fermentation step takes place followed by the distillation process (Fig. 17.5).

In regard to the fermentation process, it is possible to have a distinction between batch, fed-batch (Veloso et al. 2019), and continuous processes. The distillation shown in this case is consisting of diverse steps which include reboiler columns and

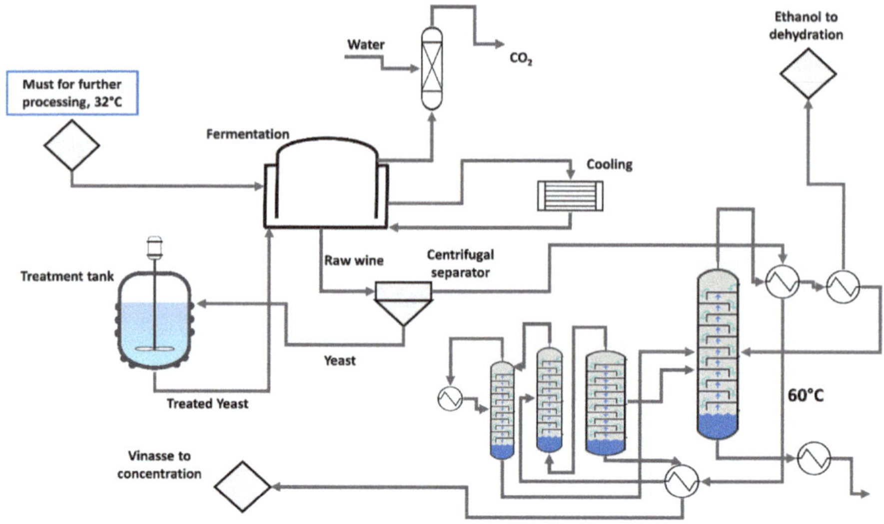


**Fig. 17.5** Fermentation and distillation steps in bioethanol production from sugarcane

condenser columns. The distillation process for the ethanol production is described in more detail at the end of this chapter with some theoretical notions.

The vinasse, produced by the fermentation and distillation process, is sent to a concentration unit while the produced ethanol is sent to a dehydration unit for further concentration to almost purity (Fig. 17.6).

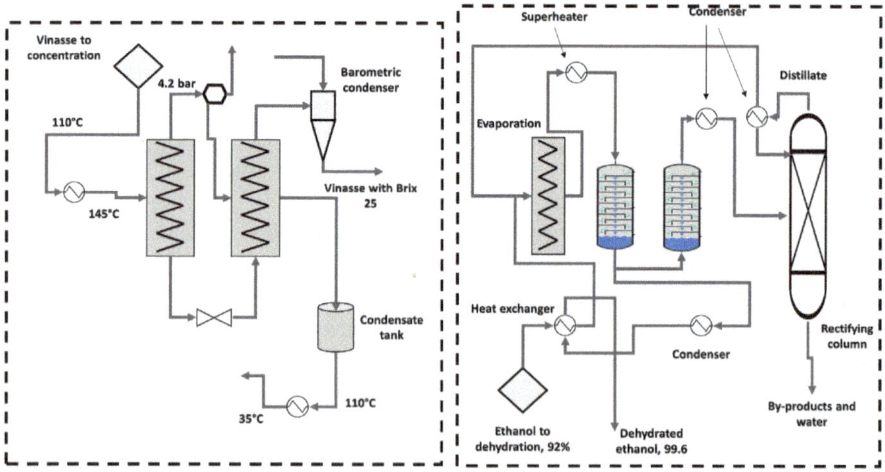


**Fig. 17.6** Vinasse concentration unit (left) and ethanol dehydration unit (right). Adapted from Kujawski and Zielinski (2006)

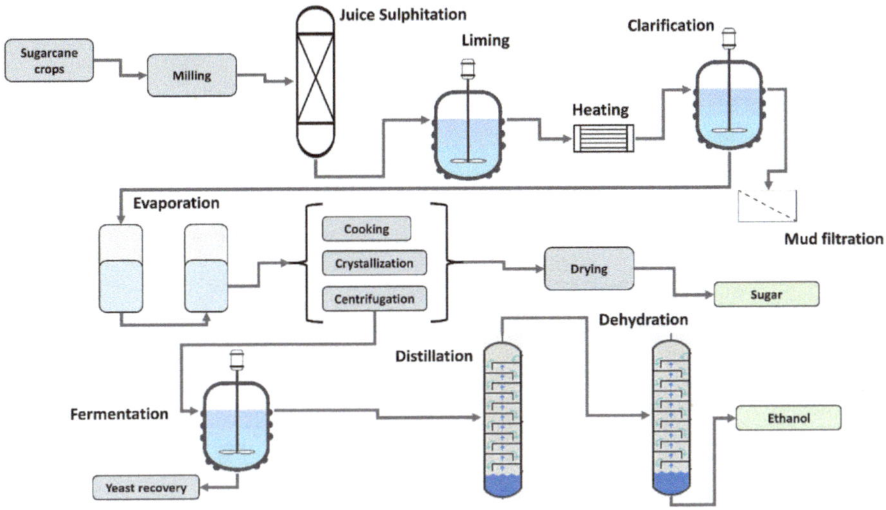


**Fig. 17.7** One example of process for bioethanol production from sugarcane. Adapted from Amores et al. (2013)

In reference to the more general representation of the process, Fig. 17.7, and with respect to the production of sugar; after the process of clarification and the evaporation of the resulted filtrate, the juice is treated further as was illustrated previously. The steps that follow are denominated as cooking, crystallization, and centrifugation. In Fig. 17.3, the evaporator depicted before the separation section is the equivalent of the crystallization shown in Figs. 17.4 and 17.7. The sugar starts to crystallize in a sugar-syrup juice, and the syrup traces are washed away from the sugar crystals by water. The crystals are then separated by centrifugation. The fermentation of the sugar allows for the production of ethanol which is then distilled and separated at the end of the process.

## 17.2 Bioethanol from Lignocellulosic Biomass

In general, the bioethanol can be produced also by feedstock which has a composition in cellulose, hemicellulose, and lignin. In this case, a pretreatment is necessary to allow for the separation of these components. The pretreatment includes enzymes and also chemicals at certain temperatures and pressures. More in detail, the lignocellulosic biomass is processed with water, acid, and steam and after a separation process (which can include centrifugation) a base is added to the produced mixture to allow for the enzymes to be used. As a matter of fact, enzymes cannot operate if the conditions are too acid. Also, the temperature conditions have to be adjusted for the enzymes. The hydrolysis step is also operated by the enzymes, these powerful

**Table 17.2** Concentration (g/100 g) of the main components in sugarcane juice and sugarcane beet juice

| Components | Sugarcane juice | Sugarcane beet juice |
|---|---|---|
| Solid | 13.7 | 17.3 |
| Sucrose | 12 | 16.5 |
| Monosaccharides | 0.63 | 0.15 |
| Polysaccharides | 0.028 | 0.019 |
| Lactate | 0.016 | |
| Acetate | 0.035 | |
| Sulfate | 0.039 | 0.02 |
| Phosphate | 0.033 | 0.047 |
| Nitrate | | 0.015 |
| Nitrite | | 0.005 |
| Aconitate | 0.09 | |
| K | 0.11 | 0.125 |
| Na | 0.005 | 0.015 |
| Cl | | 0.003 |
| Ca | 0.04 | |
| Mg | 0.028 | |
| Total-*N* | | 0.105 |
| Betaine-*N* | | 0.046 |
| Amino acid-*N* | | 0.026 |
| Ammonia-*N* | | 0.006 |
| Amide-*N* | | 0.011 |

Adapted from Drapcho et al. (2008b)

catalysts break the cellulose into glucose (which is the building block of cellulose), and the hemicellulose chains are then converted to xylose, and this sugars can be readily fermented into bioethanol. Yeast or other bacteria are then introduced into the fermentation reactor. The residence time for the ethanol to be produced is quite long, 3–5 days, and this is a little drawback for the bioethanol production since bigger reactor volumes are required to ensure a continue production. Sugars could also be converted to longer chain molecules and other kinds of biofuels. In this case, we should add hydrogen to these molecules in order to have renewable gasoline and biodiesel. The so-called stillage and the lignin residues are then separated by suitable methods (Fig. 17.8).

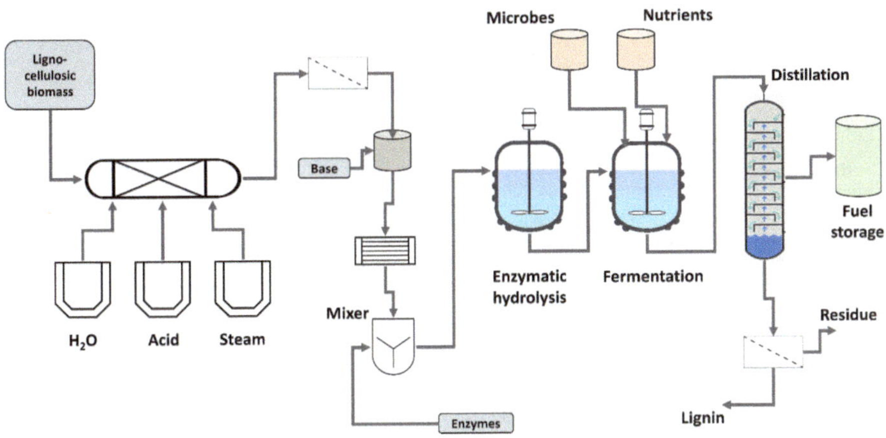


**Fig. 17.8** Production of ethanol from lignocellulosic biomass feedstock. A simplified scheme

**Table 17.3** Composition of produced molasses from cane and beets

| Item | Cane molasses | Beet molasses |
|---|---|---|
| Brix | 79.5 | 79.5 |
| Total solids (wt%) | 75.0 | 77.0 |
| Specific gravity | 1.41 | 1.41 |
| Total sugars (wt%) | 46.0 | 48.0 |
| Crude protein (wt%) | 3.0 | 6.0 |
| Nitrogen free extract (wt%) | 63.0 | 62.0 |
| Ash (wt%) | 8.1 | 8.7 |
| Ca (wt%) | 0.8 | 0.2 |
| P (wt%) | 0.08 | 0.03 |
| K (wt%) | 2.4 | 4.7 |
| Na (wt%) | 0.2 | 1.0 |
| Cl (wt%) | 1.4 | 0.9 |
| S (wt%) | 0.5 | 0.5 |
| *Trace minerals* | | |
| Cu (mg/kg) | 36 | 13 |
| Fe (mg/kg) | 249 | 117 |
| Mn (mg/kg) | 35 | 10 |
| Zn (mg/kg) | 13 | 40 |

(continued)

**Table 17.3** (continued)

| Item | Cane molasses | Beet molasses |
|---|---|---|
| *Vitamins* | | |
| Biotin (mg/kg) | 0.36 | 0.46 |
| Choline (mg/kg) | 745 | 716 |
| Pantothenic acid (mg/kg) | 21.0 | 7.0 |
| Riboflavin (mg/kg) | 1.8 | 1.4 |
| Thiamine (mg/kg) | 0.9 | – |

Adapted from Drapcho et al. (2008b)

Considering ethanol production by fermentation and from sugarcane and sugar beets, the composition of the juice produced from sugarcane and sugar beets is described in the previous table (Table 17.2).

While the produced molasses from cane and beets have the composition shown in Table 17.3.

The maximal growth of *Saccharomyces cerevisiae*, i.e., the maximal specific growth rate $\mu_{max}$, is obtained at a temperature of 30–35 °C (Zakhartsev et al. 2015), while at temperatures above 44 °C the ethanol production is reduced and the residual sugar content is increased.

## 17.3 Corn Starch as a Feedstock

When corn starch is used as a feedstock for ethanol production, the polymers forming the starch should be broken into the constituents of those polymers (sugars) and this process is called "saccharification." Starch refineries use one of the following three methods for the initial processing of corn grain. More information on these processes can be found in literature (Schwietzke et al. 2009) and also, with focus on biodiesel production, some of these processes are described (Veljković et al. 2018):

- A dry-grind process, in this case, the so-called kernels are grinded into a fine powder.
- A dry-milling process, in this case, there is the utilization of roller mills where the kernels are processed. This is followed by a separation of microorganisms and fibers by screening.
- A wet-milling process (Bothast and Schlicher 2005). In this case, sulfuric acid might be used to pretreat the kernels. In this phase, the so-called swelling of the kernels takes place and then the kernels go to the milling process. After this, the separation of microorganisms, fibers, protein, and starch is taking place by density separation and washing.

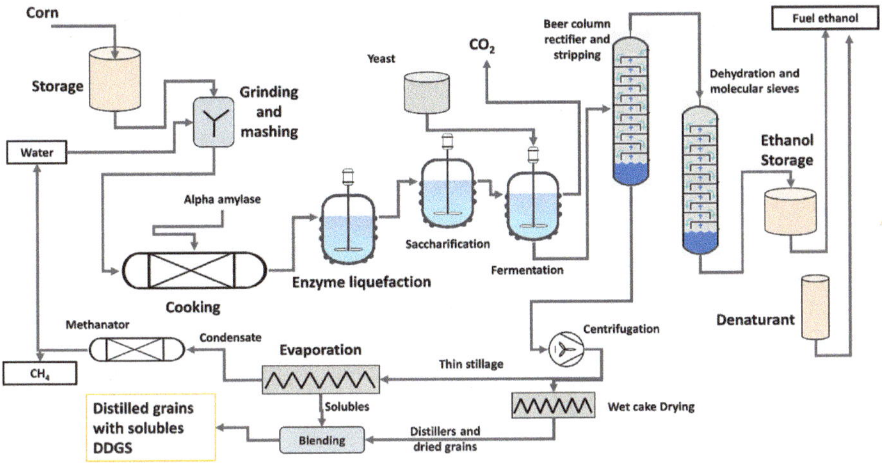


**Fig. 17.9** Schematic representation of the dry-milling process for ethanol production. Adapted from Vohra et al. (2014)

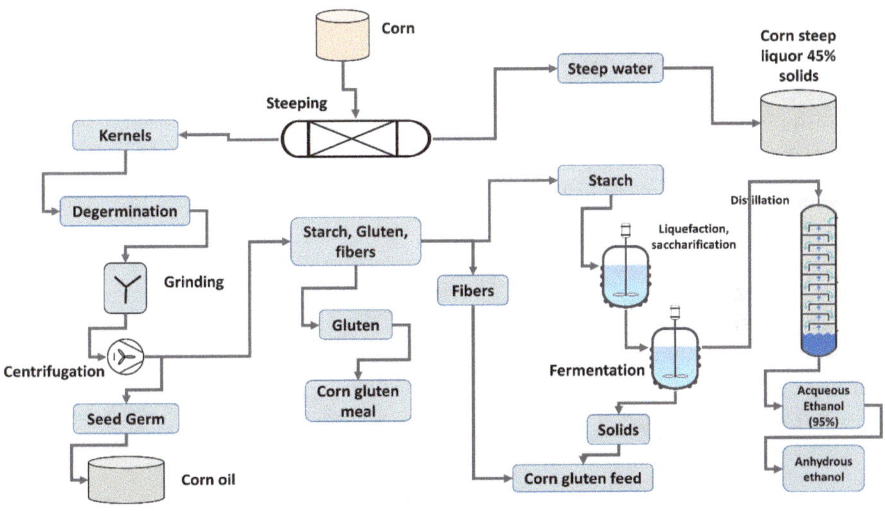


**Fig. 17.10** Schematic representation of the wet-milling process for ethanol production. Adapted from Vohra et al. (2014)

We can use the wet process or the dry-milling process for ethanol production from Fig. 17.9.

While the wet-milling process is shown in Fig. 17.10.

The reader is advised that in these representations, the connections to the distillation columns or the shown reactors are qualitative and they do not refer to the exact position within the column. In other words, if a link to a distillation column is positioned on the top of the column, this does not mean necessarily that in the real case the feeding is done at the last trays. This would require more investigation.

To be noticed from the previous figures that the utilization of enzymes in these process steps is large. To give one example, in the liquefaction of water and ground corn slurry, the $\alpha$-amylase enzyme is widely used (Kumar and Singh 2016). Figure 17.11 represents the structure of the $\alpha$-amylase enzyme.

Concerning the differences between the two methods, within the wet process, the starch is separated from the rest of the components and sent to the fermentation; while in the dry process, all the feedstock is processed first for its reduction of size and then utilized into the fermentation. In terms of heat production, approximately 29.7 MJ/kg ethanol is generated (De Jong and Van Ommen 2014). Figure 17.12 is a schematic representation of a dedicated dry-grind ethanol production plant.

Some information will be given here on the utilization of the enzymes and their process parameters. The glucoamylase enzymes exhibit maximum activity at temperatures of 60–65 °C and at a pH of about 4.5 (Hyun and Zeikus 1985). It is common that the glucoamylase is injected directly into the fermentation reactor using a process called "simultaneous saccharification and fermentation" (SSF). Large-scale industrial fermentation processes typically result in ethanol concentrations of 12–15% on a volume basis (Zhang et al. 2006). The fermentation is typically conducted in a batch process that lasts for 48–72 h (Vertes et al. 2010).

A schematic representation of a production plant, where the wet-mill ethanol production method is adopted, is demonstrated in Fig. 17.13.

In this process, the so-called steep water containing 1–2% of sulfur dioxide is added; the mixture is then held at 49–53 °C for between 22 and 50 h. After this process

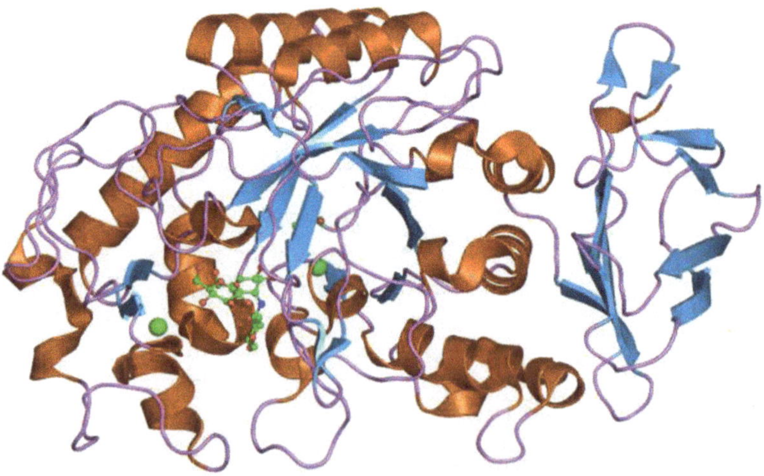

**Fig. 17.11** An example of the animal $\alpha$-amylase enzyme. The figure is produced by Jawahar Swaminathan and his colleagues at the European Bioinformatics Institute and made available for the public and free to use. The author highly acknowledge Jawahar and his collaborators

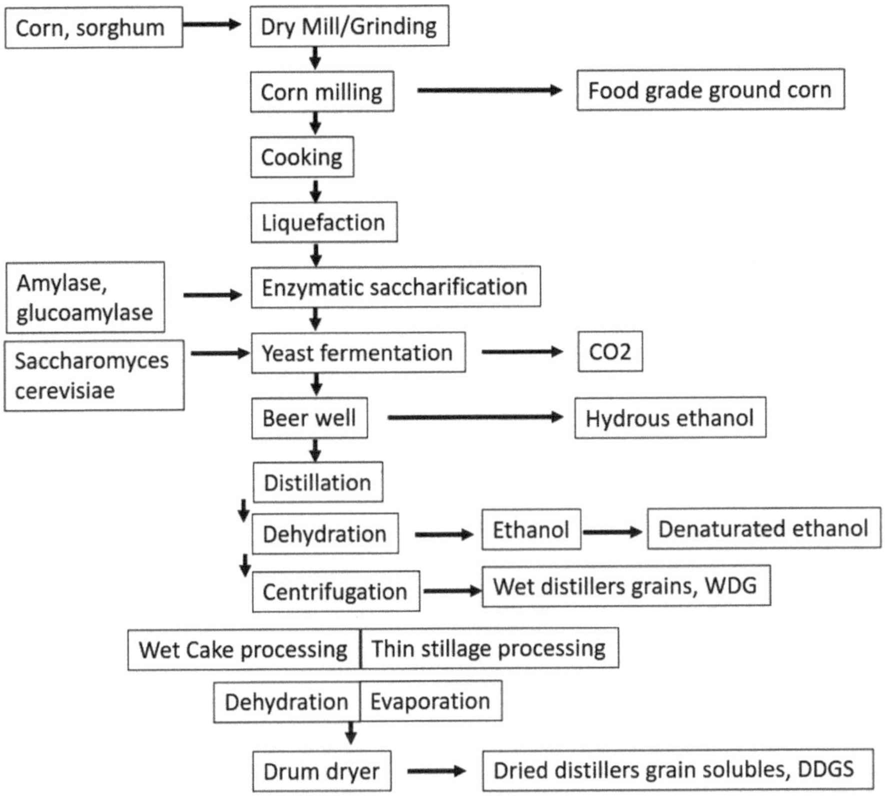


**Fig. 17.12** Schematic representation of a dedicated dry-grind ethanol production plant. Adapted from Vertes et al. (2010)

of "steeping," there is evaporation and formation of corn steep liquor and corn is converted by bacteria to lactic acid. Corn steep liquor is a particular product/feedstock which could even be utilized as fuel in microbial fuel cells, for instance (Phansroy et al. 2018). The lactic acid and sulfur dioxide cause the kernel to soften in such a way that the disulfide bonds are broken, breaking the key linkages between the starch, germ, and fiber. Finally, the fibers are separated by screens and the gluten is separated by centrifugation.

## 17.4 Ethanol Distillation

When ethanol is produced, it will be necessary to increase its concentration among the products and therefore, the distillation process is applied in this case. In this small section, only some general information is given on this process and for an

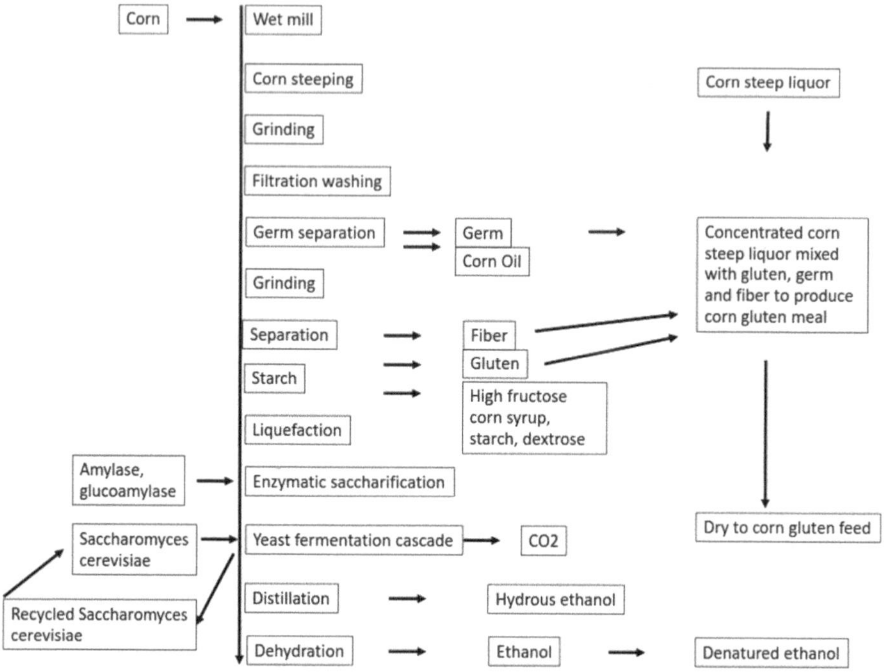


**Fig. 17.13** Schematic representation of dedicated wet-mill ethanol production plant. Adapted from Vertes et al. (2010)

accurate study of the distillation columns, the author of this manuscript suggest to read specialized textbooks available in literature, for instance (Treybal 1980). The distillation process allows to obtain a higher concentration of a component in a mixture; this is done by taking advantage of several equilibrium stages where ethanol in vapor phase is in equilibrium with its liquid phase at each of the distillation trays. If ethanol and water are mixed together, for instance, then ethanol will have a higher volatility (and lower boiling point) with respect to water and therefore the vapor produced by this mixture will contain a higher concentration of ethanol with respect to the liquid phase. Going upward, this vapor will increase the concentration of ethanol in the liquid phase of the next tray and with an higher concentration of ethanol in the liquid phase, the concentration of ethanol at this stage will be even higher than the previous stage. And, this is repeated for the successive stages in the distillation column. One of the available references on this argument is Madson (2003). A simple representation of the process is given in Fig. 17.14.

The so-called *stillage* is a liquid waste remaining after *ethanol distillation* (Choonut et al. 2015). Typical values for the ethanol in the stillage are 0.02 wt%. The feedstock to the distillation column is called Beer, and this feed has a content of 10% ethanol in volume typically. The typical number of stripping stages is eight while the stages in the rectification section are typically 14 (Madson 2003). A sim-

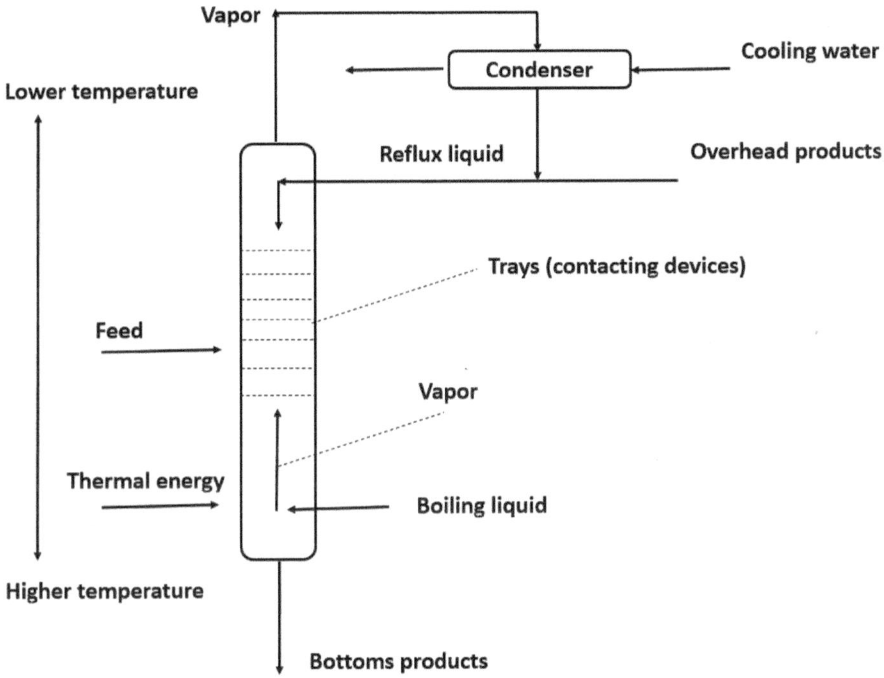


**Fig. 17.14** A simple schematics of a distillation column

ple representation of a graphical evaluation for the distillation stages is given in Fig. 17.15.

Stripping is another form of distillation. This concerns the first stage of the process when lighter components are said to be "stripped" from the heavier ones. Usually, the feed during stripping is entering at the top of the column while the heat source is at the bottom.

The distillation of two or more substances is somewhat complicated by the fact that an azeotrope can occur. An azeotrope is found when two or more components have the same concentrations in the vapor phase and the liquid phase. For this reason, this mixture cannot be altered by simple distillation. An azeotropic distillation should be performed in this case.

In a homogeneous azeotropic distillation, a solvent is used to change the volatility of the components and in this case, we are talking about extractive distillation. While in heterogeneous azeotropic distillation, we can have (Madson 2003):

- distillation using salt effects;
- pressure-swing distillation (at different pressures the equilibrium curves are translating).

Some constituents of the distillation tray are shown in Fig. 17.16.

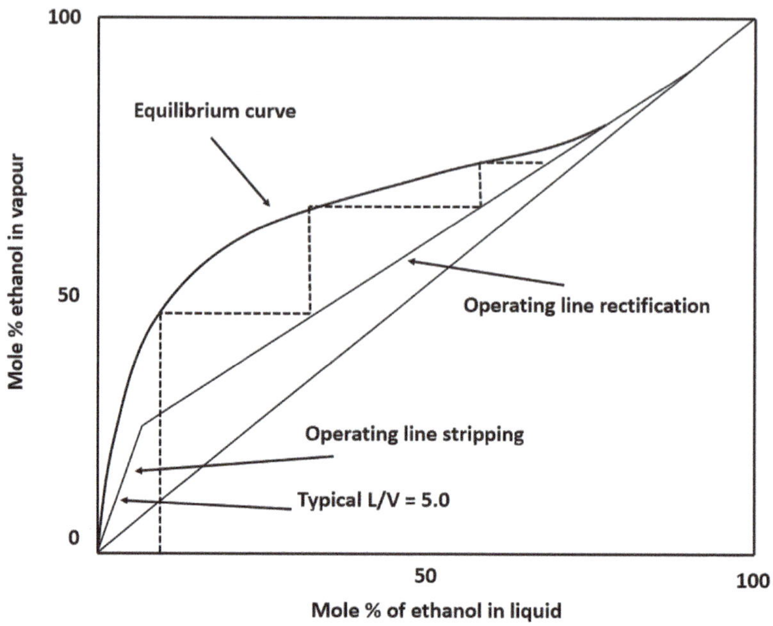


**Fig. 17.15** Graphical Evaluation of the number of stages in a distillation column

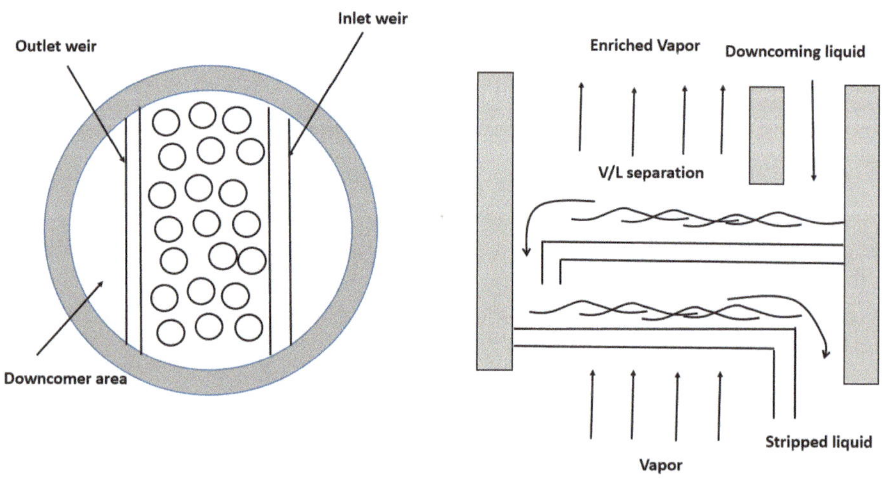


**Fig. 17.16** A top–down visual of a distillation tray with suitable passes for the vapor phase (left) and a schematic showing the passage of the liquid and gas phase over the distillation tray (right)

## 17.5 Ethanol Yields from Different Kinds of Feedstock

In literature, several data can be found on the possible ethanol yield from diverse kinds of feedstock and grains. In this last part of the chapter, some data are given as examples of yields for ethanol production. However, it should be pointed out that in literature, there are many data available and sometimes the values provided are in conflict among them.

Some values of ethanol yield are given in Table 17.4 for agricultural residues.

Biomass residues can come from different sources like corn stover, barley straw, oat straw, rice straw, sorghum and bagasse, and the provenience can be from different parts of the world. From the crop-processing by-products, some examples of the theoretical ethanol yield can be found in Table 17.5.

As for the case of ethanol yield from different feedstock kinds, the same reasoning can be done in regard to the values of the main parameters used for the production of ethanol. The literature in this field is large and a complete review of the reactors' conditions is outside the scope of the first edition of this manuscript. Some examples of values used for the main parameters in batch fermentation are given in Table 17.6.

**Table 17.4** Composition of crops and residues and theoretical ethanol yield from agricultural residues

| | Dry matter (wt%) | Carbohydrates (wt%) | Lignin (wt%) | Ethanol yield (L/kg of dry biomass) |
|---|---|---|---|---|
| Barley | 88.7 | 67.10 | 2.90 | 0.41 |
| Barley straw | 81.0 | 70.00 | 9.00 | 0.31 |
| Corn | 86.2 | 73.70 | 0.60 | 0.46 |
| Corn stover | 78.5 | 58.29 | 18.69 | 0.29 |
| Oat | 89.1 | 65.60 | 4.00 | 0.41 |
| Oat straw | 90.1 | 59.10 | 13.75 | 0.26 |
| Rice | 88.6 | 87.50 | | 0.48 |
| Rice straw | 88.0 | 49.33 | 7.13 | 0.28 |
| Sorghum | 89.0 | 71.60 | 1.40 | 0.44 |
| Sorghum straw | 88.0 | 61.00 | 15.00 | 0.27 |
| Wheat | 89.1 | 35.85 | | 0.40 |
| Wheat straw | 90.1 | 54.00 | 16.00 | 0.29 |
| Sugarcane | 26.0 | 67.00 | | 0.50 |
| Bagasse | 71.0 | 67.15 | 14.50 | 0.28 |

Adapted from Kim and Dale (2004)

**Table 17.5** Theoretical ethanol yield from crop-processing by-products

| By-products from crop processing | Starch (wt%) | Cellulose (wt%) | Lignin (wt%) | Hemicellulose (wt%) | Ethanol yield | References |
|---|---|---|---|---|---|---|
| Corn fiber | 20 | 14 | | 40 | 0.46 (L/kg dry) | Kurambhatti et al. (2018), Kálmán et al. (2006) |
| Rice husk | | 50 | 25–30 | | 14–15% (L/L) of fermented hydrolysate distilled | Onwuakor Chijioke et al. (2017), Syarif et al. (2016) |

**Table 17.6** Examples of values for the main parameters in batch fermentation

| Process conditions | Range | References |
|---|---|---|
| Cell density | 10–54 g of dry weight/L | Bunch (1994) |
| Temperature | 33–35 °C (36) | Bajpai and Margaritis (1987) |
| Ethanol concentration | Up to 14% (Vol/Vol) | Ghareib et al. (1988) |
| Residence time | Up to 10 days | Unrean and Srienc (2010) |

As we can see from the previous table, the reactors can operate at a very wide range of parameters.

## References

Alriols, M. G., Garcia, A., Chouaibi, F., Hajji, N., Labidi, J., Klemes, J., et al. (2010). Lignocellulosic biorefinery processes energy integration. *Chemical Engineering Transactions, 21,* 553–558. https://doi.org/10.3303/CET1021093.

Amores, M. J., Mele, F. D., Jiménez, L., & Castells, F. (2013). Life cycle assessment of fuel ethanol from sugarcane in Argentina. *International Journal of Life Cycle Assessment, 18,* 1344–1357. https://doi.org/10.1007/s11367.013-0584-2.

Bajpai, P., & Margaritis, A. (1987). The effect of temperature and pH on ethanol production by free and immobilized cells of *Kluyveromyces marxianus* grown on Jerusalem artichoke extract. *Biotechnology and Bioengineering, 30,* 306–313. https://doi.org/10.1002/bit.260300222.

Bird, R. B., Stewart, W. E., & Lightfoot, E. N. (2006). *Transport phenomena* (2nd ed.). Wiley.

Bothast, R. J., & Schlicher, M. A. (2005). Biotechnological processes for conversion of corn into ethanol. *Applied Microbiology and Biotechnology, 67,* 19–25. https://doi.org/10.1007/s00253-004-1819-8.

Bunch, A. W. (1994). High cell density growth of micro-organisms. *Biotechnology and Genetic Engineering Reviews, 12,* 535–562. https://doi.org/10.1080/02648725.1994.10647921.

Choonut, A., Yunu, T., Pichid, N., & Sangkharak, K. (2015). Ethanol production from reused liquid stillage. In *2015 International Conference on Alternative Energy in Developing Countries and Emerging Economies. Energy Procedia, 79,* 808–814. https://doi.org/10.1016/j.egypro.2015.11.570.

Cortes-Rodríguez, E. F., Fukushima, N. A., Palacios-Bereche, R., Ensinas, A. V., & Nebra, S. A. (2018). Vinasse concentration and juice evaporation system integrated to the conventional ethanol production process from sugarcane—Heat integration and impacts in cogeneration system. *Renewable Energy, 115,* 474–488. https://doi.org/10.1016/j.renene.2017.08.036.

De Jong, W., & Van Ommen, R. (2014). *Biomass as a sustainable energy source for the future: Fundamentals of conversion processes*. Wiley.

Drapcho, C. M., Nghiem, J., & Walker, T. (2008a). *Biofuels engineering process technology*. McGraw-Hill Education.

Drapcho, C. M., Nghim, N. P., & Walker, T. (2008b). *Biofuels, bioproducts and biorefining*. McGraw-Hill.

Ghareib, M., Youssef, K. A., & Khalil, A. A. (1988). Ethanol tolerance of *Saccharomyces cerevisiae* and its relationship to lipid content and composition. *Folia Microbiologica (Praha), 33,* 447–452. https://doi.org/10.1007/BF02925769.

Greetham, D., Zaky, A., Makanjuola, O., & Du, C. (2018). A brief review on bioethanol production using marine biomass, marine microorganism and seawater. *Current Opinion in Green Sustainable Chemistry Bioresources and Biochemicals/Biofuels and Bioenergy, 14,* 53–59. https://doi.org/10.1016/j.cogsc.2018.06.008.

Hyun, H. H., & Zeikus, J. G. (1985). General biochemical characterization of thermostable pullulanase and glucoamylase from *Clostridium thermohydrosulfuricum*. *Applied and Environment Microbiology, 49,* 1168–1173.

Kálmán, G., Recseg, K., Gáspár, M., & Réczey, K. (2006). Novel approach of corn fiber utilization. *Applied Biochemistry and Biotechnology, 131,* 738–750. https://doi.org/10.1385/ABAB:131:1:738.

Kim, S., & Dale, B. E. (2004). Global potential bioethanol production from wasted crops and crop residues. *Biomass and Bioenergy, 26,* 361–375. https://doi.org/10.1016/j.biombioe.2003.08.002.

Kujawski, W., & Zielinski, L. (2006). Bioethanol—One of the renewable energy sources. *Environment Protection Engineering, 32,* 143–149.

Kumar, D., & Singh, V. (2016). Dry-grind processing using amylase corn and superior yeast to reduce the exogenous enzyme requirements in bioethanol production. *Biotechnology for Biofuels, 9*. https://doi.org/10.1186/s13068-016-0648-1.

Kumar, D., & Singh, V. (2019). Bioethanol production from corn, Chap. 22. In S. O. Serna-Saldivar (Ed.), *Corn* (3rd ed., pp. 615–631). Oxford: AACC International Press. https://doi.org/10.1016/B978-0-12-811971-6.00022-X.

Kurambhatti, C. V., Kumar, D., Rausch, K. D., Tumbleson, M. E., & Singh, V. (2018). Increasing ethanol yield through fiber conversion in corn dry grind process. *Bioresource Technology, 270,* 742–745. https://doi.org/10.1016/j.biortech.2018.09.120.

Madson, P. W. (2003). Ethanol distillation: The fundamentals. In *The alcohol textbook*. The Nottingham University Press.

Mohd Azhar, S. H., Abdulla, R., Jambo, S. A., Marbawi, H., Gansau, J. A., Mohd Faik, A. A., et al. (2017). Yeasts in sustainable bioethanol production: A review. *Biochemistry and Biophysics Reports, 10,* 52–61. https://doi.org/10.1016/j.bbrep.2017.03.003.

Onay, M. (2018). Bioethanol production from *Nannochloropsis gaditana* in municipal wastewater. In *5th International Conference on Energy and Environment Research, ICEER 2018*, Prague, Czech Republic, July 23–27, 2018. *Energy Procedia, 153*, 253–257. https://doi.org/10.1016/j.egypro.2018.10.032.

Onwuakor Chijioke, E., Hans-Anukam Uzunma., & Uzokwe Munachi, J. (2017). Production of ethanol and biomass from rice husk using cultures of *Aspergillus flavus*, *Aspergillus eamarii* and

*Saccharomyces cerevisiae*. *American Journal of Microbiological Research, 5*, 86–90. https://doi.org/10.12691/ajmr-5-4-3.
Palacios-Bereche, R., Ensinas, A. V., Modesto, M., & Nebra, S. A. (2014). Mechanical vapour recompression incorporated to the ethanol production from sugarcane and thermal integration to the overall process applying pinch analysis. *Chemical Engineering Transactions, 39*, 397–402.
Phansroy, N., Khawdas, W., Watanabe, K., Aso, Y., & Ohara, H. (2018). Microbial fuel cells equipped with an iron-plated carbon-felt anode and *Shewanella oneidensis* MR-1 with corn steep liquor as a fuel. *Journal of Bioscience and Bioengineering, 126,* 514–521. https://doi.org/10.1016/j.jbiosc.2018.04.011.
Schwietzke, S., Kim, Y., Ximenes, E., Mosier, N. S., & Ladisch, M. (2009). Ethanol production from maize. In *Molecular genetic approaches to maize improvement*. New York: Springer.
Sutjahjo, D. H. (2018). The characteristics of bioethanol fuel made of vegetable raw materials. In *Materials science and engineering*. IOP Publishing. https://doi.org/10.1088/1757.899X/296/1/012019.
Syarif, H. U., Suriamihardja, D. A., Selintung, M., & Wahab, A. W. (2016). Analysis SEM the chemical and physics composition of used rice husks as an absorber plate. *International Journal of Engineering Science and Applications, 2,* 25–30.
Treybal, R. E. (1980). *Mass transfer operations* (3rd ed.). McGraw-Hill.
Unrean, P., & Srienc, F. (2010). Continuous production of ethanol from hexoses and pentoses using immobilized mixed cultures of *Escherichia coli* strains. *Journal of Biotechnology, 150,* 215–223. https://doi.org/10.1016/j.jbiotec.2010.08.002.
Velasquez, H. I., Mesa, C., & Giraldo, S. A. (2013). Energy and exergy analysis of the combined production process of sugar and ethanol from sugarcane (Colombia case study). In *International Conference on Efficiency, Cost, Optimisation, Simulation and Environmental Impact of Energy Systems*. Presented at the ECOS 2013.
Veljković, V. B., Biberdžić, M. O., Banković-Ilić, I. B., Djalović, I. G., Tasić, M. B., Nježić, Z. B., et al. (2018). Biodiesel production from corn oil: A review. *Renewable and Sustainable Energy Reviews, 91,* 531–548. https://doi.org/10.1016/j.rser.2018.04.024.
Veloso, I. I. K., Rodrigues, K. C. S., Sonego, J. L. S., Cruz, A. J. G., & Badino, A. C. (2019). Fed-batch ethanol fermentation at low temperature as a way to obtain highly concentrated alcoholic wines: Modeling and optimization. *Biochemical Engineering Journal, 141,* 60–70. https://doi.org/10.1016/j.bej.2018.10.005.
Vertes, A. A., Qureshi, N., Blaschek, H. P., & Yukawa, H. (Eds.). (2010). *Biomass to biofuels: Strategies for global industries*. Oxford, UK: Wiley.
Vohra, M., Manwar, J., Manmode, R., Padgilwar, S., & Patil, S. (2014). Bioethanol production: Feedstock and current technologies. *Journal of Environmental Chemical Engineering, 2,* 573–584. https://doi.org/10.1016/j.jece.2013.10.013.
Watanabe, M. (2009). *Ethanol production in Brazil: Bridging its economic and environmental aspects*. International Association for Energy Economics.
Zakhartsev, M., Yang, X., Reuss, M., & Pörtner, H. O. (2015). Metabolic efficiency in yeast *Saccharomyces cerevisiae* in relation to temperature dependent growth and biomass yield. *Journal of Thermal Biology, 52,* 117–129. https://doi.org/10.1016/j.jtherbio.2015.05.008.
Zhang, C. -H., Yang, Y., Teng, B. -T., Li, T. -Z., Zheng, H. -Y., Xiang, H. -W., et al. (2006). Study of an iron-manganese -Tropsch synthesis catalyst promoted with copper. *Journal of Catalysis, 237,* 405–415. https://doi.org/10.1016/j.jcat.2005.11.004.

# Chapter 18
# Biodiesel

*The energy in biomass doesn't come for free, you have to use fertilizers. You have the plowing of the field and even by plowing the field you will need power. I have read that for 1 hectare 100 liters of diesel will be needed. If the scope of the land is the production of biodiesel, 1/10 will be lost just for the plowing of the field.*

Michel Hartmut, Nobel Prize in Chemistry in 1988.

## 18.1 Fundamentals of Biodiesel Production

Biodiesel is sometimes considered as a renewable transportation fuel, and it is normally produced from vegetable oils and animal fats even if it can be produced from other sources such as waste cooking oils (Phan and Phan 2008). Biodiesel is consisting of fatty acid methyl esters (FAME) (Hoekman et al. 2012) and because of this composition required, it will be more understood why they are produced by vegetable oils. In the following explanations, the author of this book encourages the reader in considering also more advanced textbooks of organic chemistry (Jones and Fleming 2014) in order to understand the terminology used here.

Vegetable oils present a series of positive aspects that make them suitable for biodiesel production. More in particular:

- are rich in neutral triacylglycerols (TAG);
- can be rich in free fatty acids (FFA).

Non-conventional sources of these two categories of compounds are, for instance, non-edible vegetable oils (for instance, Jatropha oil and Polanga oil) or animal fat wastes (AFWs) (Pollardo et al. 2018) (for instance, chicken fat and white grease). These feedstocks are characterized with high free fatty acids and water.

As mentioned previously, biodiesel is constituted mainly by FAME, however, the composition can still vary quite largely when considering the process conditions and the feedstock utilized. In addition, catalysis plays a fundamental role in biodiesel production.



C. De Blasio, *Fundamentals of Biofuels Engineering and Technology*,
Green Energy and Technology, https://doi.org/10.1007/978-3-030-11599-9_18

In regard to the biodiesel composition, Table 18.1 is derived with data from Knothe (2005) and Hoekman et al. (2012) and only the main esters are reported here. This is done to demonstrate that, even if the composition of biodiesel is restricted to only some compounds, there is still a variety of them (Table 18.1).

The main reactions taking place during the biodiesel production aim at the formation of fatty acid esters (characterized by the group –OOR). If the feedstock is rich in neutral triacylglycerols (TAG), we have the following reaction:

$$\text{Triacylglycerol (TAG)} + 3R'OH \overset{\text{Catalyst}}{\longleftrightarrow} 3R'COOR + C_3H_5(OH)_3 \quad (18.1)$$

the component 3R′COOR is our fatty acid ester and in addition, glycerol is produced by this process. Notice that the terms R and R′ are referred here to carbon chains ($CH_3$–$CH_2$–$CH_2$– …).

The catalysts can be KOH, NaOH, or $H_2SO_4$. If vegetable oils are rich in free fatty acids (Gebremariam and Marchetti 2018), FFA, we have:

$$\text{RCOOH (FFA)} + R'OH \overset{\text{Acid Catalyst}}{\longleftrightarrow} RCOOR' + H_2O \quad (18.2)$$

where the catalyst is commonly HCl or $H_2SO_4$ and RCOOR′ is an alkyl ester. In transesterification, the biodiesel is obtained by the reaction of a vegetable oil with an alcohol, in the presence of a catalyst, to give monoalkyl esters.

The overall reaction is represented in Fig. 18.1.

Free fatty acids are formed when oils are treated at high temperatures where they undergo an oxidation process. Side reactions in the production of biodiesel are saponification and hydrolysis, and they are represented, respectively, by:

$$RCOOH + NaOH\ (\text{or } KOH) \leftrightarrow R'COONa^{+}\left(\text{or } R'COOK^{+}\right) + H_2O \quad (18.3)$$

$$RCOOR' + H_2O \overset{\text{Base catalyst}}{\longleftrightarrow} RCOOH + R'OH \quad (18.4)$$

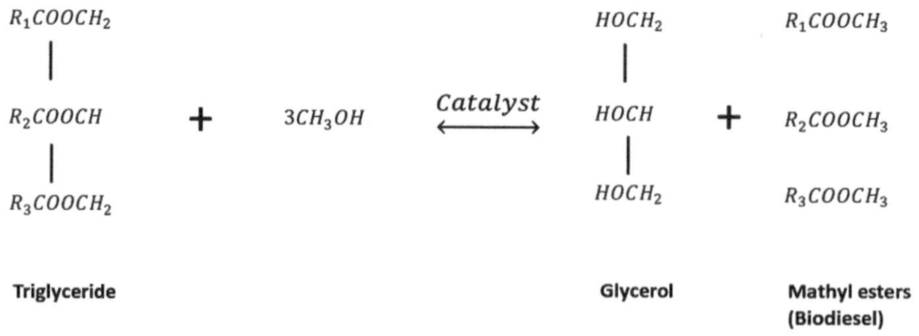


**Fig. 18.1** Reaction of a vegetable oil with an alcohol. Adapted from Lotero et al. (2005)

**Table 18.1** Main esters of biodiesel in reference to different kinds of feedstock used

| Ester | Camelina | Canola | Coconut | Corn | Jatropha | Palm | Rapeseed | Safflower | Soy | Sunflower | Tallow | Yellow grease |
|---|---|---|---|---|---|---|---|---|---|---|---|---|
| Lauric (dode-canoic); 12:0 | | | 47.7 | | | | | | | | | |
| Myristic (tetrade-canoic); 14:0 | | | 18.5 | | | | | | | | | |
| Palmitic (hexade-canoic); 16:0 | | | | 11.5 | 14.9 | 42.5 | | | 11.6 | | 24.3 | |
| Stearic (octade-canoic); 18:0 | | | | | | | | | | | 18.2 | |
| Methyl octade-cenoate (methyl oleate); 18:1 | 16.8 | 60.4 | | 26.6 | 40.4 | 41.3 | 59.5 | 14.2 | 23.7 | 21.7 | 42.2 | 44.6 |
| Methyl octadeca-dienoate (linoleate); 18:2 | 17.0 | 21.2 | | 58.7 | 36.2 | | 21.5 | 74.3 | 53.8 | 66.3 | | 25.1 |
| Methyl octadeca-trienoate (linole-nate); 18:3 | 35.6 | 9.6 | | | | | 8.4 | | | | | |

Values in % of FAME composition

It is referred in literature that when acid-catalyzed reactions occur, the saponification is greatly reduced (Gebremariam and Marchetti 2018). On the other hand, when free fatty acids (FFAs) (ROOH is the acidic group) are present in the feedstock, then there could be the formation of soaps, and this is due to a possible excess of the catalyst itself. The reaction of saponification due to an excess of base catalyst is also described in Fig. 18.2.

The formation of soaps is very negative for separation processes, the viscosity will increase for example. In waste oil, the amount of FFA suggested is not very high and around 1 wt% (Kumar Tiwari et al. 2007). While the FFA reduces the quality of the feedstock, water is eliminated by centrifugation to an acceptable level of 0.5% (wt).

Fatty acids directly affect the main parameters of the biodiesel which are:

- the cloud point (the minimum temperature at which small solid crystals can be observed when the temperature of the sample is decreased);
- the cold filter plugging point (the temperature at which the fuel filter plugs due to crystals and solid components);
- the pour point (the minimum temperature at which the fuel is actually moving sufficiently, it becomes a gel).

For the sake of comprehension, we remember here that the structure of a carboxylic acid is represented in Fig. 18.3.

In Fig. 18.3, it is possible to notice that this molecule has some parts which are nucleophile (attract protons) while, on the other hand, there are also electrophile parts (which attract electrons). It is quite important to reason on these sites of the molecules to understand and design possible reaction routes. For this reason, more information on the meaning of ionization potential and electron affinity will be given in the following sections.

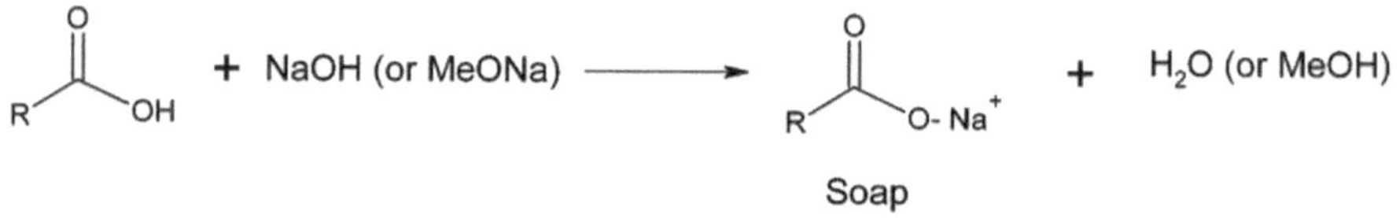


**Fig. 18.2** FFA reaction with a base catalyst and formation of soap

**Fig. 18.3** Carboxylic acid representation. Adapted from Jones and Fleming (2014)

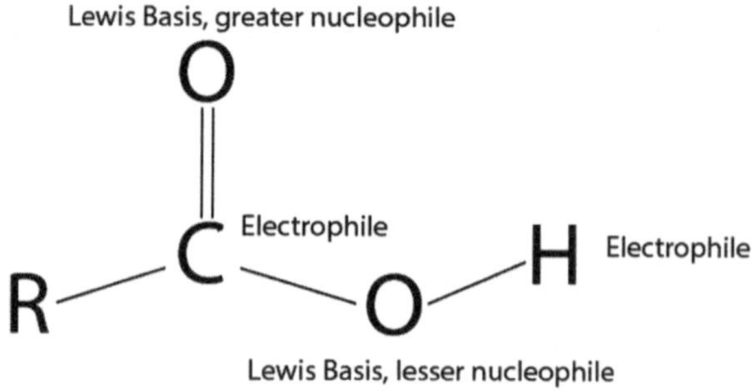

## 18.2 Ionization Potential and Electron Affinity of an Element

The ionization potential is the energy required to subtract one electron from the atom to form a cation (the easiest electron is the one situated at the upper orbital level), this is measured in electron Volt, eV, which is equivalent to $1\ V \cdot 1.6 \times 10^{-19}$ C, and this has the dimensions of Joule. Remember also that $1\ V = 1\ J/1\ C$ and that the Coulomb is actually the measure of the charge (the electrons).

The electron affinity is, on the other hand, the energy released when the electron is added to an atom. Some values for the ionization potential of some elements are given in Table 18.2 adapted from Jones and Fleming (2014).

While an example of nucleophilic substitution is given in Fig. 18.4.

To be noticed that in Fig. 18.4, the N–H group has more negative charge on the N atom, and therefore the H atom will have a more positive charge. On the other hand,

**Table 18.2** Examples of ionization potential and electron affinities (in brackets)

| | | | | | | | |
|---|---|---|---|---|---|---|---|
| H<br>13.60<br>(0.75) | | | | | | | He<br>24.59 |
| Li<br>5.39<br>(0.62) | Be<br>0.32 | B<br>8.30<br>(0.24) | C<br>11.26<br>(1.27) | N<br>14.53 | O<br>13.62<br>(1.47) | F<br>17.42<br>(3.34) | Ne<br>21.56 |
| Na<br>5.14<br>(0.55) | Mg<br>7.65 | Al<br>5.99<br>(0.46) | Si<br>8.15<br>(1.24) | P<br>10.49<br>(0.77) | S<br>10.36<br>(2.08) | Cl<br>12.97<br>(3.61 | Ar<br>15.75 |

Values are in eV

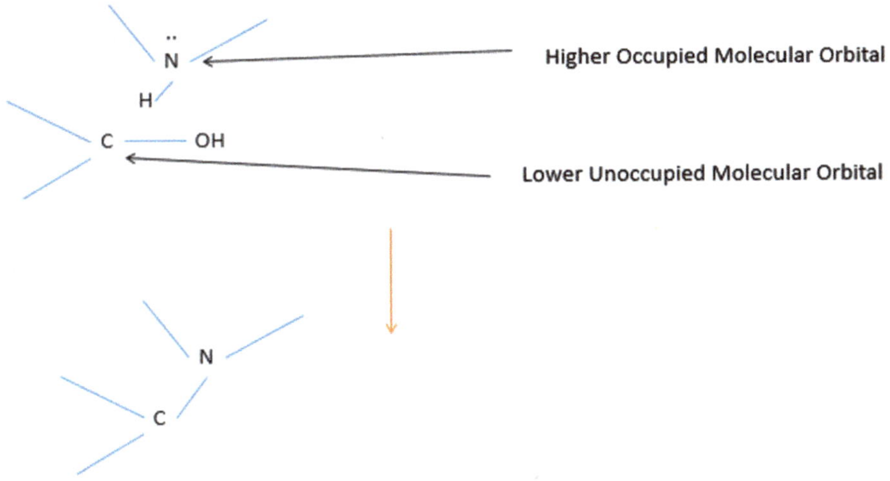


**Fig. 18.4** Nucleophilic substitution between N–H and C–OH groups

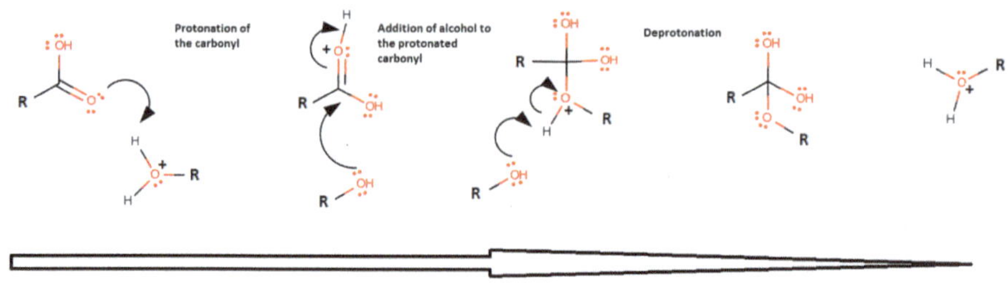


**Fig. 18.5** Formation of the tetrahedral intermediate in the esterification of hydroxyl acids. Adapted from Jones and Fleming (2014)

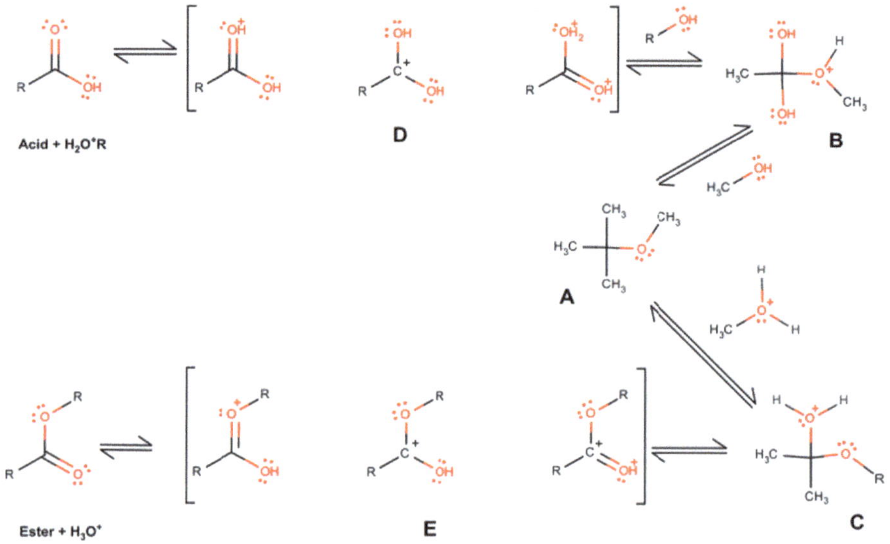


**Fig. 18.6** All steps in Fischer esterification. Adapted from Jones and Fleming (2014)

the group C–OH has a more negative charge on the OH part and therefore this part will have more possibility to react with the H atom of the N–H group.

The formation of the tetrahedral intermediate in the esterification of hydroxyl acids proceeds as shown in Fig. 18.5.

Consider the periodic table with the electron affinities, note that oxygen has an electron affinity higher than carbon; for this reason, the alcohol will bound on the carbon and not on the oxygen.

All steps in Fischer esterification are demonstrated in Fig. 18.6.

The ion $RH_2O^+$ goes to R–OH which is a stable compound. The reaction with the alkoxide group ($^-OR$) is reversible and catalytic, while the reaction of the hydroxide group ($^-OH$) is not reversible and not catalytic (Fig. 18.7).

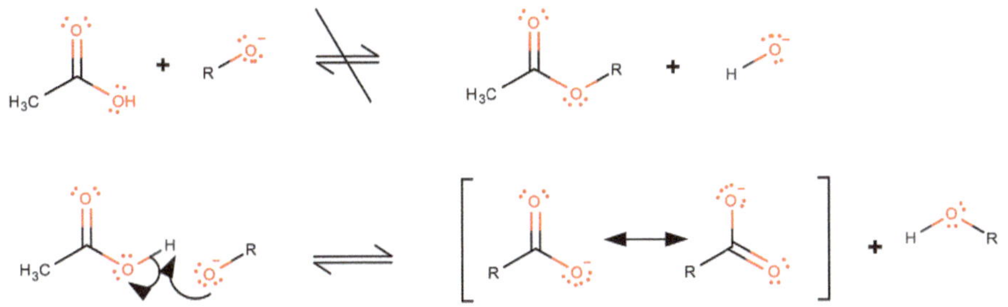


**Fig. 18.7** Reaction involving the alkoxide group ($^{-}$OR) and the hydroxide group ($^{-}$OH). Adapted from Jones and Fleming (2014)

The hydrogen from the hydroxyl acid cannot just be replaced by the R of the alkohyde; because, an acid when loses the hydrogen is actually stabilized by resonance. Remember that R is referring to the carbon chain ($CH_3$–$CH_2$ …).

As a summary and as referred previously, the catalysts used in the process of biodiesel production are mainly alkaline:

- NaOH
- KOH.

And, they are involved in the reactions:

$$CH_3O^{-}Na^{+} \rightarrow CH_3O^{-} + Na^{+} \tag{18.5}$$

$$NaOH + CH_3OH \leftrightarrow CH_3O^{-} + H_2O + Na^{+} \tag{18.6}$$

Here, we notice the formation of the methoxide ions $\left(CH_3O^{-}\right)$, and they are strong nucleophiles and attach the glyceride molecules.

## 18.3 Base-Catalyzed Production of Biodiesel

When base catalysts are used in the production of biodiesel, the reaction mechanisms are illustrated as follows. The base catalysts create an alkoxide ion, $BH^{+}$. In this way, the alkyl ion, $RO^{-}$, will attack directly the double bond of the carboxylic group.

$$ROH + B \leftrightarrow RO^{-} + BH^{+} \tag{18.7}$$

The alkyl ion attack on the double bond of the triglyceride is demonstrated in Fig. 18.8.

Notice that the RO– ion is a nucleophile. As an example, here, the formation of the methoxide ion from a base catalyst and an alcohol is demonstrated:

$$CH_3O^-Na^+ \rightarrow CH_3O^- + Na^+ \tag{18.8}$$

$$NaOH + CH_3OH \leftrightarrow CH_3O^- + H_2O + Na^+ \tag{18.9}$$

In particular, the methoxide ion (the one with the methyl group) is a strong nucleophile, which means that it is able to link at lower electrophile carbon.

## 18.4 Acid-Catalyzed Production of Biodiesel

When an acid will be adopted as catalyzer, the attack on the original triglyceride is shown in Fig. 18.9.

Acid catalysts do not show measurable susceptibility to FFA as previously mentioned; however, remember that, there has to be a limit in FFA content. In case, we

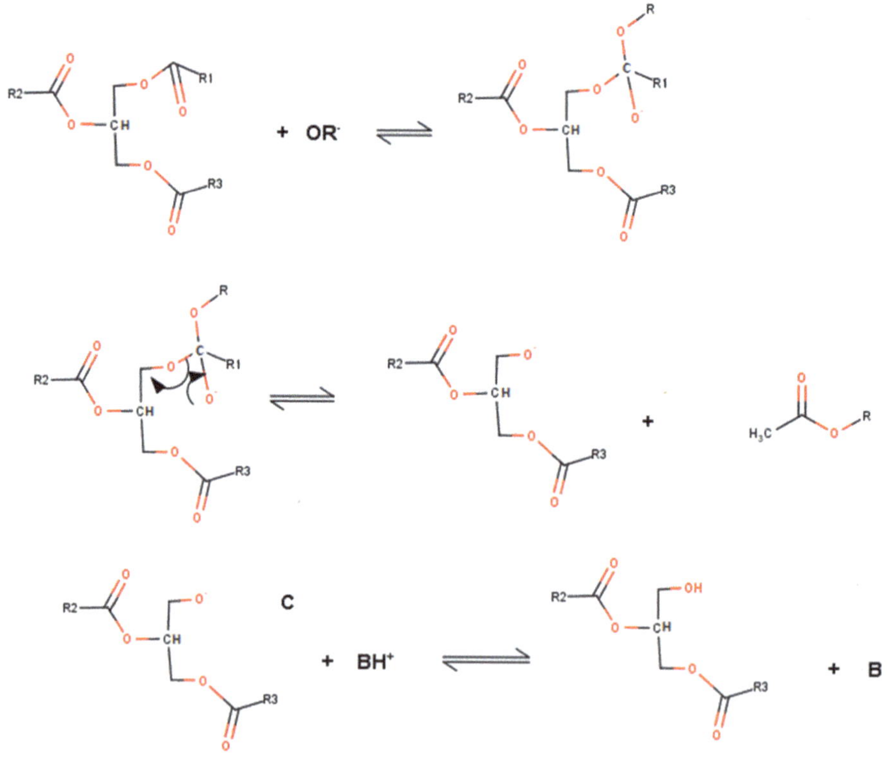


**Fig. 18.8** Catalyst alkoxide ion and attach on the carboxylic group. Adapted from Lotero et al. (2005)

consider here an alcohol described by $R_4OH$ (see Fig. 18.9), this would be a nucleophile because it can give some electrons to the cation. Nucleophiles are also called Lewis bases. If R4 is the methoxide ion, then this will be the methyl ester we want, the carbonyl oxygen becomes more electrophilic, and this is because obtains a plus charge, and this makes more easy to host the OH group of the alcohol.

An example of homogeneous acid-catalyzed process for biodiesel production is given in Fig. 18.10.

As an example of biodiesel production from waste cooking oils (WCO), it will be reported here the alkali-catalyzed transesterification of waste cooking oils. The composition of waste cooking oils is demonstrated in Table 18.3.

In this particular study, the temperature of the experiments was 30, 50, or 70 °C. While the residence time was 20, 40, 60, 90, or 120 min, the molar ratio of methanol and WCO was from 5:1 to 12:1. The amount of KOH catalyst was 0.5–1.5 wt% of the WCO.

The transesterification is commonly carried out with an extra amount of alcohol in order to shift the equilibrium to the proposed product, methyl ester.

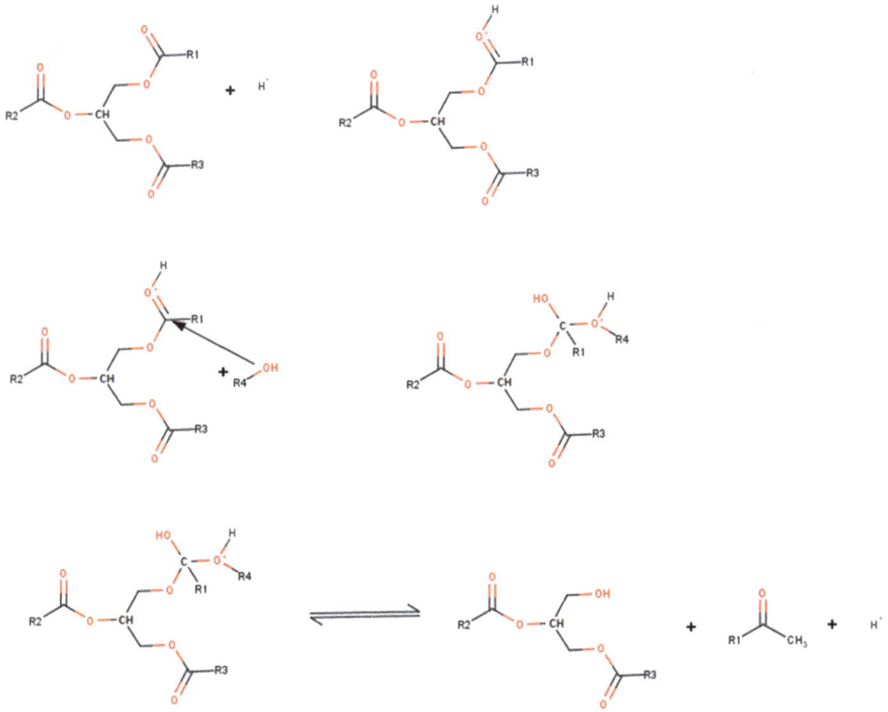

**Fig. 18.9** Acid-catalyzed reactions on the triglyceride. R1, R2, and R3 are carbon chains of the fatty acids, while R4 is an alkyl group of the alcohol. Adapted from Lotero et al. (2005)

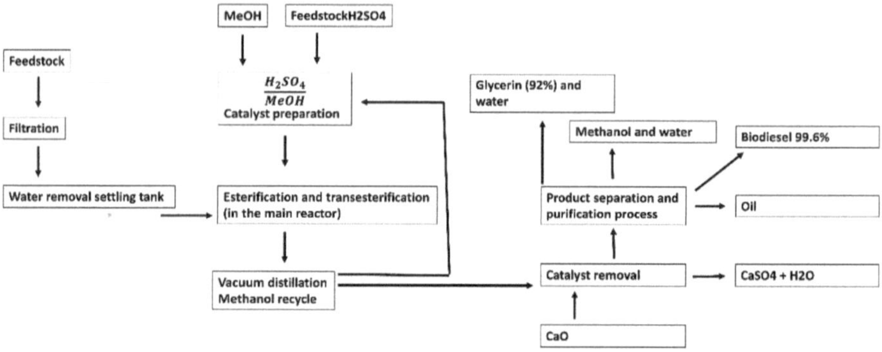


**Fig. 18.10** Example of homogeneous acid-catalyzed process for biodiesel production

**Table 18.3** Composition of waste cooking oils (WCO)

| Component | g/100 g of total fatty acid methyl ester |
|---|---|
| C6:0 | 0 |
| C8:0 | 8.82 |
| C10:0 | 6.21 |
| C12:0 | 44.65 |
| C14:0 | 16.31 |
| C15:0 | 0.00 |
| C16:0 | 10.59 |
| C16:1 | 0 |
| C17:0 | 0 |
| C18:0 | 3.29 |
| C18:1 | 8.17 |
| C18:2 | 1.96 |
| C18:3 | 0 |
| C20:0 | 0 |
| C20:1 | 0 |

Adapted from Phan and Phan (2008)

Table 18.4 reports a comparison between waste cooking oil and diesel. It is possible to notice that obviously waste cooking oils are not suitable for an engine, however, they do have good technical properties.

Methanol concentration has an effect on the conversion of biodiesel; as a matter of fact, the conversion to biodiesel can be up to 20% in case the concentration of methanol goes from five to twelve times the concentration of waste cooking oils. The temperature also affects the conversion of waste cooking oils to biodiesel; it is reported in the literature that if the concentration of KOH is 0.75 and the methanol/WCO ratio is 8:1, then the highest conversion has been noticed at 50 °C (Lotero et al. 2005).

**Table 18.4** Properties of waste cooking oils and diesel

| Parameter | WCO | Biodiesel |
|---|---|---|
| Acid number (mg KOH/g) | 2.36 | 0.11 |
| Iodine number (mg KOH/g) | 13.20 | – |
| Saponification number (mg KOH/g) | 268.22 | – |
| Density (g/cm$^3$) | 0.92 | 0.83 |
| Flash point (°C) | 269 | 69 |
| Cloud point (°C) | 21 | 0 |
| Pour point (°C) | 18 | <12 |
| Viscosity at 40 °C (mm$^2$/s) | 30.05 | 3.53 |

Adapted from Phan and Phan (2008)

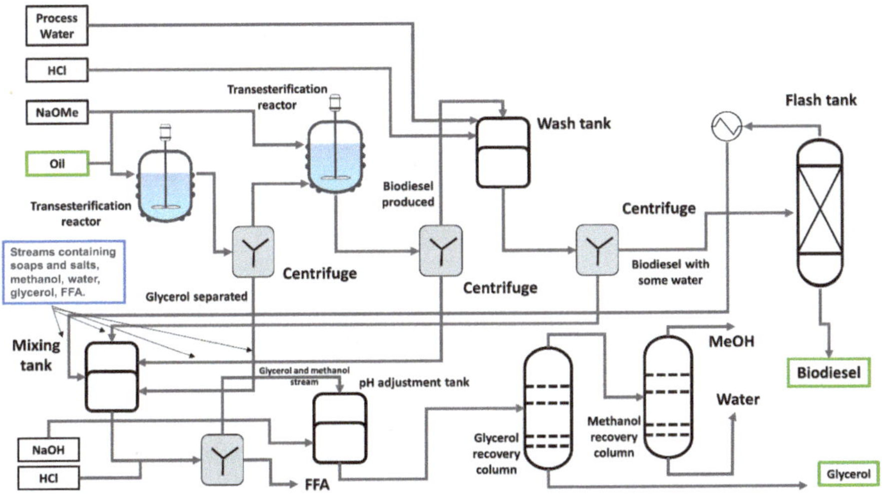


**Fig. 18.11** A more detailed flow diagram for biodiesel production by a homogeneous alkali-catalyzed process. Adapted from Apostolakou et al. (2009)

## 18.5 More Detailed Process Description for Alkali-Catalyzed Biodiesel Production

Previous studies on biodiesel production and the different alternatives available have been performed. It has been concluded that the alkali-catalyzed process for biodiesel production is the most favorable in terms of its economy (Marchetti et al. 2008) and that heterogeneous-catalyzed processes are also favorable. For this reason, the case of alkali-catalyzed biodiesel production will be described in more details here. The process is demonstrated in Fig. 18.11.

The demonstrated process is mostly suitable for oils that have a low content in FFA, like rapeseed oil. The oil and the related catalyst are sent to a transesterification reactor together with methanol. As mentioned, diverse catalysts could be used in this stage and to give one example, sodium methoxide could be used (KoohiKamali et al. 2012); however, also other types of catalysts are utilized as well. The effluent from the transesterification reactor is sent to a centrifuge where the esterificated fraction is sent to a second reactor. The glycerol is separated from this fraction and sent to a mixing tank. The same operation is basically achieved into the second transesterification reactor, and the obtained product is then sent to the washing process where the biodiesel fraction is processed with water and hydrochloric acid. The fraction containing soaps or salts, methanol, water, glycerol, and FFA is sent to a mixing tank, while the remaining biodiesel (which contains a small fraction of water) is sent to a flash tank unit where the refined biodiesel is then separated.

The fraction with the highest amount of glycerol and coming from the transesterification reactors is sent, together with the fraction coming from the washing process, to be mixed and the separation of FFA occurs by centrifugation with the help of HCl solution. The derived glycerol and methanol stream is then adjusted for its pH values with NaOH and sent to a glycerol recovery column where the glycerol is finally separated. The effluent stream with the highest amount of methanol is sent to a methanol recovery column where it is separated.

## References


Apostolakou, A. A., Kookos, I. K., Marazioti, C., & Angelopoulos, K. C. (2009). Techno-economic analysis of a biodiesel production process from vegetable oils. *Fuel Processing Technology, 90,* 1023–1031. https://doi.org/10.1016/j.fuproc.2009.04.017.

Gebremariam, S. N., & Marchetti, J. M. (2018). Biodiesel production through sulfuric acid catalyzed transesterification of acidic oil: Techno economic feasibility of different process alternatives. *Energy Conversion and Management, 174,* 639–648. https://doi.org/10.1016/j.enconman.2018.08.078.

Hoekman, S. K., Broch, A., Robbins, C., Ceniceros, E., & Natarajan, M. (2012). Review of biodiesel composition, properties, and specifications. *Renewable and Sustainable Energy Reviews, 16,* 143–169. https://doi.org/10.1016/j.rser.2011.07.143.

Jones, M., & Fleming, S. A. (2014). *Organic chemistry* (5th ed.). W. W. Norton & Company.

Knothe, G. (2005). Dependence of biodiesel fuel properties on the structure of fatty acid alkyl esters. *Fuel Processing Technology Biodiesel Processing and Production, 86,* 1059–1070. https://doi.org/10.1016/j.fuproc.2004.11.002.

KoohiKamali, S., Tan, C. P., & Ling, T. C. (2012). Optimization of sunflower oil transesterification process using sodium methoxide. *The Scientific World Journal, 2012,* 8 pages. https://doi.org/10.1100/2012/475027.

Kumar Tiwari, A., Kumar, A., & Raheman, H. (2007). Biodiesel production from jatropha oil (*Jatropha curcas*) with high free fatty acids: An optimized process. *Biomass and Bioenergy, 31,* 569–575. https://doi.org/10.1016/j.biombioe.2007.03.003.

Lotero, E., Liu, Y., Lopez, D. E., Suwannakarn, K., Bruce, D. A., & Goodwin, J. G. (2005). Synthesis of biodiesel via acid catalysis. *Industrial and Engineering Chemistry Research, 44,* 5353–5363.

Marchetti, J. M., Miguel, V. U., & Errazu, A. F. (2008). Techno-economic study of different alternatives for biodiesel production. *Fuel Processing Technology, 89,* 740–748. https://doi.org/10.1016/j.fuproc.2008.01.007.

Phan, A. N., & Phan, T. M. (2008). Biodiesel production from waste cooking oils. *Fuel, 87,* 3490–3496. https://doi.org/10.1016/j.fuel.2008.07.008.

Pollardo, A. A., Lee, H., Lee, D., Kim, S., & Kim, J. (2018). Solvent effect on the enzymatic production of biodiesel from waste animal fat. *Journal of Cleaner Production, 185,* 382–388. https://doi.org/10.1016/j.jclepro.2018.02.210.

# Chapter 19
# Some Chemical Analyses in Biodiesel Production and Biofuel Characteristics

*If I don't see it, I don't believe it.*
Sentence normally attributed to St. Thomas.

## 19.1 Amount of Free Fatty Acid

Biodiesel has a series of positive and negative aspects if compared with normal diesel. The main positive aspects of biodiesel production are listed as (Lin and Lin 2012):

- higher biodegradability;
- pollutant reduction;
- higher flash point;
- biodiesel increases lubricity;
- biodiesel is considered to not contribute to global warming since it is produced by renewable biomass. However, this is not a suitable conclusion for large-scale productions, as it was referred in other parts of this book.

As an example of reduction in emissions, Fig. 19.1 shows that when biodiesel is blended to normal diesel, we have actually a reduction of pollution.

On the other hand, negative aspects with biodiesel production are listed as:

- higher $NO_x$ production;
- higher viscosity;
- lower volatility;
- presence of unreacted glycerols.

All the properties listed are highly influenced by the composition of the fuel itself; therefore, it is beneficial to have some knowledge of some analyses done on the fuel and the feedstock.

In the previous chapter, it was mentioned that the production of biodiesel is influenced by the feedstock composition and, because of this composition, also the production method will be influenced. It is important to understand, ones the product



C. De Blasio, *Fundamentals of Biofuels Engineering and Technology*,
Green Energy and Technology, https://doi.org/10.1007/978-3-030-11599-9_19

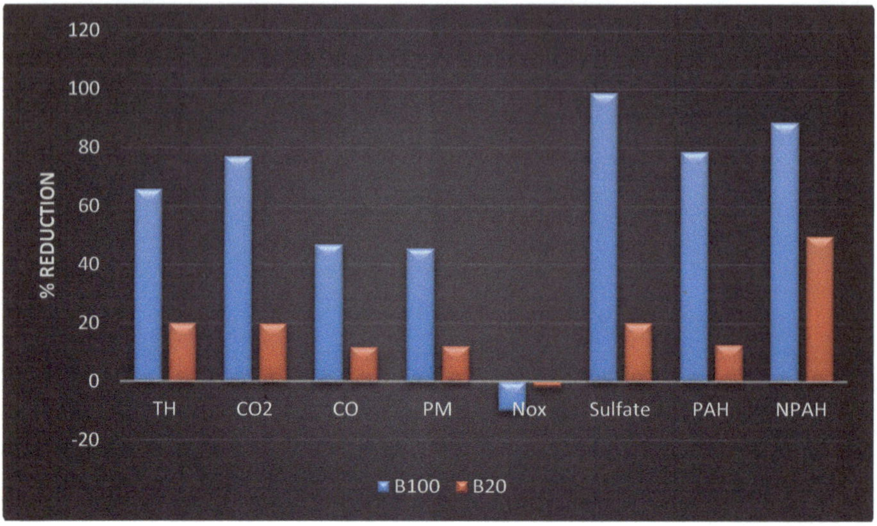


**Fig. 19.1** Percentage of pollutants reduction as a function of the fuel blend. B20 means "fuel blend," in this case, 20% of biodiesel is mixed with 80% of normal diesel. Data from Drapcho et al. (2008b)

is obtained, what are the physical and chemical analyses necessary to assess our product. Among the main kind of analyses, we have the determination of the amount of free fatty cid (FFA).

The procedures for the FFA or in general the total acid number (TAN) determination follow diverse standards. For instance, we can recall here the ASTM D664-11a, the ASTM D974-14, the ASTM D3339-12, and the AOCS Cd 3d 63 (Park et al. 2017). And, there are also several variations of these procedures. The basic principle behind these types of analyses is that we measure the sample at hand, then the sample is dissolved into a solution and a particular solvent is used to do this. For instance, as solvents, we have mixtures of toluene, 2-propanol, and water (Baig and Ng 2011) or a mixture of ethanol and water (Aricetti and Tubino 2012); or only propanol can be used. Isopropyl alcohol or isopropanol is also very common (Patel and Shah 2015, p. 11). The dissolved sample is then titrated with a suitable base, for instance, KOH and the change in pH values will determine the typical change in color. The change in color at a precise pH value is determined by particular chemical compounds, for example, phenolphthalein. When the molecule is in its acidic form, there is no detectable change in color; however, when the solution is titrated with a base, KOH, for instance, the pH of the solution will change and at some point, the phenolphthalein will be converted in its pink base form. The structure of the molecule in the two states is depicted in the following Fig. 19.2.

A complete procedure will be given in this case with respect to the product obtained. We assume here that we have produced biodiesel and we want to know if in our biodiesel there is still a quantity of FFA, however, this kind of analyses can

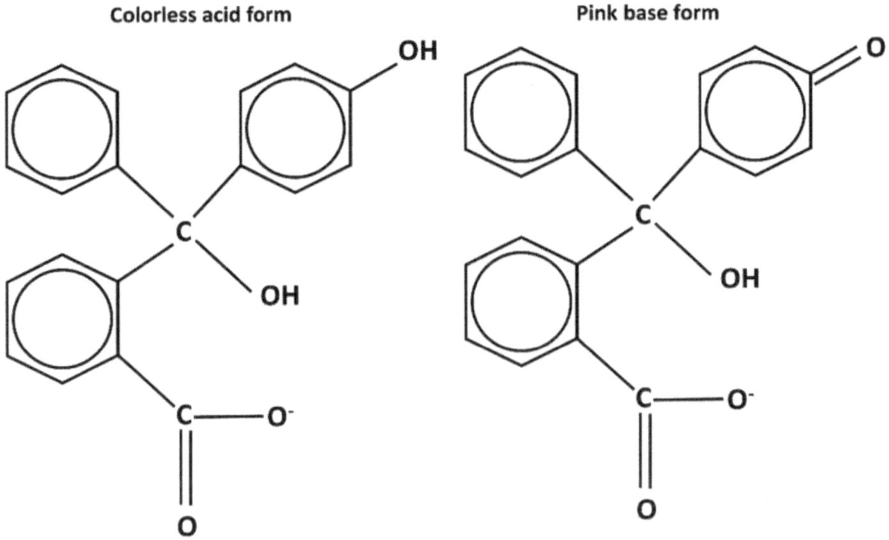


**Fig. 19.2** Structure of the phenolphthalein molecule in its acidic for and base form. Adapted from Zumdahl and Zumdahl (2014)

be performed for any type of feedstock oil or produced bio-oil. The procedure is as follows:

- The first step is to weigh a quantity of sample (for instance, biodiesel). Let us assume that we have 15–20 g. Make sure that the scale used is a quite precise one and reports the exact weight up to the second decimals.
- At this point, we need a suitable solvent to solubilize the sample and one of the possible solvents available is the isopropyl alcohol. Take a sufficient amount of isopropyl alcohol (for instance, 120–140 mL) with some drops of phenolphthalein (for instance, 7–9 drops).
- The sample should be mixed at this point with the titration solvent solution made previously.
- At this stage, you can titrate with KOH solution (with known concentration of the basis, for instance, 0.1 N).
- Titrate until the color is purple and the color is stable for a sufficient time (for instance 20–30 s).

The acid number is calculated by the following:

$$\text{Acid number} = \frac{[(A - B) \cdot N \cdot MW]}{W_s} \tag{19.1}$$

where

*A* volume (L) of KOH solution necessary for the titration;

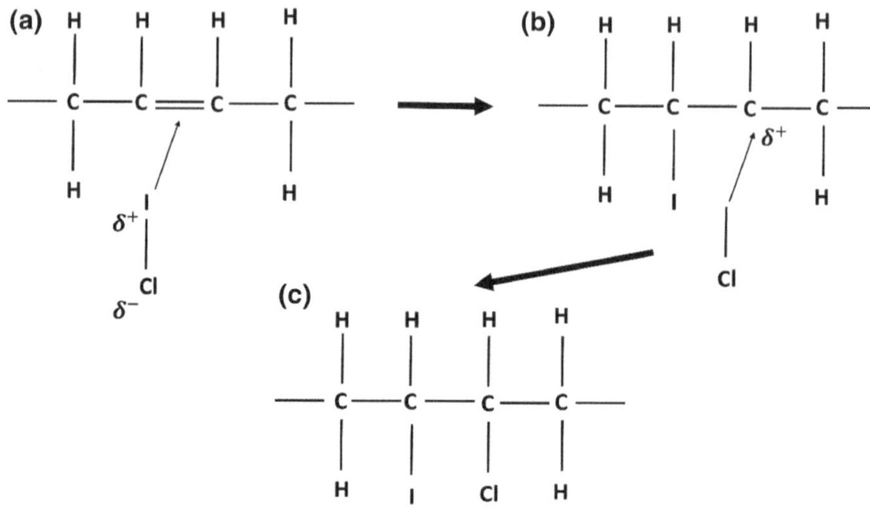


**Fig. 19.3** Mechanism of iodine saturation of the double bonds in unsaturated molecules

$B$ volume (L) of titrant used in the blank trial (no biodiesel used);
$N$ normality of titrant (0.1 in this case);
$MW$ molecular weight of the base used for the titration, in this case KOH, 56.1;
$W_s$ weight of sample in grams.

## 19.2 Iodine Number Determination

The iodine number is commonly measured to have an indication of the saturation degree of oils or in this case biofuels like biodiesel. The degree of saturation reflects the amount of double bonds present within the sample molecule. There is a high correlation between the degree of saturation and the iodine number; for instance, if we consider the average number of double bonds in the sample, the iodine number is higher if the amount of double bonds is larger (Giakoumis 2018). Unsaturated oils' examples are the canola oil, corn oil, linseed oil, and safflower oil. The result from this kind of analysis will, in practice, give the amount of iodine required for performing the saturation of one hundred grams of the sample.

The principle on the basis of this kind of analysis is that the molecules of iodine monochloride will eventually saturate the double bonds present in the molecules. This is possible because the iodine monochloride molecule is quite polar with a positive and negative part, Fig. 19.3, and the double bond (where there is an overcharge of electrons) will be attacked by the positive part of the iodine monochloride molecule. The next step would be the formation of a positive site within the oil molecule where the negatively charged chlorine will attach the remaining "free" space.

The titration, in this case, is concerning the presence of the remaining iodine monochloride left into the solution. The method is also denominated as Andrews' titration after Launcelot W. Andrews in 1903 (Andrews 1903).

The commercially available solution for the iodine monochloride molecules is called Wij's solution. For the determination of the iodine number of a particular sample, the procedure is as follows:

- It is necessary to measure the mass of the sample (for instance, oil or biodiesel), in this case, a suitable amount would be 0.1–0.2 g.
- At this point, we need a sufficient amount of iodine monochloride. This is available as Wij's solution. Add a sufficient amount, for instance, 25–30 mL so we are sure that this amount is enough to titrate the sample.
- At this stage, we need to add the solvent which, in this case, is the trichloroethene. A quantity of 10–20 mL will be sufficient in this case.
- The procedure continues by mixing the test flask. Then, some waiting time is required for the dissolution to take place; therefore, it is good to close the flask and take it in the dark for half hour at room temperature.
- It is also required to do the same procedure as for the steps written previously, but in this case, there should be no sample (oil or biodiesel) in the flask. This is called Blank, we need this titration as a reference.
- After keeping the flask in the dark, it is necessary to add to the solution potassium iodide. For instance, 15–20 mL of 10% potassium iodide solution (this is an excess to ensure that all the I-Cl is used). In this way, the leftover of the Wij solution will oxidize the Iodine in the KI solution, $2I^- = I_2$.
- At this point, it is necessary to titrate the $I_2$ with sodium thiosulphate solution. This is done in the presence of fresh starch.
  - A suitable titration medium would be sodium thiosulphate solution with a known concentration (for instance, 0.1 M). In this step, the $I_2$ is reduced back to iodide, $I^-$ ions. The titration is continuing until a yellow color appears (straw yellow).
  - Then, fresh starch is added (some drops of starch solution), and the color of the solution turns blue.
  - Finally, the titration is continued with sodium thiosulphate until the color turns white.

This procedure is similar to other procedures found in literature (Yasuda 1931). The iodine value is calculated as follows:

$$\text{Iodine value} = \frac{(V_{\text{blank}} - V_{\text{oil}}) \cdot [\text{Thio}] \cdot MW_{\text{t}}\left(\frac{\text{g}}{\text{mol}}\right)}{\text{grams of oil}} \frac{100\ \text{g}}{1000\ \text{cc/L}} \quad (19.2)$$

where

$V_{(\text{blank})}$ and $V_{(\text{oil})}$ is the volume of sodium thiosulphate used in titration for the sample and the blank;

[Thio] is the concentration of the thiosulphate;

$MW_{\text{t}}$, is the molecular weight of thiosulphate, 126.9 g/mol.

The titration of the $I_2$ with sodium thiosulphate is shown as:

$$2Na_2S_2O_3 + I_2 = Na_2S_2O_4 + 2NaI \quad (19.3)$$

## 19.3 Saponification Value of Fats and Oils

As we have seen previously within the explanation of the production process for biodiesel, soaps can be formed by the reaction of the fatty acids and an excess of homogeneous catalyst, like KOH. The evaluation of the saponification value utilizes just this reaction. By definition, with saponification value, we consider the milligrams of KOH needed to neutralize the fatty acids formed by the complete hydrolysis of one gram of fat or oil. The main reactions in this case are as follows:

$$\text{Fat} + \text{water} + \text{KOH} = \text{fatty acids} + \text{alcohol} \quad (19.4)$$

$$\text{Fatty acid} + \text{KOH} = \text{Salt of Fatty acid} + \text{alcohol} \quad (19.5)$$

In these reactions, a certain quantity of KOH is consumed from a known quantity put into the solution and the principle is to titrate the remaining quantity of KOH. A suitable example for the second reaction reported above is the reaction between tristearin and KOH:

$$\text{Tristearin} + \text{KOH} \rightarrow \text{glycerol} + \text{potassium stearate } (3KOOCC_{17}H_{35}) \quad (19.6)$$

The structure of the tristearin is reported in Fig. 19.4.

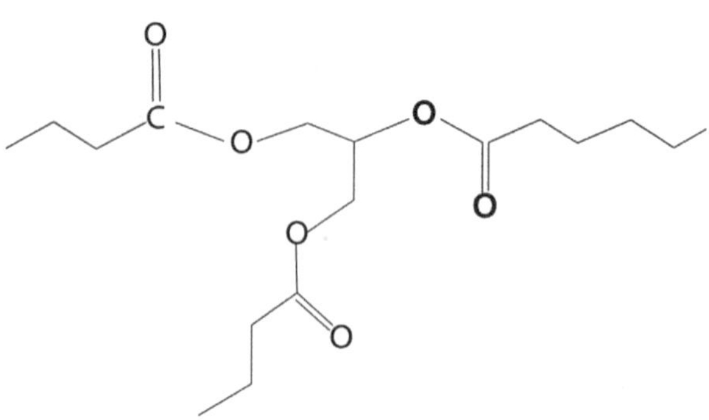


**Fig. 19.4** Tristearin structure

The procedure for the estimation of the saponification value is as follows:

- Also, in this case, it is necessary to weight a sufficient amount of sample. For instance, 0.5 g or 1 g of oil or biodiesel.
- We have then to dissolve the sample in KOH, and this is normally done by boiling under reflux with excess KOH (for example, 40 mL or 50 mL of KOH 0.1 M).
- We find the moles of KOH left by titration with HCl (for instance, 0.1 M HCl and some drops of phenolphthalein). The reaction in this case is:

$$KOH + HCl = KCl + H_2O \tag{19.7}$$

- At the end of the titration, the color goes from pink to white.

The calculations to be performed are:

- Number of moles used (left) = (concentration) – (volume used);
- Moles of KOH reacted = (initial moles) – (moles left);
- Ml of KOH = 1000 – (moles reacted) – (molecular mass, 56);
- Divide this amount by the grams of oil.

## 19.4 Biodiesel Characteristics and Properties

When biodiesel is produced, it is necessary to evaluate also some physical characteristics of the fuel. This is a very important assessment since, in order to be utilized in an engine, biodiesel has to behave in a good manner with respect to:

- The cloud point, which is the temperature at which the crystals of solid biodiesel first become visible, in case we decrease the temperature.
- The pour point also called gel point, this is the temperature where biodiesel becomes solid and can no longer be pumped.
- The flash point, this is the minimum temperature at which the vapors of the fuel ignite using an ignition source.
- The autoignition temperature, this is the temperature at which the vapors ignite with no source of ignition.
- The fire-point, this is the minimum temperature at which the flame is still burning with no source of ignition.

Table 19.1 shows some characteristics of biodiesel. In regard to the table, the nomenclature is as follows: viscosity (vis), acid value (AV), cloud point (CP), pour point (PP), flash point (FP), density (Den), and water content (WC).

While as a comparison, some properties of diesel and biodiesel are shown in Table 19.2.

Equation (19.8) (Satyanarayana and Rao 1992) is an example of flash point calculation. Notice that there are also other methods to calculate this value.

**Table 19.1** Biodiesel characteristics (adapted from Phan and Phan 2008)

| Parameter | EN 14214 | Cetinkaya and Karaosmanoglu (2004) | Georgogianni et al. (2007) | Encinar et al. (2005) | Phan and Phan (2008) |
|---|---|---|---|---|---|
| Vis 40 °C, $mm^2/s$ | 3.5–5.0 | 5.29–6.46 | 4.76–4.45 | 4.8 | 4.89 |
| AV, max mg KOH/g | 0.5 | 0.289 | 0.8–0.5 | – | 0.43 |
| CP, °C | – | 9 | (−3) to (−4) | 4.7 | 3 |
| PP, °C | – | −3 | (−4) to (−5) | −3.9 | 0 |
| FP, °C | >1.1 | 176 | 67–83 | 177 | 120 |
| Den (g/mL) | 0.86–0.90 | 0.88–0.89 | 0.85–0.82 | 0.89 | 0.88 |
| WC, max, ppm | 500 | 480.07 | – | – | Trace |

The original references are also reported within the table

**Table 19.2** Properties of diesel and biodiesel

| Property | Diesel | Biodiesel |
|---|---|---|
| Oxygen (wt%) | 0 | 11 |
| Composition | HC (10–C21) | FAME (C12–C22) |
| Kinetic viscosity ($mm^2/s$) at 40 °C | 1.9–4.1 | 1.9–6.0 |
| Water (vol.%) | 0.05 | 0.05 |
| Specific gravity (g/mL) | 0.85 | 0.88 |
| Flash point, °C | 60–80 | 100–170 |
| Cloud point, °C | (−15) to (5) | (−3) to (12) |
| Pour point, °C | (−35) to (−15) | (−15) to (16) |
| Carbon (wt%) | 87 | 77 |
| Hydrogen (wt%) | 13 | 12 |
| BOCLE scuff (g) | 3600 | >7000 |
| Sulfur (wt%) | 0.05 | 0.05 |
| Cetane number | 40–55 | 48–60 |
| HFRR (μm) | 685 | 314 |
| Standard | ASTM D975 | ASTM D6751 |

*HC* hydrocarbons, *FAME* fatty acid methyl esters, *HFRR* high-frequency reciprocating ring, *BOCLE* ball on cylinder lubricity evaluator
Adapted from Lotero et al. (2005)

**Table 19.3** *a, b, c* coefficients for different groups

| Group | Number of compounds | *A* | *B* | *C* | AAE% |
|---|---|---|---|---|---|
| Hydrocarbons | 230 | 225.10 | 537.60 | 2217.00 | 0.410 |
| Alcohols | 150 | 230.80 | 390.50 | 1780.00 | 0.186 |
| Amines | 70 | 222.40 | 416.60 | 1900.00 | 0.2 |
| Acids | 40 | 323.20 | 600.10 | 2970.00 | 0.531 |
| Ethers | 80 | 275.90 | 700.00 | 2879.00 | 0.343 |
| Sulfur | 40 | 238.00 | 577.90 | 2297.00 | 0.484 |
| Esters | 120 | 260.80 | 449.20 | 2217.00 | 0.186 |
| Ketones | 80 | 260.50 | 296.00 | 1908.00 | 0.722 |
| Halogens | 200 | 262.10 | 414.00 | 2154.00 | 0.672 |
| Aldehydes | 45 | 264.50 | 293.00 | 1970.00 | 0.771 |
| Phosphorus | 20 | 201.70 | 416.10 | 1666.00 | 0.815 |
| Nitrogens | 125 | 185.70 | 432.00 | 1645.00 | 0.622 |
| Petroleum fractions | 21 | 237.90 | 334.40 | 1807.00 | 0.294 |

Data taken from Dean (1985), Aldrich Chemical Company, Inc (1990), Weiss (1980)
AAE% is the percentage of average absolute error in predicted flash point

$$T_{\mathrm{f}} = a + \frac{b\left(\frac{c}{T_{\mathrm{b}}}\right)^{2} e^{-\frac{c}{T_{\mathrm{b}}}}}{\left(1 - e^{-\frac{c}{T_{\mathrm{b}}}}\right)^{2}} \tag{19.8}$$

where $T_{\mathrm{f}}$ and $T_{\mathrm{b}}$, are the flash and the normal boiling point in Kelvin, while the parameters *a*, *b*, *c* are given in Table 19.3.

The letters *a*, *b*, *c* refer to the formula for the calculation of the flash point given previously.

Additional methods and more accurate explanation of the procedures utilized are available in literature (Rand 2010); however, some more examples will be given here. For instance, the cetane number (CN) is among the most important parameters for biodiesel assessment, this value is calculated according to Date (2012):

$$\mathrm{CN} = \%\,\mathrm{by\ \ volume\ \ of}\, n\_\mathrm{hexadecane} + 0.15 \cdot (\%\ by\ \ \mathrm{volume\ \ of\,HMN}) \tag{19.9}$$

where HMN indicates the 2,2,4,4,6,8,8-heptamethylnonane.

The CN is a measure of the lowest compression ratio that would produce autoignition.

In the cetane test, we measure the time when the first detectable increase of pressure is taking in place. When this is happening, it means that the fuel is ignited. Figure 19.5 describes the structures of the cetane and the isocetane. In the figure, the white circles represent the hydrogen atoms, while the line crossings have carbon

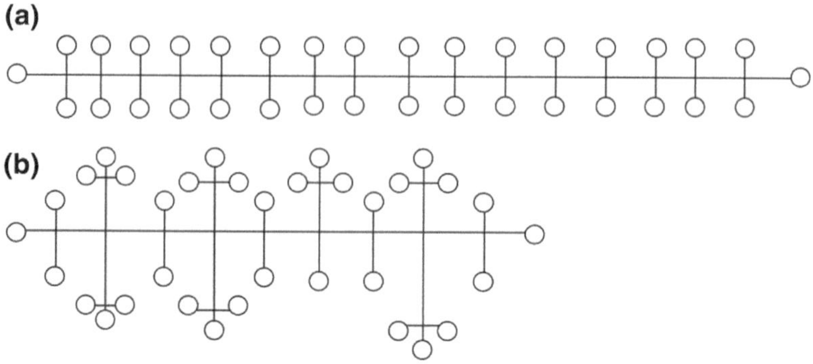


**Fig. 19.5** **a** Cetane (normal hexadecane) C16H34, cetane number = 100. **b** Isocetane C16H34, 2,2,4,4,6,8,8-heptamethylnonane with cetane number = 15

atoms, therefore each of the crossings represents a carbon atom (Mollenhauer and Tschöke 2010) (Fig. 19.5).

In biodiesel, it is also possible to add additives. For instance, a quantity of ethanol can be commonly added to biodiesel. Ethanol has 60% of the energy of diesel, but the octane number is higher. Therefore, the energy amount can be very similar to the one of diesel.

Concerning the definition of octane number: an engine which has the same self-ignition properties of a mixture of 2.2,4-trimethylpentane, the so-called iso-octane, has the octane number equal to the percentage of iso-octane in an iso-octane–heptane mixture.

The heptane formula is: $C_7H_{16}$, while the iso-octane is represented in Fig. 19.6.

**Fig. 19.6** Structure of Iso-octane. Adapted from Morrison and Boyd (1992)

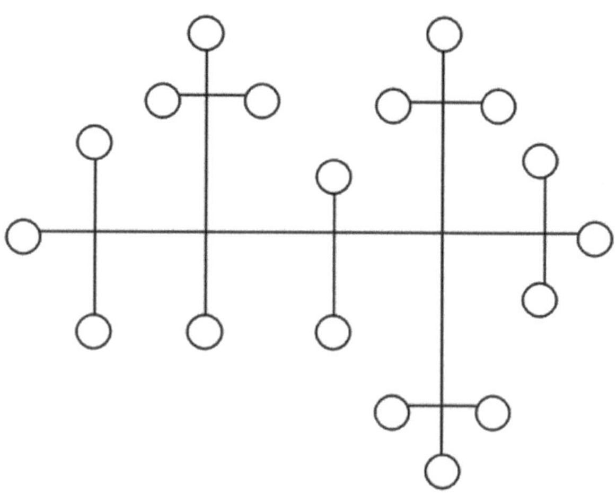

In diesel engines, a high cetane number and low octane number is favorable, this is because the injection of the fuel is done after the compression of the air; therefore, the fuel should ignite soon after an increase of pressure.

On the other hand, in gasoline engines, the fuel is compressed together with the air and when the piston reaches the highest pressure, a spark is released to ignite the fuel and the air mixture. In gasoline engines, we do not want to have ignition before the spark and therefore we want high octane number and low cetane number.

Table 19.4 shows diverse kinds of feedstock with values for their level of unsaturation. As described elsewhere in this manuscript, the iodine number represents the degree of unsaturation, while the cetane number is a measure of the facility of ignition of the diesel. The more the cetane number the less time takes for the fuel to ignite.

In Table 19.4, the MUFA refers to the monounsaturated fatty acids, PUFA stands for the polyunsaturated acids. While the other abbreviations refer to safflower (SAF), grape (GRP), silybum marianum (SIL), hemp (HMP), sunflower (SFL), wheat germ (WHG), pumpkin seed (PMS), sesame (SES), rice bran (RB), almond (ALM), rapeseed (RPS), peanut (PNT), olive (OL), and coconut oil (COC).

Data on the characteristics of different kinds of feedstock and fuels are reported in Table 19.5. Among the reported properties, we have the density, the kinematic viscosity (Kin-VIS), the cetane number (CN), the higher heating value (HHV), the flash point (FP), the carbon residue (CR), the iodine value (IV), and the pour point (PP).

Biodiesel has been also produced by utilizing microorganisms, and there is large amount of research on this topic at the moment. Some microbial-based source of biodiesel is given in Table 19.6.

As additional data, in Table 19.7, some properties of diesel, biodiesel, and different kinds of fuels are given. It is useful to do this kind of comparisons when assessing the suitability of fuels and feedstock. In regard to the table: diesel (D), biodiesel (BD), sunflower oil (SO), rapeseed oil (RO), soy (S), sunflower oil (SFO), gasoline (G), ethanol (E), methanol (M), viscosity (Vis), density (Den), flash point (FP), boiling point (BP), cloud point (CD), pour point (PP), combustion point (ComP), autoignition temperature (AT), distillation (Dis), cetane index (CI), octane number (ONum), lubricity (Lub), and heat of vaporization (HOV).

Among the latest trends in biodiesel production, enzyme-catalyzed enzymatic transesterification is one of the techniques that could be applied as novel technology.

The fatty acid methyl ester (FAME) can be produced from waste choice white grease (CWG) biocatalyzed by immobilized lipase (Adewale et al. 2015). In this case, the enzyme could be, for instance, the Candida Antarctica lipase B (CALB).

The yield of FAME is given by:

$$\text{Yield(FAME, \%)} = \frac{\sum \text{FAME}}{\sum(\text{TG} + \text{DG} + \text{MG} + \text{FFA} + \sum \text{FAME})} \tag{19.10}$$

TG, DG, MG, and FFA are triacylglycerides, diacylglycerides, monoacylglycerides, and free fatty acids, respectively.

**Table 19.4** Diverse kinds of feedstock with values for their level of unsaturation

| FA [%] | SAF | GRP | SIL | HMP | SFL | WHG | PMS | SES | RB | ALM | RPS | PNT | OL | COC |
|---|---|---|---|---|---|---|---|---|---|---|---|---|---|---|
| C6:0 | | | | | | | | | | | | | | 0.52 |
| C8:0 | | 0.01 | | | | | | | | | | | | 7.6 |
| C10:0 | | | | | | | | | | | 0.01 | | | 5.5 |
| C12:0 | | 0.01 | 0.01 | | 0.02 | 0.07 | | | | 0.09 | | | | 47.7 |
| C14:0 | 0.10 | 0.05 | 0.09 | 0.07 | 0.09 | | 0.17 | | 0.39 | 0.07 | | 0.04 | | 19.9 |
| C15:0 | | 0.01 | 0.02 | | | 0.04 | | | | | 0.02 | | | |
| C16:0 | 6.7 | 6.6 | 7.9 | 6.4 | 6.2 | 17.4 | 13.1 | 9.7 | 20.0 | 6.8 | 4.6 | 7.5 | 16.5 | |
| C17:0 | 0.04 | 0.06 | 0.06 | 0.05 | 0.02 | 0.03 | 0.13 | | | 0.05 | 0.04 | 0.07 | | |
| C18:0 | 2.4 | 3.5 | 4.5 | 2.6 | 2.8 | 0.7 | 5.7 | 6.5 | 2.1 | 2.3 | 1.7 | 2.1 | 2.3 | 2.7 |
| C20:0 | | 0.16 | 2.6 | | 0.21 | | 0.47 | 0.63 | | 0.09 | | 1.01 | 0.43 | |
| C22:0 | | | | | | | | 0.14 | | | | | 0.15 | |
| C16:1 ($n-7$) | 0.08 | 0.08 | 0.05 | 0.11 | 0.12 | 0.21 | 0.12 | 0.11 | 0.19 | 0.53 | 0.21 | 0.07 | 1.8 | |
| C17:1 ($n-7$) | | | 0.03 | | | | | | | | | | | |
| C18:1cis ($n-9$) | 11.5 | 14.3 | 20.4 | 11.5 | 28.0 | 12.7 | 24.9 | 41.5 | 42.7 | 67.2 | 63.3 | 71.1 | 66.4 | 6.2 |
| C18:1trans ($n-9$) | | | | | | | | | | | 0.14 | | | |
| C20:1 ($n-9$) | | 0.40 | 0.15 | 16.5 | 0.18 | 7.91 | 1.08 | 0.32 | 1.11 | 0.16 | 9.1 | | 0.30 | |
| C18:2cis ($n-6$) | 79.0 | 74.7 | 63.3 | 59.4 | 62.2 | 59.7 | 54.2 | 40.9 | 33.1 | 22.8 | 19.6 | 18.2 | 16.4 | 1.6 |
| C18:3 ($n-3$) | 0.15 | 0.15 | 0.88 | 0.36 | 0.16 | 1.2 | 0.12 | 0.21 | 0.45 | | 1.2 | | 1.6 | |
| C18:3 ($n-6$) | | | | 3.0 | | | | | | | | | | |
| SFAs | 9.3 | 10.4 | 15.1 | 9.2 | 9.4 | 18.2 | 19.6 | 16.9 | 22.5 | 9.3 | 6.3 | 10.7 | 19.4 | 92.1 |
| MUFAs | 11.6 | 14.8 | 20.7 | 28.1 | 28.3 | 20.9 | 26.1 | 42.0 | 44.0 | 67.9 | 72.8 | 71.1 | 68.2 | 6.2 |
| PUFAs | 79.1 | 74.9 | 64.2 | 62.8 | 62.4 | 61.0 | 54.3 | 41.2 | 33.6 | 22.8 | 20.9 | 18.2 | 18.0 | 1.6 |
| $n-3$ PUFAs | 0.2 | 0.2 | 0.9 | 0.4 | 0.2 | 1.2 | 0.1 | 0.2 | 0.5 | 0.0 | 1.2 | 0.0 | 1.6 | 0.0 |
| $n-6$ PUFAs | 79.0 | 74.7 | 63.3 | 62.4 | 62.2 | 59.7 | 54.2 | 40.9 | 33.1 | 22.8 | 19.6 | 18.2 | 16.4 | 1.6 |

Adapted from Orsavova et al. (2015)

**Table 19.5** Fuel properties of different feedstock kinds and biofuels

| Feed | Density ($kg/m^3$) | Kin-VIS (cS) | CN | HHV (MJ/kg) | FP (°C) | CR (wt%) | IV (gI2/g) | PP (°C) | S (wt%) |
|---|---|---|---|---|---|---|---|---|---|
| Diesel | 820–890 | 1.6–5.9 | >50 | >45 | >61 | <0.15 | | | <1.0 |
| Fuel-oil | 920–990 | <180 | >30 | >43 | >66 | <12 | | <15 | <3 |
| Biodiesel | 860–900 | 3.5–5 | >51 | | 120 | 0.3 | 120 | | 1 |
| Babassu | 946 | 30 | 38 | | 150 | | 16 | | |
| Castor | 955 | 251 | 42 | 37.4 | | | 83–86 | | |
| Coconut | 918 | 27 | 40–42 | 37.1 | | | 8–11 | | 0.01 |
| Corn | 910 | 31–35 | 38 | 39.5 | 277 | 0.24 | 103–128 | −40 | 0.01 |
| Cottonseed | 915 | 34 | 42 | 38.7–39.5 | 234 | 0.24 | 103–115 | −15 to 12 | 0.01 |
| Crambe | 905 | 54 | 45 | 40.5 | 274 | 0.23 | | | |
| Jatropha | 940 | 34 | 39 | 38.8 | 225 | | 82–98 | −15 | 0.01 |
| Linseed | 924 | 26–27 | 35 | 39.3–39.5 | 241 | 0.22 | 180 | | |
| Mahua | 960 | 25 | | 36 | 232 | | 58–70 | | |
| Neem | 919 | 50 | | | | | 65–80 | | |
| Palm | 918 | 40–45 | 42 | 39.5 | 267 | 0.23 | 48–58 | −7 | 0.01 |
| Peanut | 903 | 40 | 42 | 39.8 | 271 | 0.24 | 84–100 | −32 | 0.01 |
| Rapeseed | 912 | 35–37 | 41 | 39.7 | 246 | 0.3 | 105 | −9 | 0.01 |
| Sesame | 913 | 36 | 40–42 | 39.4 | 260 | 0.24 | 103–106 | −12 | 0.01 |
| Soybean | 914 | 29–33 | 38 | 39.6 | 254 | 0.25 | 128–143 | −7 | 0.01 |
| Safflower | 914 | 31 | 41 | 39.5 | 260 | 0.25 | 145 | −15 | |
| Sunflower | 916 | 34–36 | 37 | 39.6 | 274 | 0.27 | 125–140 | | 0.01 |

Adapted from Blin et al. (2013)

**Table 19.6** Examples of microbial-based source of Biodiesel

| Microbial-based oils | Oil content (% dw) | Lauric 14:0 | Palmitic 16:0 | 16:1 | 18:0 | Olein 18:1 | Linoleic 18:2 | Linolenic 18:3 | ArA 20:4 ω − 6 | EpA 20:5 ω − 3 | DhA 22:6 ω − 3 |
|---|---|---|---|---|---|---|---|---|---|---|---|
| Algae | | | | | | | | | | | |
| *Amphidinium carterae* | – | 8.0 | 15.0 | 5.0 | – | 5.0 | 6.0 | 17.0 | – | 4.0 | 2.0 |
| *Botryococcus braunii* | 50 | – | 15.4 | 10.6 | 28.2 | 13.3 | 22.12 | – | – | – | – |
| *Chlorella* spp | 30 | – | 25.0 | 2.0 | 0.9 | 5.0 | 20 | 19.0 | – | – | – |
| *Chlorella pyrenoidosa* | 45 | – | 22.0 | 3.0 | 0.9 | 6.5 | 18 | 27.0 | – | – | – |
| *Chlorella vulgaris* | 38 | – | 26.0 | 2.0 | 0.8 | 16.0 | 24 | 20.0 | – | – | – |
| *Chlorella protothecoides* | 55 | 1.3 | 12.9 | – | 2.7 | 60.8 | 17.3 | – | – | – | – |
| *Crypthecodinium cohnii* | 15 | 17.0 | 17.0 | 1.0 | 3.0 | 10.0 | – | – | – | – | 44.0 |
| *Cylindrotheca fusiformis* | 14 | – | 25.1 | 27.0 | 1.6 | 3.8 | 0.5 | 0.5 | 9.1 | 12 | 0.7 |
| *Dunaliella bardawil* | 8 | – | 1.7 | 1.3 | – | 2.5 | 2.5 | 0.5 | 1.5 | 1.02 | – |
| *Isochrysis galbana* | 29 | 12.0 | 10.0 | 11.0 | 0.7 | 3.0 | 2.0 | – | – | 25.0 | 11.0 |
| *Nannochloropsis* spp | 50 | – | 27.9 | 32.4 | 2.1 | 10.4 | 1.9 | – | 2.1 | 20.1 | 0.5 |

(continued)

**Table 19.6** (continued)

| Microbial-based oils | Oil content (% dw) | Lauric 14:0 | Palmitic 16:0 | 16:1 | 18:0 | Olein 18:1 | Linoleic 18:2 | Linolenic 18:3 | ArA 20:4 $\omega-6$ | EpA 20:5 $\omega-3$ | DhA 22:6 $\omega-3$ |
|---|---|---|---|---|---|---|---|---|---|---|---|
| *Neochloris oleoabundans* | 45 | – | – | – | – | – | – | – | – | – | – |
| *Mortiella alpine* 1S-4 | 56 | 2.4 | 15.0 | – | 2.3 | 10.0 | 7.2 | 4.0 | 40.3 | – | – |
| *Mortiella elongate* | 43 | – | 9.4 | – | 3.5 | 50.9 | 8.2 | 3.5 | 16.5 | – | – |
| *Mucor circinelloides* | 20 | – | 23 | 1 | – | 40 | 11 | 16 | – | – | – |
| *Penicillium spinulosum* | 64 | – | 18 | – | 12 | 12 | 43 | 21 | – | – | – |
| *Pythium irregular* | 43 | 16.8 | 18.6 | 4.1 | 2.8 | 17.3 | 16.0 | 1.2 | 8.2 | 10.5 | – |
| *Rhizopus arrhizus* | 57 | – | – | – | 6 | – | – | – | – | – | – |
| *Yeast* | | | | | | | | | | | |
| *Cryptococcus albidus* | 65 | – | 12.0 | 1.0 | 3 | 73 | 12 | – | – | – | – |
| *Lipomyces starkeyi* | 63 | – | 34 | 6 | 5 | 51 | 3 | – | – | – | – |
| *Rhodotorula glutinis* | 72 | – | 37 | 1 | 3 | 47 | 8 | – | – | – | – |
| *Trichosporon pullulans* | 65 | – | 15 | – | 2 | 57 | 24 | 1 | – | – | – |

Adapted from Drapcho et al. (2008b)

**Table 19.7** Properties of diesel, biodiesel, and different kinds of fuels

| Property | Units | D | BD | G | SO | RO | S | SFO | S | E | M |
|---|---|---|---|---|---|---|---|---|---|---|---|
| Vis (cS at 40 °C) | V (40 °C). (cSt) | 1.9–4.1 | 1.9–6 | 0.8 | 4.2–4.6 | 4.8 | 4.1–4.5 | 4.5 | 4.4 | 1.5 | 0.75 |
| Den (g/ml) | g/mL | 0.85 | 0.88 | 0.72–0.78 | 0.86 | 0.89 | 0.89 | 0.87 | 0.88 | 0.8 | 0.8 |
| FP (°C) | °C | 60–80 | 100–170 | −43 | 164–183 | 153 | 141–188 | 187 | 195 | 13 | 11 |
| BP (°C) | °C | 210–235 | – | 30–225 | – | – | 339 | – | – | 78 | 65 |
| CP (°C) | °C | −19 to 5 | −3 to 12 | – | −3 | −3 | 2 | −1 | 1 | – | – |
| PP (°C) | (°C) | −35 to −15 | – 15 to 16 | – | – 6 | – | – 2 | – 8 | – 4 | – | – |
| ComP (°C) | °C | 92 | – | – | 183 | – | 171 | 192 | – | – | – |
| AT (°C) | °C | 254 | – | 370 | – | – | – | – | – | 423 | 464 |
| Dis (°C) | 50% (°C) | – | – | – | – | – | – | 353 | 344 | – | – |
| CI [–] | [–] | 40–55 | 48–60 | <15 | 49 | 52 | 45–47 | 49.4 | 48.2 | <15 | <15 |
| ONum [–] | [–] | – | – | 82–92 | – | – | – | – | – | 90–102 | 89–102 |
| Lub (HFRR μm) | HFRR (μm) | 685 | 314 | – | – | – | – | – | – | – | – |
| Lub (BOCLE g) | BOCLE (g) | 3600 | >7000 | – | – | – | – | – | – | – | – |
| HHV (MJ/kg) | | 45.2 | – | 43.5 | 40.1 | 40.0 | 38.8 | 38.9 | 40.0 | 27 | 20.1 |
| HOV (kJ/kg) | kJ/kg | 375 | – | 380 | – | – | – | – | – | 920 | 1185 |
| $H_2O$ (%vol) | %vol | 0.05 | 0.05 | – | – | – | – | – | – | – | – |
| *Elemental analysis* | | | | | | | | | | | |
| C (wt%) | wt% | 87 | 77 | – | – | – | – | – | – | – | – |
| H (wt%) | wt% | 13 | 12 | – | – | – | – | – | – | – | – |
| O (wt%) | wt% | 0 | 11 | – | – | – | – | – | | – | – |
| S (wt%) | wt% | 0.05 | 0.05 | – | 0.01 | <0.01 | 0.01 | <0.01 | 0.1 | – | – |

Adapted from Drapcho et al. (2008a)

In enzyme-catalyzed enzymatic transesterification for biodiesel production, the main parameters are:

- the type of biocatalysts used;
- the different solvents used;
- the effect of speed of agitation;
- the biocatalyst amount;
- effects of different alcohols;
- temperature.

The enzymatic transesterification follows the so-called Ping Pong Bi Bi, mechanism (El Rassy et al. 2004). The main reactions in this mechanism are:

$$TG + M \overset{E}{\longleftrightarrow} F + G \tag{19.11}$$

$$E + TG \overset{k1,k2}{\longleftrightarrow} [ETG] \overset{k3}{\rightarrow} E' + F \tag{19.12}$$

$$E' + M \overset{k4,k5}{\longleftrightarrow} \left[E'M\right] \overset{k3}{\rightarrow} E + G \tag{19.13}$$

where $k_1$ and $k_4$ are the forward reaction rate constants, while $k_2$ and $k_5$ are the backward reaction rate constants. Triacylglycerides (TG) and methanol (M) are the substrates, while FAME (F) and glycerol (G) are the products. In this mechanism, the enzyme (E) must bind the first substrate (TG) to form an enzyme–substrate complex [ETG], then release the first product (F), and transform intermediate enzyme (E′). This is followed by binding of a second substrate (M) to E′ and the breakdown of the transitory complex [E′M] to free enzyme (E) and the second product (G).

The production/consumption rates for the main compounds in enzyme transesterification are expressed as:

$$\frac{d(TG)}{dt} = -k_1[TG][E] + k_2[ETG] \tag{19.14}$$

$$\frac{d(M)}{dt} = -k_4[M]\left[E'\right] + k_5\left[E'M\right] \tag{19.15}$$

$$\frac{d(F)}{dt} = k_3[ETG] \tag{19.16}$$

$$\frac{d(G)}{dt} = k_6\left[E'M\right] \tag{19.17}$$

$$\frac{d(E)}{dt} = -k_1[TG][E] + k_2[ETG] + k_6\left[E'M\right] \tag{19.18}$$

$$\frac{d(E')}{dt} = -k_4[M]\left[E'\right] + k_5\left[E'M\right] + k_3[ETG] \tag{19.19}$$

$$\frac{d(ETG)}{dt} = k_1[TG][E] - k_2[ETG] - k_3[ETG] \tag{19.20}$$

$$\frac{d(E'M)}{dt} = k_4[M][E'] - k_5[E'M] - k_6[E'M] \tag{19.21}$$

where (TG) are triacylglycerides, (M) is methanol, (F) is FAME, (G) is glycerol, (E) is enzyme, (TG) is the first substrate, [ETG] is the enzyme–substrate complex, (F) is the first product, (E′) is the intermediate enzyme, (M) is the second substrate, and [E′M] is the transitory complex.

## References


Adewale, P., Dumont, M.-J., & Ngadi, M. (2015). *Enzyme-catalyzed synthesis and kinetics of ultrasonic-assisted methanolysis of waste choice white grease for fatty acid methyl ester production* [WWW Document]. https://doi.org/10.1021/acs.energyfuels.5b00849.

Aldrich Chemical Company, Inc. (1990). *Aldrich catalogue handbook of fine chemicals*. Aldrich Chemical Company, Inc.

Andrews, L. W. (1903). Titrations with potassium iodate. *Journal of the American Chemical Society, 25,* 756–761. https://doi.org/10.1021/ja02009a012.

Aricetti, J. A., & Tubino, M. (2012). A green and simple visual method for the determination of the acid-number of biodiesel. *Fuel, 95,* 659–661. https://doi.org/10.1016/j.fuel.2011.10.058.

Baig, A., & Ng, F. T. T. (2011). Determination of acid number of biodiesel and biodiesel blends. *Journal of the American Oil Chemists Society, 88,* 243–253. https://doi.org/10.1007/s11746-010-1667-x.

Blin, J., Brunschwig, C., Chapuis, A., Changotade, O., Sidibe, S. S., Noumi, E. S., et al. (2013). Characteristics of vegetable oils for use as fuel in stationary diesel engines—Towards specifications for a standard in West Africa. *Renewable and Sustainable Energy Reviews, 22,* 580–597. https://doi.org/10.1016/j.rser.2013.02.018.

Cetinkaya, M., & Karaosmanoglu, F. (2004). Optimisation of base-catalysed transesterification reaction of used cooking oil. *Energy & Fuels, 18,* 1888–1895.

Date, A. W. (2012). *Analytic combustion with thermodynamics, chemical kinetics and mass transfer*. Cambridge: Cambridge University Press. https://doi.org/10.1017/CBO9780511976759.

Dean, J. A. (1985). *Lange's handbook of chemistry* (13th ed.). McGraw-Hill.

Drapcho, C. M., Nghiem, J., & Walker, T. (2008a). *Biofuels engineering process technology*. McGraw-Hill Education.

Drapcho, C. M., Nghim, N. P., & Walker, T. (2008b). *Biofuels, bioproducts and biorefining*. McGraw-Hill.

El Rassy, H., Perrard, A., & Pierre, A. C. (2004). Application of lipase encapsulated in silica aerogels to a transesterification reaction in hydrophobic and hydrophilic solvents: Bi-Bi Ping-Pong kinetics. *Journal of Molecular Catalysis. B, Enzymatic, 30,* 137–150. https://doi.org/10.1016/j.molcatb.2004.03.013.

Encinar, J. M., González, J. F., & Rodríguez-Reinares, A. (2005). Biodiesel from used frying oil. Variables affecting the yields and characteristics of the biodiesel. *Industrial and Engineering Chemistry Research, 44,* 5491–5499. https://doi.org/10.1021/ie040214f.

Georgogianni, K. G., Kontominas, M. G., Tegou, E., Avlonitis, D., & Vergis, V. (2007). Biodiesel production: Reaction and process parameters of alkali-catalysed transesterification of waste frying-oils. *Energy & Fuels, 21,* 3023–3027.

Giakoumis, E. G. (2018). Analysis of 22 vegetable oils' physico-chemical properties and fatty acid composition on a statistical basis, and correlation with the degree of unsaturation. *Renewable Energy, 126,* 403–419. https://doi.org/10.1016/j.renene.2018.03.057.

Lin, C.-Y., & Lin, Y.-W. (2012). Fuel characteristics of biodiesel produced from a high-acid oil from soybean soapstock by supercritical-methanol transesterification. *Energies, 5,* 2370–2380. https://doi.org/10.3390/en5072370.

Lotero, E., Liu, Y., Lopez, D. E., Suwannakarn, K., Bruce, D. A., & Goodwin, J. G. (2005). Synthesis of biodiesel via acid catalysis. *Industrial and Engineering Chemistry Research, 44,* 5353–5363.

Mollenhauer, K., & Tschöke, H. (2010). *Handbook of diesel engines*. Berlin: Springer.

Morrison, R. T., & Boyd, R. N. (1992). *Organic chemistry* (6th ed.). Prentice Hall.

Orsavova, J., Misurcova, L., Vavra Ambrozova, J., Vicha, R., & Mlcek, J. (2015). Fatty acids composition of vegetable oils and its contribution to dietary energy intake and dependence of cardiovascular mortality on dietary intake of fatty acids. *International Journal of Molecular Sciences, 16,* 12871–12890. https://doi.org/10.3390/ijms160612871.

Park, L. K.-E., Liu, J., Yiacoumi, S., Borole, A. P., & Tsouris, C. (2017). Contribution of acidic components to the total acid number (TAN) of bio-oil. *Fuel, 200,* 171–181. https://doi.org/10.1016/j.fuel.2017.03.022.

Patel, N. K., & Shah, S. N. (2015). Biodiesel from plant oils, Chap. 11. In S. Ahuja (Ed.), *Food, energy, and water* (pp. 277–307). Boston: Elsevier. https://doi.org/10.1016/B978-0-12-800211-7.00011-9.

Phan, A. N., & Phan, T. M. (2008). Biodiesel production from waste cooking oils. *Fuel, 87,* 3490–3496. https://doi.org/10.1016/j.fuel.2008.07.008.

Rand, S. (2010). *Significance of tests for petroleum products* (8th ed.). United States: American Society for Testing and Materials, ASTM.

Satyanarayana, K., & Rao, P. G. (1992). Improved equation to estimate flash points of organic compounds. *Journal of Hazardous Materials, 32,* 81–85. https://doi.org/10.1016/0304-3894(92)85106-B.

Weiss, G. (1980). *Hazardous chemicals data book.* Park Ridge, HJ: Noyes Data Corporation.

Yasuda, M. (1931). The determination of the iodine number of lipids. *Journal of Biological Chemistry, 94,* 401–409.

Zumdahl, S. S., & Zumdahl, S. A. (2014). *Chemistry* (9th ed.). Brooks Cole Cengage Learning.

# Chapter 20
# Fischer–Tropsch (FT) Synthesis to Biofuels (BtL Process)

*Nature is like a painting of Monet, very often when you get close to part of nature, it's like walking up to an impressionist painting: you see less and less. All you see is little dots. In order to understand the meaning and the importance and the law contained in physics, you must step back and back to see the all and the perfection.*

Robert B. Laughlin, Nobel Prize in Physics 1998.

## 20.1 Background Information

The FT reaction produces long-chain hydrocarbons from synthetic gas, which is a mixture of $H_2$ and CO, over CO- or Fe-based catalysts (Triantafyllidis et al. 2013)

$$nCO + 2nH_2 \rightarrow (-CH_2-)n + nH_2O \tag{20.1}$$

A short history of this technology is given in (Schulz 1999). In the case of BtL (Biomass to Liquid) technology, the biomass is converted to synthetic gas by the gasification process, a thermochemical conversion technology operating at 500–1200 °C in the presence of a gasifying agent (air, oxygen, steam, $CO_2$, or a mixture of these components) (Triantafyllidis et al. 2013).

A schematic view of BtL technology is given in Fig. 20.1.

A diagram showing the gasification of biomass process is given in Fig. 20.2. In this case, the steam gasification is shown here.

The process shown in Fig. 20.2 demonstrates the gasification of a general biomass feedstock. The feedstock is processed in a bubbling bed reactor at temperatures of 800–900 °C, and the produced char could be sent to a combustion unit, where it will be burned to produce energy. The main reactions taking place during this process are:

$$C + CO_2 \rightarrow 2CO \tag{20.2}$$



C. De Blasio, *Fundamentals of Biofuels Engineering and Technology*, Green Energy and Technology, https://doi.org/10.1007/978-3-030-11599-9_20

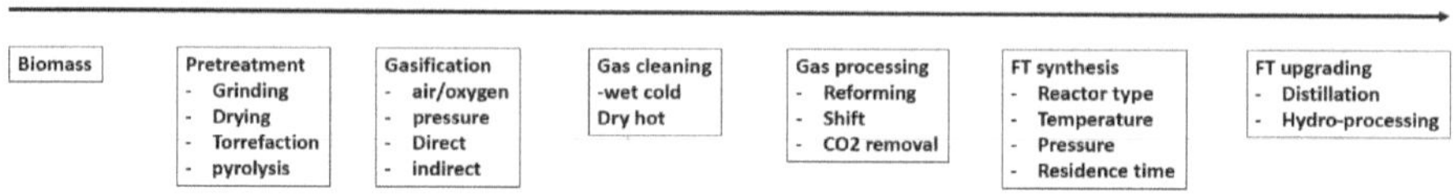


**Fig. 20.1** A schematic view of BtL technology

$$C + H_2O \rightarrow CO + H_2 \tag{20.3}$$

$$CO + H_2O \leftrightarrow CO_2 + H_2 \tag{20.4}$$

$$C + 2H_2 \rightarrow CH_4 \tag{20.5}$$

On the other hand, a GtL plant includes the conversion of natural gas to synthesis gas through steam reforming, dry reforming, partial oxidation (POX), or oxidative steam reforming (autothermal reforming (ATR)).

The steam reforming could be applied to diverse kinds of biomass, for instance, glycerol, with the presence of catalytic agents (Huang et al. 2018). The reaction of steam reforming is:

$$CH_4 + H_2O = CO + 3H_2; \tag{20.6}$$

while the dry reforming is represented by:

$$CO_2 + CH_4 \rightarrow 2H_2 + 2CO \tag{20.7}$$

Partial oxidation (POX) of biomass is an interesting method to derive energy from a feedstock with not necessity to destroy it completely (Özdenkçi et al. 2017):

$$CH_4 + 0.5O_2 = 2H_2 + CO; \tag{20.8}$$

while the autothermal reforming reactions (ATR) are demonstrated as (Triantafyllidis et al. 2013):

$$CH_4 + H_2O \rightarrow CO + 2H_2O, \quad \Delta H_R = 206\ \text{kJ/mol} \tag{20.9}$$

$$CH_4 + \frac{3}{2}O_2 \rightarrow CO + 2H_2O, \quad \Delta H_R = -520\ \text{kJ/mol} \tag{20.10}$$

$$CO + H_2O \rightarrow CO_2 + H_2, \quad \Delta H_R = -41\ \text{kJ/mol} \tag{20.11}$$

Notice that sometimes in literature the reactions are not balanced. The enthalpies are calculated on one mole of methane. A more general representation of the reactions in Fisher–Tropsch synthesis is given in Eqs. (20.12–20.21)

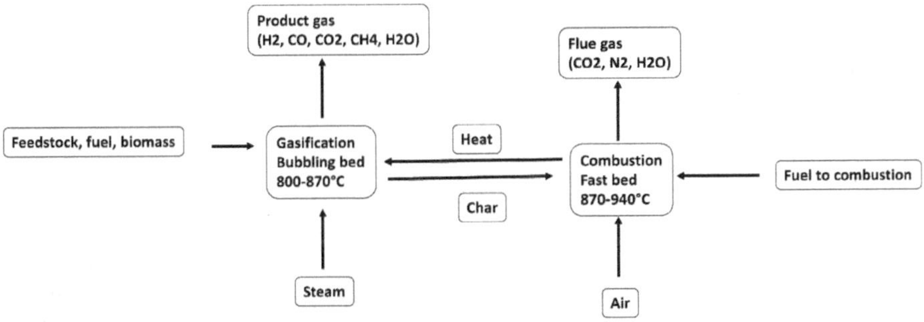


**Fig. 20.2** A diagram showing the steam gasification of biomass process

$$nCO + 2nH_2 \rightarrow (-CH_2-)n + nH_2O \tag{20.12}$$

$$2CO + H_2 \rightarrow (-CH_2-) + CO_2, \quad \Delta H_{300K} = -204 \text{ kJ/mol} \tag{20.13}$$

$$3CO + H_2 \rightarrow (-CH_2-) + 2CO_2, \quad \Delta H_{300K} = -244.5 \text{ kJ/mol} \tag{20.14}$$

$$CO_2 + H_2 \rightarrow (-CH_2-) + H_2O, \quad \Delta H_{300K} = -125.2 \text{ kJ/mol} \tag{20.15}$$

$$CO + H_2O \rightarrow H_2 + CO_2, \quad \Delta H_{300K} = -39.8 \text{ kJ/mol} \tag{20.16}$$

$$\text{Alkanes} : nCO + (2n+1)H_2 \rightarrow C_nH_{2n+2} + nH_2O \tag{20.17}$$

$$\text{Alkenes} : nCO + (2n)H_2 \rightarrow C_nH_{2n} + nH_2O \tag{20.18}$$

$$\text{Alcohols} : nCO + 2nH_2 \rightarrow C_nH_{2n+2}O + (n-1)H_2O \tag{20.19}$$

$$\text{Aldehydes} - \text{ketones} : nCO + (2n-1)H_2 \rightarrow C_nH_{2n}O + (n-1)H_2O \tag{20.20}$$

$$\text{Carboxylic acids} : nCO + (2n-2)H_2 \rightarrow C_nH_{2n}O_2 + (n-2)H_2O \tag{20.21}$$

Concerning the catalyst type and $H_2/CO$ ratio in FT synthesis, a Fe-based catalyst has higher Water-Gas Shift activity in general (Ma et al. 2015) and also compared to a Co-based catalyst increasing the $H_2/CO$ ratio. In the case of Cobalt-based catalysis, the products might be shifted from long-chain hydrocarbons to mostly methane (Riedel et al. 1999). This ratio has a significant impact on CO conversion, $C_{5+}$ selectivity, product distribution, and chain growth probability. A low inlet $H_2/CO$ ratio leads to selectivity to longer hydrocarbons (increase of $C_{5+}$ selectivity) and a drop in the hydrocarbon formation rate due to the decrease in productivity of the catalyst. A reduction of the hydrogen in the feed leads to a lower selectivity to $CH_4$, but an increase in the olefin/paraffin ratio; while CO conversion increases at high $H_2/CO$.

Activities of different metal catalysts on different supports and reaction rates are given in literature (Demitras and Muetterties 1977):

$$\text{Ru} > \text{Fe} > \text{Ni} > \text{Co} > \text{Rh on Alumina} \tag{20.22}$$

$$\text{Co} > \text{Fe} > \text{Ru} > \text{Ni} > \text{Rh on silica} \tag{20.23}$$

We have that, on alumina, the Ru has higher activity than Fe and so on. The kind of support used for the catalyst metal influences the activity of the metal itself. In this way, we have that the selectivity of the $C_{5+}$ hydrocarbons is higher going from $Co/TiO_2$ to Co/MgO as follows (Jahangiri et al. 2014):

$$Co/TiO_2 > Co/SiO_2 > Co/Al_2O_3 > Co/C > Co/MgO \tag{20.24}$$

In conventional FT processes, four types of reactors are used (Triantafyllidis et al. 2013):

- fixed bed multitubular reactors;
- fluidized bed reactors;
- slurry-bed reactors;
- microchannel reactors.

However, we can find also the circulating fluidized bed reactor among the suitable options (Steynberg et al. 2004, p. 2).

As reported by the same references, fluidized bed reactors have:

- superior quality in heat transfer and the temperature control could be performed in a better way during highly exothermic FT reactions;
- there is the possibility of using smaller catalyst particles, in this way, we can avoid intra-particle diffusion. This phenomenon can limit the reaction rate and the pressure drop;
- as also commonly referred to fluidized bed reactors, we have better mixing of the catalyst particles due to fluidization and a higher gas–solid mass transfer efficiency;
- the catalyst in this case can be replaced in a shorter time and there is the possibility of loading fresh catalyst during the run;
- high production capacity due to the possibility of using more gas flows.

As an example of slurry-bed reactor, the SASOL slurry-bed reactor is demonstrated in Fig. 20.3.

An important drawback of the slurry reactor is the separation of the catalyst from the waxes. In Table 20.1, the main reactors parameters in FT synthesis are given while in Table 20.2 also the $H_2/CO$ ratio is reported. The original references are given within the two tables.

Some effects of the main parameters on FT synthesis are described as: (Triantafyllidis et al. 2013):

**Table 20.1** Main reactor parameters in FT synthesis

| Catalyst | Reactor | Reducing agent | Temperature (°C) | Temperature rise (°C/min) | Pressure (bar) | Space velocity ($l \cdot g^{-1} h^{-1}$ | Hours (h) |
|---|---|---|---|---|---|---|---|
| Fe/Cu/k$SiO_2$ (Hao et al. 2009) | Stirred, Slurry | Syngas $H_2$/CO = 0.67 | 270 | 0.17 | 10–25 | 1 | 13 |
| Fe/Cu/La/$SiO_2$ (Pour et al. 2008) | Fixed bed | 5% $H_2$/$N_2$ gas mixture | 400 | 5 | Atm | 15.1 | 1 |
| Fe–Mn ultrafine catalyst (Bai et al. 2002) | Stirred, slurry | Syngas $H_2$/CO = 2 | 275 | – | 35 | 1 | 32 |
| Fe–Mn catalyst (Ji et al. 2001) | Stirred, slurry | Syngas $H_2$/CO = 1 | 280 | – | 15–20 | 0.23 | 48 |
| FeCrAlY foam (Chin et al. 2005) | Microchannel | 5% $H_2$/He | 350 | – | Atm | – | 12 |
| Fe–Mn catalyst (Liu et al. 2007) | Fixed bed | Syngas $H_2$/CO = 2 | 400 | – | 25–30 | 1 | 32 |
| Cu–Ru/$Al_2O_3$ (Sari et al. 2009) | Stirred tank, slurry | $H_2$ | 400 | 2 | Atm | 1.2 | 24 |
| Co–Ru/$Al_2O_3$ (Tavasoli et al. 2010) | Fixed bed | $H_2$ | 400 | 1 | Atm | – | 12 |
| Co–Re/$Al_2O_3$ (Borg et al. 2007) | Fixed bed | $H_2$ | 350 | 1 | 1 | – | 16 |

(continued)

**Table 20.1** (continued)

| Catalyst | Reactor | Reducing agent | Temperature (°C) | Temperature rise (°C/min) | Pressure (bar) | Space velocity $(l \cdot g^{-1} h^{-1}$ | Hours (h) |
|---|---|---|---|---|---|---|---|
| Co–Re/$Al_2O_3$ (Oukaci et al. 1999) | Fixed bed | $H_2$ | 250-350 | – | Atm | – | 8 |
| Co–Re/$Al_2O_3$ (Das et al. 2003) | Fixed bed | $H_2/He = 0.5$ | Step 1: 100<br>Step 2: 350 | Step 1: 2<br>Step 2: 1 | Atm | – | Step 1: 1<br>Step 2: 10 |
| Co–Pt/$Al_2O_3$ (Xu et al. 2005) | Fixed bed | $H_2$ | 350 | – | Atm | – | 8 |

**Table 20.2** Main reactor parameters in FT synthesis with conversion and $H_2/CO$ ratio

| Catalyst | Reactor | Temperature (°C) | Pressure (bar) | $H_2/CO$ | Space velocity ($l \cdot g^{-1}h^{-1}$ | CO conversion (%) |
|---|---|---|---|---|---|---|
| Fe–Mn ultrafine catalyst (Bai et al. 2002) | Stirred tank, slurry | 260–300 | 15–31 | 0.65–2 | 1–2.5 | 91–95 |
| Fe–Mn catalyst (Ji et al. 2001) | Stirred tank, slurry | 260–290 | 9.3–25.3 | 0.8–2.5 | 1–6.6 | 60–80 |
| Fe–Mn catalyst (Liu et al. 2007) | Fixed bed | 280–340 | 22.5 | 1.01–2.74 | 0.8–5.6 | 64–88 |
| Fe–Mn–K/$SiO_2$ (Zhang et al. 2006) | Slurry reactor | 250 | 15 | 1.35–1.5 | – | 59–74 |
| Fe–Mn–Cu–K/$SiO_2$ (Zhang et al. 2006) | Slurry reactor | 250 | 15 | 1.35–1.5 | – | 62–76 |
| Co–Ru/$Al_2O_3$ (Dalai and Davis 2008) | Stirred tank, slurry | 210–240 | 20–35 | 1–2.5 | 0.5–1.5 | |
| Co–Ru/$Al_2O_3$ (de Klerk and Furimsky 2010) | Fixed bed | 210–240 | 25 | 0.5–2 | 0.448 | |
| Co–Re/$Al_2O_3$ (Borg et al. 2007) | Fixed bed | 210 | 20 | 2.1 | – | 40–50 |
| Co–Re/$Al_2O_3$ (Das et al. 2003) | Fixed bed | 210 | 19.7 | 2 | 2 | 70–75 |
| Co–Pt/$Al_2O_3$ (Xu et al. 2005) | Slurry | 230 | 20 | 2 | – | 61.3 |

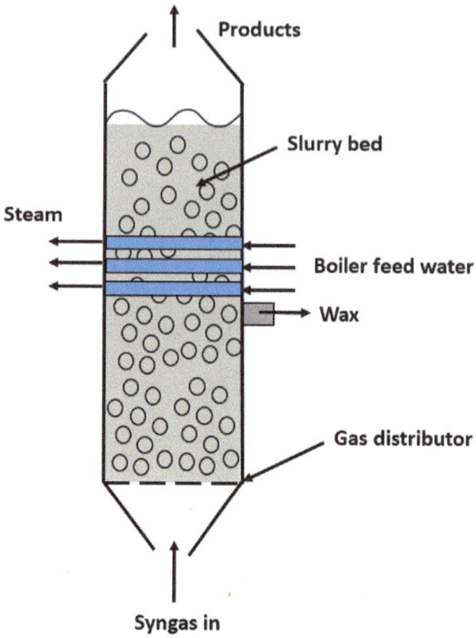


**Fig. 20.3** SASOL slurry-bed reactor. Adapted from Steynberg et al. (2004, p. 2)

- higher temperature increases CO conversion, $CH_4$ and $C_2$–$C_4$ selectivity. The selectivity for olefins and oxygenates is also increased. There is, however, a decrease in the selectivity toward carbon chains with more than five carbon atoms.
- On the other hand, the increase in pressure increases the CO conversion and the selectivity for chains with more than five carbon atoms. This means that also the chain growth probability, α will be increased. More explanations on the meaning for the chain growth probability will be given later at the end of the chapter.
- Increasing the $H_2$/CO ratio results in a higher CO conversion. It can be noticed also an increase in the selectivity of alkenes, and a decrease in the selectivity for carbon chains with more than five carbon atoms due to the enhancement of $H_2$ species.
- A higher space velocity (1/residence time) results in a sharp decrease in CO conversion, due to a decrease in the residence time of the reactants and products.

While some of the reaction kinetics proposed in literature are given in Table 20.3.

More details are given in Table 20.4 where the kind of catalyst, reactor type, temperature, and pressure are given together with the $H_2$/CO ratio.

The intrinsic kinetic expressions presented by the references given in Table 20.4 are, respectively, as follows:

$$-R_{H_2+CO} = \frac{aP_{H_2}^2 P_{CO}}{\left(1 + bP_{H_2}^2 P_{CO}\right)} \tag{20.25}$$

$$-R_{H_2+CO} = aP_{H_2} P_{CO}^{-0.5} \tag{20.26}$$

**Table 20.3** Intrinsic kinetic expressions for the reaction between carbon monoxide and hydrogen

| Intrinsic kinetic expression | Catalyst used | |
|---|---|---|
| $-R_{H_2+CO} = \frac{ap_{H_2}^2}{p_{CO}}$ | $Co/MgO/ThO_2$/Diatomite | Brötz (1949) |
| $-R_{H_2+CO} = \frac{ap_{H_2}^{0.55}}{p_{CO}^{0.33}}$ | $Co/La_2O_3/Al_2O_3$ | Pannell et al. (1981) |
| $-R_{H_2+CO} = \frac{ap_{H_2}^2 p_{CO}}{\left(1+bp_{H_2}^2 p_{CO}\right)}$ | $Co/ThO_2$/Diatomite | Anderson (1956a) |
| $-R_{H_2+CO} = \frac{ap_{H_2}}{p_{CO}^{0.5}}$ | $Co/CuO/Al_2O_3$ | Yang et al. (1979) |
| $-R_{H_2+CO} = \frac{ap_{H_2} p_{CO}}{p_{CO}+bp_{H_2O}}$ | Fused Fe/K | Atwood and Bennett (1979) |
| $-R_{CO} = \frac{ap_{H_2} p_{CO}^{0.5}}{(1+bp_{CO}^{0.5})^3}$ | $Co/Al_2O_3$ | Outi et al. (1981) |
| $-R_{CO} = \frac{ap_{H_2}^{0.5} p_{CO}^{0.5}}{\left(1+bp_{CO}^{0.5}+cp_{H_2}^{0.5}+dp_{CO}\right)^2}$ | Co/Diatomite | Sarup and Wojciechowski (1989) |
| $-R_{CO} = \frac{ap_{H_2}^{0.5} p_{CO}}{\left(1+bp_{CO}+cp_{H_2}^{0.5}\right)^2}$ | Co/Diatomite | Wojciechowski (1988) |
| $-R_{CO} = \frac{ap_{H_2} p_{CO}}{(1+bp_{CO})^2}$ | $Co/MgO/SiO_2$ | Chanenchuk et al. (1991) |
| $-R_{H_2+CO} = \frac{ap_{H_2}^2 p_{CO}}{p_{CO} p_{H_2}+bp_{H_2O}}$ | Precipitated Fe | Deckwer et al. (1986) |

**Table 20.4** Adapted from Yates and Satterfield (1991)

| | Catalyst | Reactor type | T (°C) | P (MPa) | $H_2/CO$ |
|---|---|---|---|---|---|
| Anderson (1956b) | $Co/ThO_2$/Diatomaceous earth | Fixed bed | 186–207 | 0.1 | 0.9–3.5 |
| Yang et al. (1979) | $Co/CuO/Al_2O_3$ | Fixed bed | 235–270 | 0.17–5.5 | 1.0–3.0 |
| Outi et al. (1981) | $Co/Al_2O_3$ | Fixed bed | 250 | 0.1 | 0.2–4.0 |
| Sarup and Woj-ciechowski (1989) | Co/Diatomaceous earth | Berty | 190 | 0.2–1.5 | 0.5–8.3 |

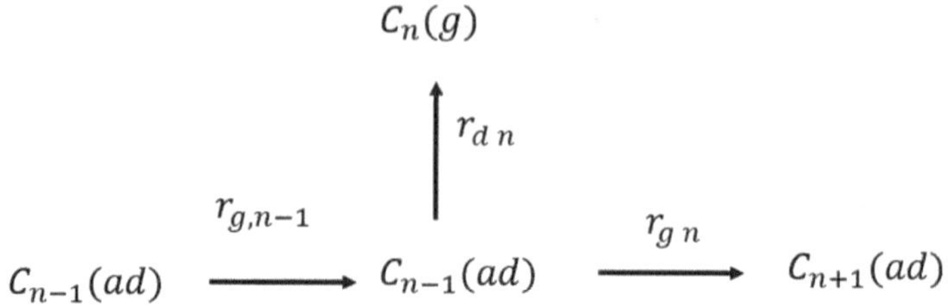


**Fig. 20.4** Chain growth for compounds with *n* size on the catalyst. Adapted from Cheng et al. (2008). The (ad) means adsorbed on the catalyst, while (g) refers to the release of compounds in the gas phase

$$-R_{CO} = a P_{H_2} P_{CO}^{0.5} / \left(1 + b P_{CO}^{0.5}\right)^3 \tag{20.27}$$

$$-R_{CO} = a P_{CO}^{0.5} P_{H_2}^{0.5} / \left(1 + b P_{CO}^{0.5} + c P_{H_2}^{0.5} + d P_{CO}\right)^2 \tag{20.28}$$

## 20.2 Chain Growth Probability (Products Distribution) for the FT Synthesis

The chain growth probability, α, related to an element with *n* carbon number, is defined as:

$$\alpha_n = \frac{\sum_{i=n+1}^{\infty} m_i}{\sum_{i=n}^{\infty} m_i}; \tag{20.29}$$

where $m_n$ is the molar fraction of the component with *n* carbon atoms. The chain growth probability can be also written as (Cheng et al. 2008):

$$\alpha_n = \frac{r_{g,n}}{r_{g,n} + r_{d,n}}; \tag{20.30}$$

where *n* is the chain length, $\boldsymbol{r}_{g,n}$ is the chain growth rate and $\boldsymbol{r}_{d,n}$ is the chain termination rate. These reaction rates refer to Fig. 20.4:

Taking into consideration the chain growth probability, the Anderson–Schulz–Flory (ASF) distribution is given by:

$$M_n = M_1 \cdot \alpha^{n-1}; \tag{20.31}$$

where $M_n$ refers to mole fraction and $M_1$ is a constant. Making the assumption that α is not dependent on hydrocarbon chain length, an equation may be derived in the following equation (Tavakoli et al. 2008),

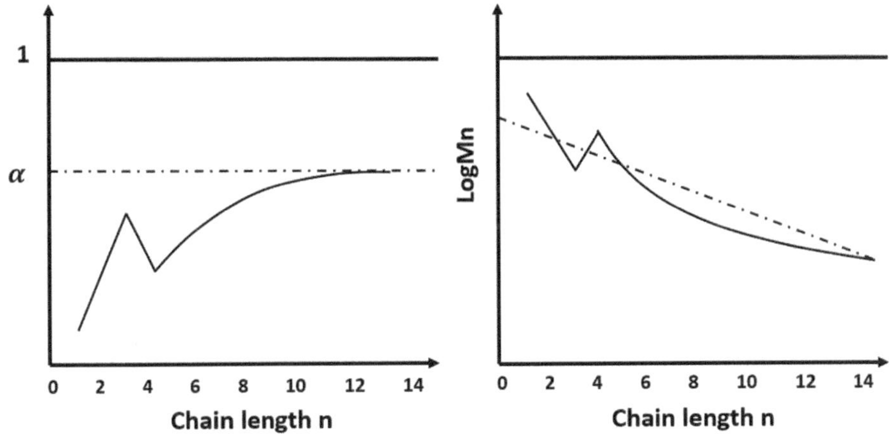


**Fig. 20.5** The α value and product molar fraction against the chain length. The dashed lines represent the ideal ASF distribution while the solid curves are the deviated ASF distribution values. Adapted from Cheng et al. (2008)

$$\log\left(\frac{W_n}{n}\right) = n \log \alpha + \text{const} \tag{20.32}$$

where $W_n$ is the weight fraction for the component with $n$ carbons. Notice that in the simplified ASF distribution the chain growth probability is constant, however, in reality this value is a function of the chain length. The linearized version of the distribution can provide us with a straight line where the chain growth probability can be derived. Figure 20.5 demonstrated that the values of $\alpha$ and $M_n$ indeed are different from the ideal case.

The mass fraction for different components as a function of their chain growth probability is reported in Fig. 20.6 as a qualitative estimation. It is possible to notice that the chain growth probability is just the distribution function for each of these compounds.

The main process parameters such as temperature, pressure, and $H_2$/CO ratio have an effect on the product distribution for the different components, the olefin/paraffin ratio, carbon deposition, and methane selectivity. These are positively or negatively affected by the process parameters. This is demonstrated in Table 20.5.

More information on the Sasol process and how the FT synthesis is performed at low temperatures is given in (Espinoza et al. 1999), while the commercial process development is described in (Geerlings et al. 1999). On the other hand, the high-temperature FT synthesis is described in (Steynberg et al. 1999). Information on the selection and advances in reactor design for FT synthesis is given in (Sie and Krishna 1999) while more information on the reactions' kinetics and product distributions are given in (Patzlaff et al. 1999; Schulz and Claeys 1999). Some patented catalysts used in FT synthesis are compared in bubble bed column reactors by (Oukaci et al. 1999).

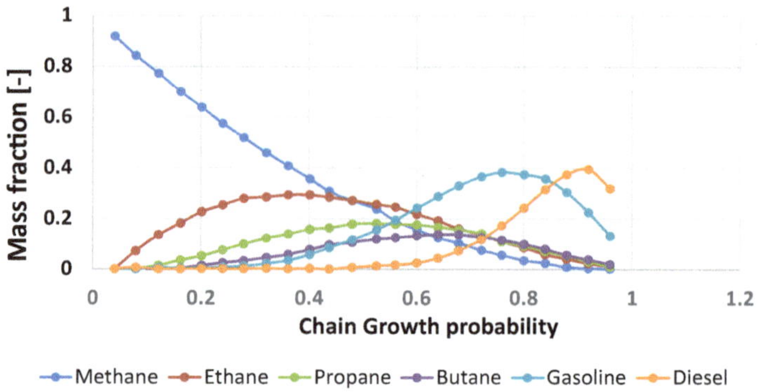


**Fig. 20.6** Distribution functions for diverse components and their mass fraction. Qualitative estimation

**Table 20.5** Effect of FT process parameters on the chain growth probability, olefin/paraffin ratio, carbon deposition, and methane selectivity

| Parameter | Chain growth probability | Olefin/paraffin ratio | Carbon deposition | Methane selectivity |
|---|---|---|---|---|
| Temperature increase | Down | Down | Up | Up |
| Pressure increase | Up | Complex | Complex | Down |
| $H_2/CO$ ratio increase | Down | Down | Down | Up |
| Conversion increase | Complex | Down | Up | Up |
| Space velocity increase | Complex | Up | Complex | Down |

Adapted from De Deugd (2004)

The catalyst preparation is also a very important part of the development of this technology, and some details are given in (Ernst et al. 1999). The FT synthesis not necessarily should be performed by using CO and $H_2$ only, but it can be performed also by using $CO_2$ and $H_2$ syngas (Riedel et al. 1999); in this case, the catalysts are made on the basis of Iron and Cobalt.

## 20.3 Syngas Treatment for FT Synthesis

In the FT synthesis, the used syngas is usually treated to obtain the necessary purity and the minimal requirements to carry on the process. The wet-cold gas cleaning proceeds according to the scheme presented in Fig. 20.7.

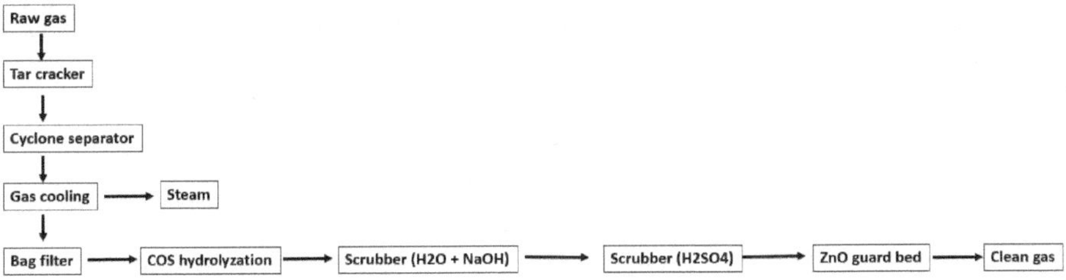


**Fig. 20.7** Wet-cold gas cleaning process. Adapted from Triantafyllidis et al. (2013)

The so-called acid scrubbing is also utilized. In this process, the main components used are: HCl, $H_2S$, HF, HCN, HBr, $NH_3$, and the main reactions taking place for each of these constituents are given as follows:

$$HCl + NaOH \rightarrow NaCl + H_2O \tag{20.33}$$

$$HCN + NaOH \rightarrow NaCN + H_2O \tag{20.34}$$

$$2NaOH + Cl_2 \rightarrow NaOCl + NaCl + H_2O \tag{20.35}$$

$$NaCN + (2NaOH + Cl_2) \rightarrow NaCNO + 2NaCl + H_2O \tag{20.36}$$

$$H_2S + 2NaOH \rightarrow Na_2S + 2H_2O \tag{20.37}$$

It should be noticed that the compound NaCNO is less toxic than NaCN and can be removed easily.

When Halogen vapors ($Cl_2$, $F_2$, $Br_2$) are involved, the following reactions are taking place:

$$Cl_2 + 2XOH \rightarrow XCl + XOCl + H_2O \quad (\text{at} < 20\ ^\circ\text{C}) \tag{20.38}$$

$$3Cl_2 + 6XOH \rightarrow 5XCl + XClO_3 + 3H_2O \quad (\text{at} \cong 75\ ^\circ\text{C}) \tag{20.39}$$

In the previous Eq. (20.38 and 20.39), the term $X$ can be Na or K.

The scrubbing of $NH_3$ is usually performed by utilizing sulfuric acid according to:

$$2NH_3 + H_2SO_4 \rightarrow (NH_4)_2SO_4 \tag{20.40}$$

while a $CO_2$ Scrubber involves the following reactions:

$$2NaOH(aq) + CO_2(g) \rightarrow Na_2CO_3(aq) + H_2O(l) \tag{20.41}$$

**Table 20.6** Main reactions involved in a wet flue gas desulfurization (WFGD) process

| Rate-determining steps | Reactions |
|---|---|
| Absorption of gaseous $SO_2$ in liquid water | $SO_2 + H_2O \rightleftarrows H^+ + HSO_3^-$ <br> $HSO_3^- \rightleftarrows H^+ + SO_3^{2-}$ |
| Oxidation of $HSO_3^-$ (liquid phase) | $HSO_3^- + \frac{1}{2}O_2 \rightleftarrows H^+ + SO_4^{2-}$ |
| Solid limestone is dissolving in acidic environment (pH 5.5, Industrial process) | $HSO_4^{2-} \rightleftarrows SO_4^{2-} + H^+$ <br> $CaCO_3 \rightleftarrows Ca^{2+} + CO_3^{2-}$ <br> $CO_2 + H_2O \rightleftarrows HCO_3^- + H^+$ <br> $HCO_3^- \rightleftarrows CO_3^{2-} + H^+$ <br> $H_2O \rightleftarrows H^+ + OH^-$ |
| Crystallization of gypsum | $Ca^{2+} + SO_4^{2-} + 2H_2O \rightleftarrows CaSO_4 \cdot 2H_2O$ |

Adapted from De Blasio et al. (2013)

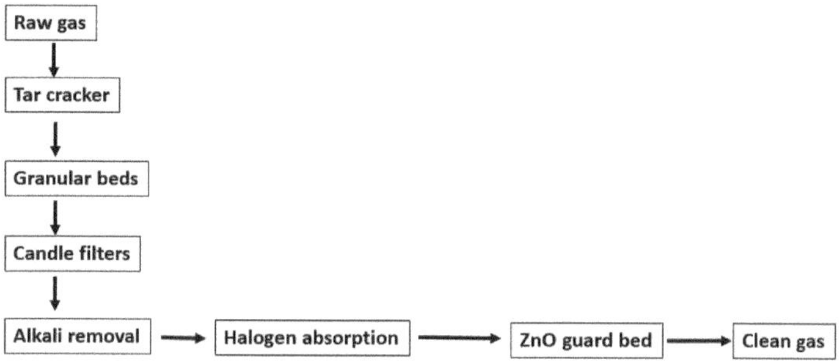


**Fig. 20.8** Dry-hot gas cleaning process for removing tar components from syngas

$$Na_2CO_3(aq) + Ca(OH)_2(s) \rightarrow 2NaOH(aq) + CaCO_3(s) \quad \Delta H^\circ = -5.3\ \text{kJ/mol} \tag{20.42}$$

Concerning the separation of $SO_2$ from the gaseous stream, wet scrubbing is usually applied. The main reactions involved in a wet flue gas desulfurization (WFGD) process are given in Table 20.6.

While Table 20.7 gives the gas cleaning requirements for the FT synthesis on the basis of diverse kinds of contaminants and cleaning methods.

The dry-hot gas cleaning process is described in Fig. 20.8; this process is commonly applied to tar and particulate removal from syngas produced, for example, from gasification of biomass (Zwart et al. 2010).

Two well-known tar cracking catalysts are naturally occurring minerals: dolomite and olivine (Soomro et al. 2018). The first method is with catalyst mixed with the feed biomass in so-called catalytic gasification or pyrolysis (in situ). In this case, tar is removed in the gasifier itself (usually in a fluidized bed gasifier). In the second

**Table 20.7** FT synthesis gas cleaning requirements for diverse contaminant and cleaning methods

| Contaminant | Poplar (wt%) | FT cleaning requirements (ppb) | Cleaning efficiency required (%) | Wet-cold cleaning | Dry-hot cleaning |
|---|---|---|---|---|---|
| Particulate | 1.33 | 0 | >99.9 | Cyclone<br>Bag filter<br>Scrubber | Granular bed<br>Metallic filter<br>Ceramic candle filter<br>Cyclone |
| $HCN + NH_3$ | 0.47 | 20 | >99.9 | Scrubber (H2SO4)<br>Chemical and physical acid removal (rectisol, Selexol, MDEA) | ZnO, CuO guard bed |
| $H_2S + COS$ | 0.01 | 10 | >99.9 | Scrubber<br>ZnO, CuO guard bed<br>Activated charcoal<br>COS hydrolyzation | Safeguard filter<br>ZnO, CuO guard bed<br>Sorption bed of $\alpha$−Fe |
| Alkali | 0.1 | 10 | >99.9 | Condensation on particles made by cooling | Adsorption and chemisorption |
| HCl | 0.1 | 10 | >99.9 | Absorption by dolomite (in tar cracking)<br>Bag filter (reaction with particulates)<br>Scrubber (NaOH) | Dry fly ash |
| Tars | – | 0 | >99.9 | Scrubber (organic oil-RME) | Catalytic, thermal reforming |

Adapted from Triantafyllidis et al. (2013)

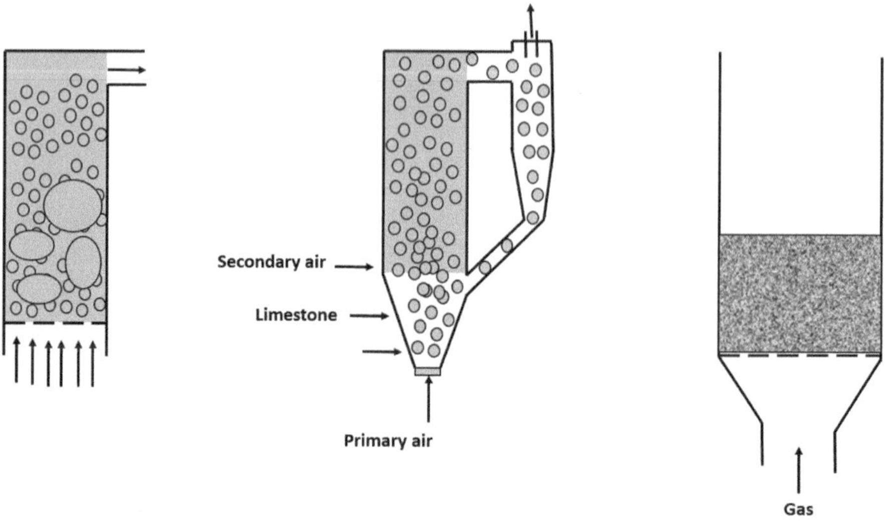


**Fig. 20.9** Bubbling fluidized bed (BFB) (left), circulating fluidized bed (CFB) (center), fixed bed reactor (right). Adapted from Kunii and Levenspiel (1991)

method, tar is treated downstream of the gasifier in a secondary reactor, outside of the gasifier (fixed bed catalytic reactor).

Figure 20.9 describes some of the available tar cracking solutions. Reactors are mainly bubbling fluidized bed (BFB), circulating fluidized bed (CFB), and fixed bed reactors.

These are configurations used as well to perform the gasification of biomass. It is today always more common to use integrated gasification combined cycle (IGCC) power plant and in these configurations, the air separation unit (ASU) is one of the most important operations which improves the efficiency of the combined cycle itself (Jones et al. 2011).

A simple representation of a IGCC plant is reported in Fig. 20.10.

Gasification of biomass can be coupled with processes aimed at the production of biofuels via the FT synthesis. The integration can also involve different feedstock kinds, for instance, black liquor, denominated as BL in Fig. 20.11. To be noticed that, in the case gasification is applied, it is required to dry the fed biomass before gasifying it. This is a very energy-demanding process, if we consider that water has a quite large heat capacity and a latent heat of vaporization which is reported as 2260 kJ/kg or 40 kJ/mol (Datt 2011).

In the next section, some notions will be given on the gasification process with a focus for the supercritical water gasification process.

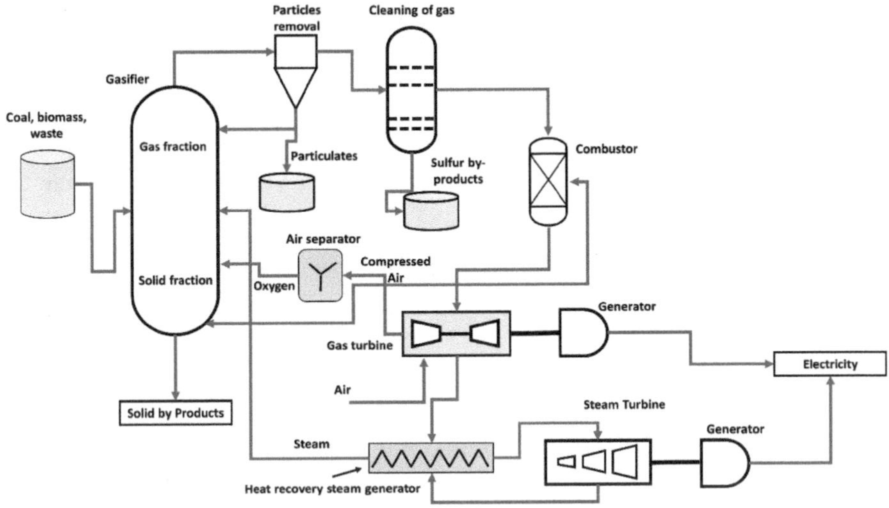


**Fig. 20.10** Integrated gasification combined cycle (IGCC) power plant. Adapted from Shadle and Breault (2012)

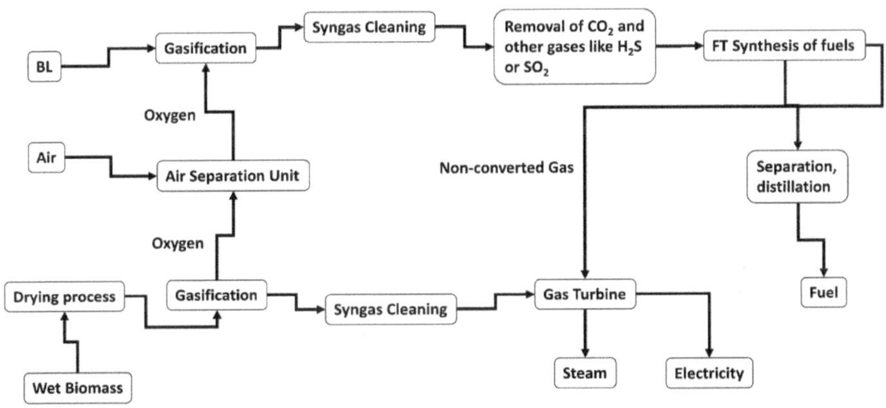


**Fig. 20.11** Integration of gasification with FT synthesis and production of power

# References

Anderson, R. B. (1956a). *Catalysis Vol IV: Hydrocarbon synthesis, hydrogenation and Cyclization.* New York, USA: Reinhold Publishing Corporation.

Anderson, R. B. (1956b). *Catalysis Vol IV: Hydrocarbon synthesis, hydrogenation and cyclization.* New York, USA: Reinhold Publishing Corporation.

Atwood, H. E., & Bennett, C. O. (1979). Kinetics of the Fischer-Tropsch reaction over iron. *Industrial and Engineering Chemistry Product Research and Development, 18,* 163–170.

Bai, L., Xiang, H.-W., Li, Y.-W., Han, Y.-Z., & Zhong, B. (2002). Slurry phase Fischer–Tropsch synthesis over manganese-promoted iron ultrafine particle catalyst. *Fuel, 7th China-Japan Symposium on C1 Chemistry, 81*, 1577–1581. https://doi.org/10.1016/S0016-2361(02)00089-3.

Borg, Ø., Eri, S., Blekkan, E. A., Storsæter, S., Wigum, H., Rytter, E., et al. (2007). Fischer-Tropsch synthesis over γ-alumina-supported cobalt catalysts: Effect of support variables. *Journal of Catalysis, 248*, 89–100. https://doi.org/10.1016/j.jcat.2007.03.008.

Brötz, W. (1949). Zur Systematik der Fischer-Tropsch catalyze. *Zeitscrift für Elektrochemie, 53*, 301.

Chanenchuk, C. A., Yates, I. C., & Satterfield, C. N. (1991). The Fischer-Tropsch synthesis with a mechanical mixture of a cobalt catalyst and a copper-based water gas shift catalyst. *Energy & Fuels, 5*, 847–855.

Cheng, J., Hu, P., Ellis, P., French, S., Kelly, G., & Lok, C. M. (2008). A DFT study of the chain growth probability in Fischer-Tropsch synthesis. *Journal of Catalysis, 257*, 221–228. https://doi.org/10.1016/j.jcat.2008.05.006.

Chin, Y., Hu, J., Cao, C., Gao, Y., & Wang, Y. (2005). Preparation of a novel structured catalyst based on aligned carbon nanotube arrays for a microchannel Fischer-Tropsch synthesis reactor. *Catalyst Today, Catalytic Microstructured Reactors Catalytic Microstructured Reactors, 110*, 47–52. https://doi.org/10.1016/j.cattod.2005.09.007.

Dalai, A. K., & Davis, B. H. (2008). Fischer-Tropsch synthesis: A review of water effects on the performances of unsupported and supported Co catalysts. *Applied Catalysis General, 348*, 1–15. https://doi.org/10.1016/j.apcata.2008.06.021.

Das, T. K., Jacobs, G., Patterson, P. M., Conner, W. A., Li, J., & Davis, B. H. (2003). Fischer-Tropsch synthesis: characterization and catalytic properties of rhenium promoted cobalt alumina catalysts☆. *Fuel, 82*, 805–815. https://doi.org/10.1016/S0016-2361(02)00361-7.

Datt, P. (2011). Latent heat of vaporization/condensation. In V. P. Singh, P. Singh & U. K. Haritashya (Eds.), *Encyclopedia of snow, ice and glaciers* (pp. 703–703). Springer Netherlands, Dordrecht. https://doi.org/10.1007/978-90-481-2642-2_327.

De Blasio, C., Carletti, C., Westerlund, T., & Järvinen, M. (2013). On modeling the dissolution of sedimentary rocks in acidic environments. An overview of selected mathematical methods with presentation of a case study. *Journal of Mathematical Chemistry, 51*(8), 2120–2143. https://doi.org/10.1007/s10910-013-0202-3.

De Deugd, R. M. (2004). *Fischer-Tropsch synthesis revisited; efficiency and selectivity benefits from imposing temporal and/or spatial structure in the reactor*. Ponsen & Looijen B.V, Wageningen, The Netherlands.

de Klerk, A., & Furimsky, E. (2010). *Catalysis in the refining of Fischer-Tropsch Syncrude*. RSC Catalysts Series 4. RCS Publishing.

Deckwer, W., Kokuun, R., Sanders, E., & Ledakowicz, S. (1986). Kinetic studies of Fischer-Tropsch synthesis on suspended iron/potassium catalyst—Rate inhibition by carbon dioxide and water. *Industrial and Engineering Chemistry Process Design and Development, 25*, 643–649.

Demitras, G. C., & Muetterties, E. L. (1977). Metal clusters in catalysis. A new fischer-tropsch synthesis. *Journal of the American Chemical Society, 99*, 2796–2797.

Ernst, B., Libs, S., Chaumette, P., & Kiennemann, A. (1999). Preparation and characterization of Fischer-Tropsch active Co/$SiO_2$ catalysts. *Applied Catalysis General, 186*, 145–168. https://doi.org/10.1016/S0926-860X(99)00170-2.

Espinoza, R. L., Steynberg, A. P., Jager, B., & Vosloo, A. C. (1999). Low temperature Fischer-Tropsch synthesis from a Sasol perspective. *Applied Catalysis General, 186*, 13–26. https://doi.org/10.1016/S0926-860X(99)00161-1.

Geerlings, J. J. C., Wilson, J. H., Kramer, G. J., Kuipers, H. P. C. E., Hoek, A., & Huisman, H. M. (1999). Fischer-Tropsch technology—From active site to commercial process. *Applied Catalysis General, 186*, 27–40. https://doi.org/10.1016/S0926-860X(99)00162-3.

Hao, Q., Bai, L., Xiang, H., & Li, Y. (2009). Activation pressure studies with an iron-based catalyst for slurry Fischer-Tropsch synthesis. *Journal of Natural Gas Chemistry, 18*, 429–435. https://doi.org/10.1016/S1003-9953(08)60134-6.

Huang, C., Xu, C., Wang, B., Hu, X., Li, J., Liu, J., et al. (2018). High production of syngas from catalytic steam reforming of biomass glycerol in the presence of methane. *Biomass and Bioenergy, 119,* 173–178. https://doi.org/10.1016/j.biombioe.2018.05.006.

Jahangiri, H., Bennett, J., Mahjoubi, P., Wilson, K., & Gu, S. (2014). A review of advanced catalyst development for Fischer-Tropsch synthesis of hydrocarbons from biomass derived syn-gas. *Catalysis Science & Technology, 4,* 2210–2229. https://doi.org/10.1039/C4CY00327F.

Ji, Y.-Y., Xiang, H.-W., Yang, J.-L., Xu, Y.-Y., Li, Y.-W., & Zhong, B. (2001). Effect of reaction conditions on the product distribution during Fischer-Tropsch synthesis over an industrial Fe-Mn catalyst. *Applied Catalysis General, 214,* 77–86. https://doi.org/10.1016/S0926-860X(01)00480-X.

Jones, D., Bhattacharyya, D., Turton, R., & Zitney, S. E. (2011). Optimal design and integration of an air separation unit (ASU) for an integrated gasification combined cycle (IGCC) power plant with $CO_2$ capture. *Fuel Processing Technology, 92,* 1685–1695. https://doi.org/10.1016/j.fuproc.2011.04.018.

Kunii, D., & Levenspiel, O. (1991). *Fluidization engineering* (2nd ed.). Elsevier.

Liu, Y., Teng, B.-T., Guo, X.-H., Li, Y., Chang, J., Tian, L., et al. (2007). Effect of reaction conditions on the catalytic performance of Fe-Mn catalyst for Fischer-Tropsch synthesis. *Journal of Molecular Catalysis Chemical, 272,* 182–190. https://doi.org/10.1016/j.molcata.2007.03.046.

Ma, T., Imai, H., Shige, T., Sugio, T., & Li, X. (2015). Synthesis of hydrocarbons from $H_2$-deficient syngas in fischer-tropsch synthesis over co-based catalyst coupled with Fe-based catalyst as water-gas shift reaction. *Journal Nanomaterial.* [WWW Document].. https://doi.org/10.1155/2015/268121.

Oukaci, R., Singleton, A. H., & Goodwin, J. G., Jr. (1999). Comparison of patented Co F-T catalysts using fixed-bed and slurry bubble column reactors. *Applied Catalysis General, 186,* 129–144. https://doi.org/10.1016/S0926-860X(99)00169-6.

Outi, A., Rautavuoma, I., & van der Baan, H. S. (1981). Kinetics and mechanism of the fischer tropsch hydrocarbon synthesis on a cobalt on alumina catalyst. *Applied Catalysis, 1,* 247–272.

Özdenkçi, K., De Blasio, C., Muddassar, H. R., Melin, K., Oinas, P., Koskinen, J., et al. (2017). A novel biorefinery integration concept for lignocellulosic biomass. *Energy Conversion and Management, 149,* 974–987. https://doi.org/10.1016/j.enconman.2017.04.034.

Pannell, R. B., Kibby, C. L., Kobylinski, T. P. (1981). A steady-state study of Fischer-Tropsch product distributions over cobalt, iron and ruthenium. *Studies in Surface Science and Catalysis, 7*(A), 447–459.

Patzlaff, J., Liu, Y., Graffmann, C., & Gaube, J. (1999). Studies on product distributions of iron and cobalt catalyzed Fischer-Tropsch synthesis. *Applied Catalysis General, 186,* 109–119. https://doi.org/10.1016/S0926-860X(99)00167-2.

Pour, A. N., Zamani, Y., Tavasoli, A., Kamali Shahri, S. M., & Taheri, S. A. (2008). Study on products distribution of iron and iron–zeolite catalysts in Fischer-Tropsch synthesis. *Fuel, 87,* 2004–2012. https://doi.org/10.1016/j.fuel.2007.10.014.

Riedel, T., Claeys, M., Schulz, H., Schaub, G., Nam, S.-S., Jun, K.-W., et al. (1999). Comparative study of Fischer-Tropsch synthesis with $H_2/CO$ and $H_2/CO_2$ syngas using Fe- and Co-based catalysts. *Applied Catalysis General, 186,* 201–213. https://doi.org/10.1016/S0926-860X(99)00173-8.

Sari, A., Zamani, Y., & Taheri, S. A. (2009). Intrinsic kinetics of Fischer-Tropsch reactions over an industrial Co–Ru/γ-Al2O3 catalyst in slurry phase reactor. *Fuel Processing Technology, 90,* 1305–1313. https://doi.org/10.1016/j.fuproc.2009.06.024.

Sarup, B., & Wojciechowski, B. W. (1989). Studies of the fischer-tropsch synthesis on a cobalt catalyst II. Kinetics of carbon monoxide conversion to methane and to higher hydrocarbons. *Canadian Journal of Chemical Engineering, 67,* 62–74.

Schulz, H. (1999). Short history and present trends of Fischer-Tropsch synthesis. *Applied Catalysis General, 186,* 3–12. https://doi.org/10.1016/S0926-860X(99)00160-X.

Schulz, H., & Claeys, M. (1999). Kinetic modelling of Fischer-Tropsch product distributions. *Applied Catalysis General, 186,* 91–107. https://doi.org/10.1016/S0926-860X(99)00166-0.

Shadle, L. J., & Breault, R. W. (2012). Integrated gasification combined cycle (IGCC). In W.-Y. Chen, J. Seiner, T. Suzuki & M. Lackner (Eds.), *Handbook of climate change mitigation* (pp. 1545–1604). Springer US, New York, NY. https://doi.org/10.1007/978-1-4419-7991-9_40.

Sie, S. T., & Krishna, R. (1999). Fundamentals and selection of advanced Fischer-Tropsch reactors. *Applied Catalysis General, 186,* 55–70. https://doi.org/10.1016/S0926-860X(99)00164-7.

Soomro, A., Chen, S., Ma, S., & Xiang, W. (2018). Catalytic activities of nickel, dolomite, and olivine for tar removal and $H_2$-enriched gas production in biomass gasification process, Catalytic activities of nickel, dolomite, and olivine for tar removal and $H_2$-enriched gas production in biomass gasification process. *Energy Environment* 0958305X18767848. https://doi.org/10.1177/0958305X18767848.

Steynberg, A. P., Dry, M. E., Davis, B. H., & Breman, B. B. (2004). Fischer-Tropsch reactors, Chap. 2. In A. Steynberg & M. Dry (Eds.), *Studies in surface science and catalysis, Fischer-Tropsch Technology* (pp. 64–195). Elsevier. https://doi.org/10.1016/S0167-2991(04)80459-2.

Steynberg, A. P., Espinoza, R. L., Jager, B., & Vosloo, A. C. (1999). High temperature Fischer-Tropsch synthesis in commercial practice. *Applied Catalysis General, 186,* 41–54. https://doi.org/10.1016/S0926-860X(99)00163-5.

Tavakoli, A., Sohrabi, M., & Kargari, A. (2008). Application of Anderson–Schulz–Flory (ASF) equation in the product distribution of slurry phase FT synthesis with nanosized iron catalysts. *Chemical Engineering Journal, 136,* 358–363. https://doi.org/10.1016/j.cej.2007.04.017.

Tavasoli, A., Pour, A. N., & Ahangari, M. G. (2010). Kinetics and product distribution studies on ruthenium-promoted cobalt/alumina Fischer-Tropsch synthesis catalyst. *Journal of Natural Gas Chemistry, 19,* 653–659. https://doi.org/10.1016/S1003-9953(09)60133-X.

Triantafyllidis, K., Lappas, A., & Stöcker, M. (2013). *The role of catalysis for the sustainable production of bio-fuels and bio-chemicals*. Elsevier.

Wojciechowski, B. W. (1988). The kinetics of Fischer-Tropsch synthesis. *Catalysis Reviews–Science and Engineering, 30,* 629.

Xu, D., Li, W., Duan, H., Ge, Q., & Xu, H. (2005). Reaction performance and characterization of Co/Al2O3 Fischer-Tropsch catalysts promoted with Pt, Pd and Ru. *Catalysis Letters, 102,* 229–235. https://doi.org/10.1007/s10562-005-5861-7.

Yang, C., Massoth, F. E., Oblad, A. G. (1979). In E. L. Kugler & F. W. Steffgen (Eds.), *Hydrocarbon synthesis from carbon monoxide and hydrogen*. Washington, D.C.: ACS.

Yates, I. C., & Satterfield, C. N. (1991). Intrinsic kinetics of the Fischer-Tropsch synthesis on a cobalt catalyst. *Energy & Fuels, 5,* 168–173. https://doi.org/10.1021/ef00025a029.

Zhang, C.-H., Yang, Y., Teng, B.-T., Li, T.-Z., Zheng, H.-Y., Xiang, H.-W., et al. (2006). Study of an iron-manganese Fischer-Tropsch synthesis catalyst promoted with copper. *Journal of Catalysis, 237,* 405–415. https://doi.org/10.1016/j.jcat.2005.11.004.

Zwart, R., Van der Heijden, S., Emmen, R., Dall Bentzen, J., Ahrenfeldt, J., Stoholm, P., et al. (2010). *Tar removal from low-temperature gasifiers*. Risø DTU, DFBT and Anhydro, Denmark: ECN and Dahlman from the Netherlands and Dall Energy.

# Chapter 21
# Notions of Biomass Gasification

*If I don't do something really new, I get bored*
Susumu Tonegawa, Nobel Prize in Medicine in 1087.

## 21.1 Preliminary Notions

Some information on the gasification of biomass process was given previously in the second chapter of this part of the book, and here, only a preliminary introduction will be provided about the general process with more focus on one particular kind of gasification which considers water as the medium at supercritical conditions. Gasification of biomass can be considered to follow different stages. The first step would be the drying of the feedstock followed by pyrolysis. The pyrolysis can be then followed by oxidation and reduction, depending on where air or oxygen is utilized during this process. In pyrolysis, which is reported to take place in a wide range of temperatures between 200 and 600 °C (Zhang et al. 2015), the biomass feedstock is having reactions of decomposition and as the last result, the char is then left as the part not containing volatiles. The formed char is then reacting within the section where we have higher temperatures (Fig. 21.1).

Reactions taking place during gasification of biomass can be exothermic and endothermic. The reactions written below are mainly taken from (De Blasio et al. 2015; De Blasio and Järvinen 2017; Özdenkçi et al. 2017). The exothermic reactions are written as follows.

Combustion:

$$CH_3{-}CH_2{-}\cdots + O_2 \rightarrow CO_2 \tag{21.1}$$

Partial oxidation:

$$CH_3{-}CH_2{-}\cdots + O_2 \rightarrow CO \tag{21.2}$$



C. De Blasio, *Fundamentals of Biofuels Engineering and Technology*,
Green Energy and Technology, https://doi.org/10.1007/978-3-030-11599-9_21

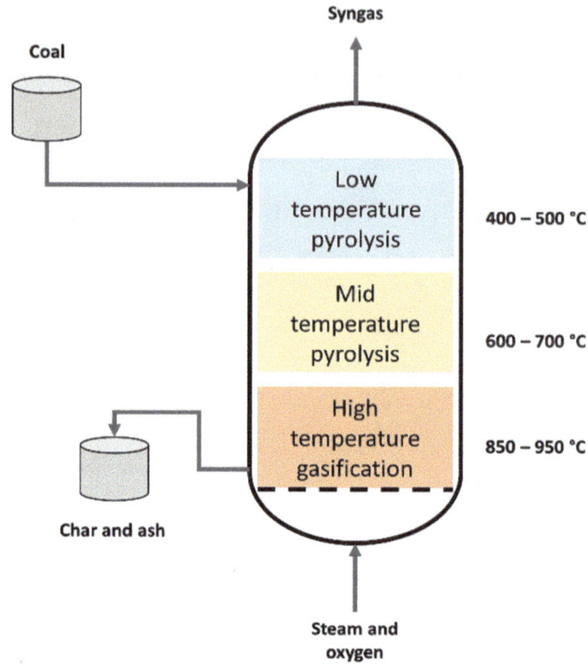


**Fig. 21.1** Stages of the gasification process. Adapted from Chen et al. (2018)

Methanation:

$$CH_3{-}CH_2{-}\cdots + H_2 \rightarrow CH_4 \tag{21.3}$$

where with $CH_3$–$CH_2$–... it is referred to the general biomass and char. While the endothermic reactions are written as:

Steam–carbon reaction:

$$CH_3{-}CH_2{-}\cdots + H_2O \rightarrow CO + H_2 \tag{21.4}$$

Boudouard reaction:

$$CH_3{-}CH_2{-}\cdots + CO_2 \rightarrow 2CO \tag{21.5}$$

Particular examples and additional reactions are given here with data on their standard enthalpy of reaction. For the water-gas shift (WGS) and CO methanation reactions, we have (Stangeland et al. 2017):

$$CO_{(g)} + H_2O_{(g)} \leftrightarrow CO_{2(g)} + H_{2(g)} \quad \Delta\, H^0_{298.15K} = -41.2\ \text{kJ/mol} \tag{21.6}$$

$$CO_{(g)} + 3H_{2(g)} \leftrightarrow CH_{4(g)} + H_2O_{(g)} \quad \Delta H^0_{298.15K} = -206.1\ \text{kJ/mol} \tag{21.7}$$

The $CO_2$ methanation reaction is reported as (Stangeland et al. 2017):

$$CO_2 + 4H_2 \rightarrow CH_4 + 2H_2O \quad (\Delta H = -165 \text{ kJ/mol}) \tag{21.8}$$

For steam reforming of acetic acid:

$$C_2H_4O_2(g) \rightarrow 2CO(g) + 2H_2(g) \quad \Delta H^0_{298.15K} = 221.2 \text{ kJ/mol} \tag{21.9}$$

The decarboxylation of formic acid and acetic acid is instead written as:

$$HCOOH(g) \rightarrow CO_{2(g)} + H_{2(g)} \quad \Delta H^0_{298.15K} = -14.9 \text{ kJ/mol} \tag{21.10}$$

$$CH_3COOH(g) \rightarrow CH_{4(g)} + CO_{2(g)} \quad \Delta H^0_{298.15K} = -36.1 \text{ kJ/mol} \tag{21.11}$$

In addition, the following reaction can also occur:

$$C_{(s)} + 2H_{2(g)} \leftrightarrow CH_{4(g)} \quad \Delta H^0_{298.15K} = -74.8 \text{ kJ/mol}; \tag{21.12}$$

and the coke gasification reactions:

$$CO(g) + H_{2(g)} \leftrightarrow C(s) + H_2O(g) \quad \Delta H^0_{298.15K} = -131.3 \text{ kJ/mol}; \tag{21.13}$$

$$CO_{2(g)} + 2H_{2(g)} \leftrightarrow C(s) + 2H_2O(g) \quad \Delta H^0_{298.15K} = -90.1 \text{ kJ/mol}. \tag{21.14}$$

The most common types of gasifiers are generally classified as (Sikarwar et al. 2016)

- fluidized;
- entrained flow gasifiers.

And, on the other hand, they also are subject to further classification. In fact, the fluidized kind can be divided into:

- bubbling fluidized bed;
- circulating fluidized bed;
- dual fluidized bed.

The entrained flow gasifiers can be classified as:

- top-fed gasifiers;
- side-fed gasifiers.

Regarding the nomenclature of these kinds of reactors, their names are quite descriptive of their functioning. As a matter of fact, the fluidized kind of reactor has the so-called fluidized bed, and, depending on the particle size distribution of the bed and the feeding flow (gas), they behave differently. In this section, only some information will be given regarding the basic principles of these kinds of reactors, and for a more detailed treatise, the reader can refer to the available literature (Kunii and Levenspiel 1991). The main difference between the reactors was depicted in Fig. 10.9; the circulating fluidized bed reactor differs from the others, since there is a "recirculation" of the solid particles. This makes it different in comparison with the bubbling kind since in the last case the solid particle are less fluidized but they are staying in place within the same area of the reactor while this is not happening in the case when a recirculation is taking place by means of a cyclone or suitable devices.

These reactors have a series of advantages and disadvantages. For instance, the fixed bed reactors are commonly referred to be more simple and also more economic. It is also referred that the tar, which could have been formed during the gasification process, is completely degraded in this kind of reactors, and this is because we have very high temperatures (around or more than 1000 °C), (Sikarwar et al. 2016). This is an important feature of these types of gasifier, but it also gives more costs. On the other hand, the circulating fluidized bed reactors and the dual fluidized bed ones have a more complicated design, and therefore, the costs are also higher. Nevertheless, it is also clear that the heat transfer mechanisms and the temperature gradients within the reactors are much more favorable.

Without going into detail with the physical principles behind the utilization of a particular bed, the inert mass is needed to influence the heat capacity inside our system (the inner volume of the reactor). For this reason, the temperature gradients between the reacting particles and the spaces where there is no reaction will be significantly decreased with beneficial outcomes.

An additional factor to be taken into account is the mixing of the particles; fluidization increases the degree of mixing of the particles and increases the reactions' efficiency. Nevertheless, the mixing and the transport phenomena are indeed increased when the particles are smaller and, on the other hand, a recirculation is then necessary.

Some examples of commercially available and operating gasifiers are as follows, and they will be illustrated with their basic principles of functioning (Breault 2010):

- The GE Energy Gasifier;
- ConocoPhillips E-Gas;
- Shell;
- Siemens;
- KBR Transport;
- British Gas Lurgi;
- Lurgi Multi-Purpose Gasifier;
- Lurgi Mark IV Gasifier;
- U-Gas, Synthesis Energy Systems;
- High Temperature Winkler Gasifier;
- PRENFLO™ Gasifier/Boiler.

The GE gasifier is a type of reactor produced by the company General Electric and can be utilized for coal gasification. As a representative illustration, Fig. 21.2 gives some information about the process.

The process involves a cocurrent flow of the feedstock (a coal slurry in this case) and the oxygen which is coming from an air separation plant unit. As gasifying

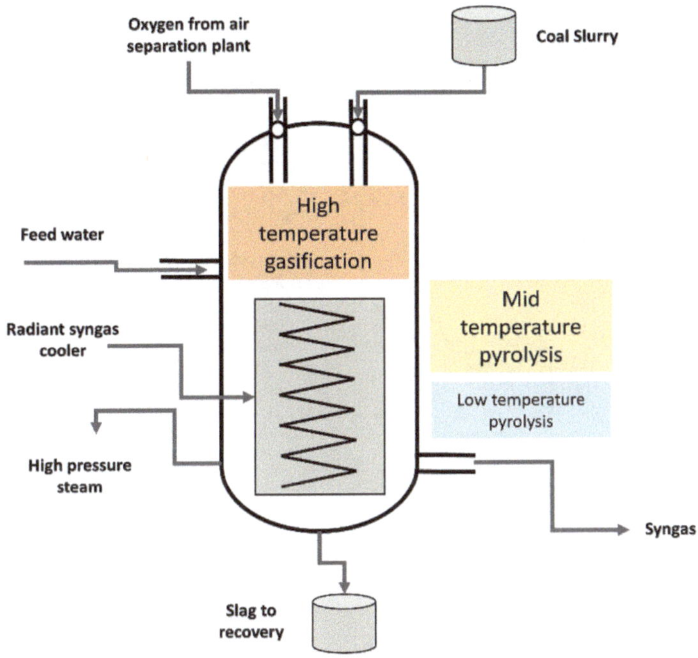


**Fig. 21.2** A simple schematics of the GE Energy Gasifier. Adapted from De Agarwal et al. (2018)

agent, we have steam in this case, steam has a large importance since it functions as transport agent, it helps with the heat transfer-related phenomena, and it is involved in the gasification reactions which were previously described (water-gas shift, steam reforming and related). The steam participates actively within the heterogeneous reactions between steam and carbon. The feeding system, in this case, is on the top of the reactor, while at the bottom, the low-temperature pyrolysis takes place and the syngas is released for its use and collection.

The ConocoPhillips process is depicted in Fig. 21.3, and in this case, we have some small difference in comparison with the previous described process. The slurry of feedstock and the oxygen are inserted at the bottom of the reactor where we have the first stage of the process. Eventually, char from other processes or recovered from the same system is inserted also within the first stage, the high-temperature gasification process.

The Shell gasifier is illustrated in Fig. 21.4. This gasifier can be also classified as a "Water wall" gasifier, and this is for the fact that at the wall, there is a series of water tubes as it is represented in Fig. 21.4.

A water-walled reactor offers additional possibilities in regard to the maintenance of the reactor and also concerning the control of the operating conditions. As a matter of fact, by having this kind of arrangement, it is possible to control in a better way the inner temperature of the reactor. The durability is also increased, since there is

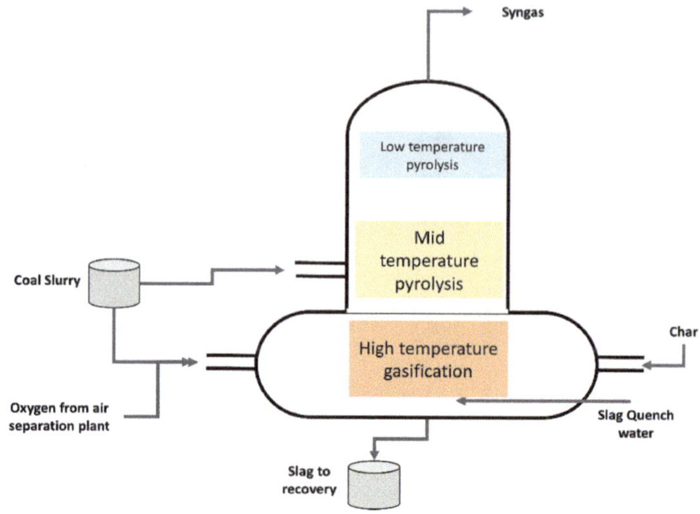


**Fig. 21.3** Drawing of the ConocoPhillips E-gas gasifier

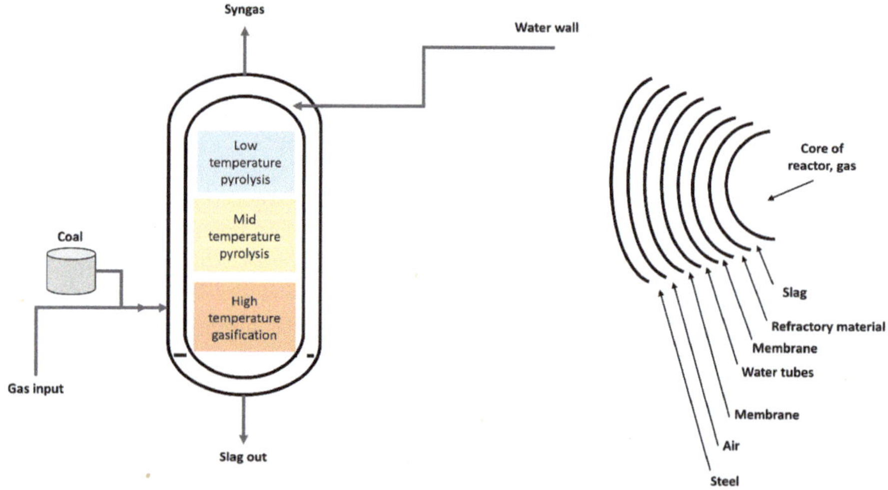


**Fig. 21.4** Shell gasifier with a description of the constituting parts of the reactor wall. Adapted from Lisandy et al. (2016)

less temperature gradients related to the refractory material. However, also this kind of arrangements is subject to failure.

Siemens has been also active within the development of the gasification technology, and in Fig. 21.5 there is a schematics of the system adopted in this case.

In reference to Fig. 21.5, the feedstock and the gas are inserted from the same side and the first stage of the process is the development of a high-temperature section; in this case, we have indeed high temperature, in the range of 1300–1800 °C, (Görsch et al. 2011).

It should be noted that, in gasification, there is a difference if the water is injected in one section or another within the reactor. In this case, the water is injected at the beginning of the second section, and we inject here the quench water.

One of the main reactions driven by the injected water is the water-gas shift reaction, WGSR, for which a simple schematics is reported in Fig. 21.6.

This indeed has a contribution especially in the case the ultimate goal is to a have a syngas rich in hydrogen.

The KBR Transport Gasifier (TRIG$^{TM}$) operating principles are similar to the circulating fluidized bed combustion systems. The steam, feedstock (for instance coal), air, and sorbent are mixed into the mixing zone, while oxygen is inserted at the bottom of the main part of the gasifier. The products are elevated up into a

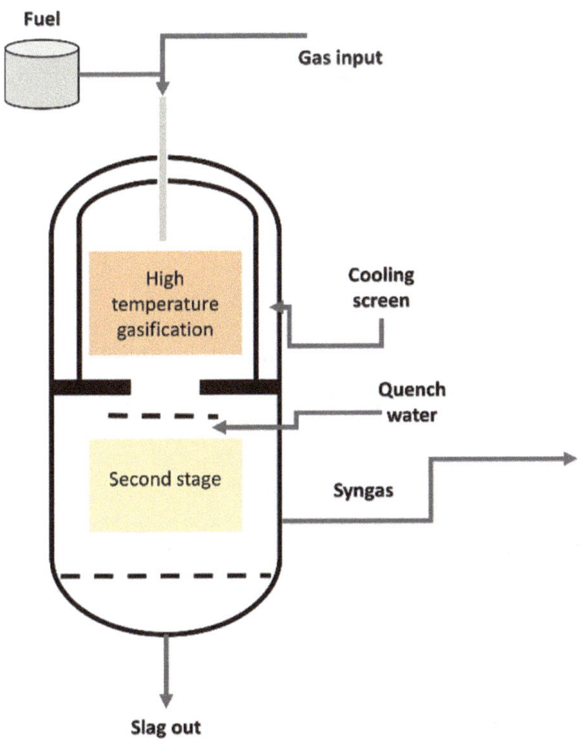


**Fig. 21.5** Siemens gasification reactor with quench water system

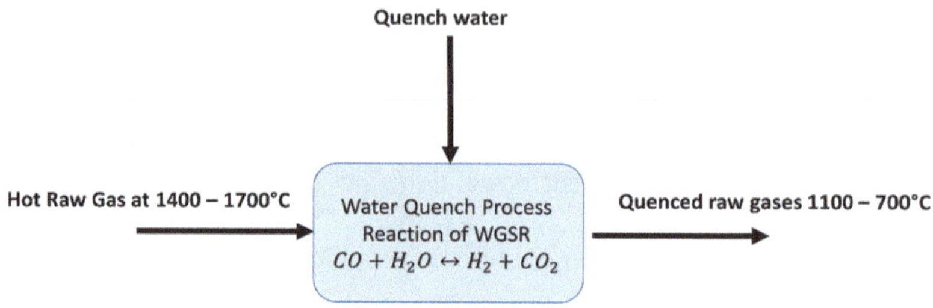


**Fig. 21.6** Schematics of the quench process and its contribution to the WGSR

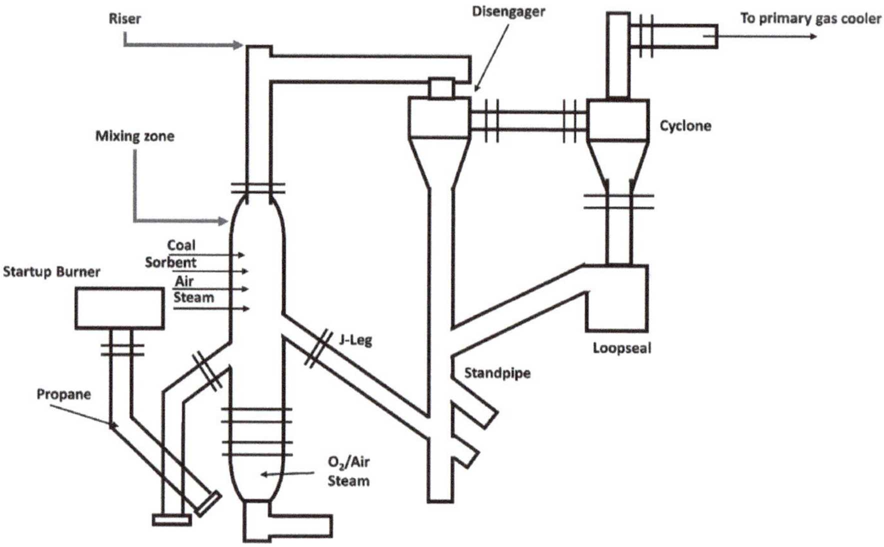


**Fig. 21.7** Representation of the KBR Transport Gasifier (TRIG$^{TM}$). Adapted from Ariyapadi et al. (2008)

conduct called *riser*, and after this, the bigger particles are firstly removed into a *disengage* unit and after this, into a cyclone. The particles collected are sent back to the gasification zone by going first into a stand pipe and then a J-leg. The produced gases are processed further with separation and cleaning processes (Fig. 21.7).

The British Gas Lurgi Gasifier is described in a simple way in Fig. 21.8.

This case differs from the others since we have here a different direction of the flow for the feedstock and for the injected steam/air/oxygen. It is also beneficial to remember that the inlet of the reactor contains still the bed material and in this case the walls are constituted by refractory material.

The company Lurgi AG has also developed a Multi-Purpose Gasifier, MPG, for which a simple schematics is reported here (Fig. 21.9).

The simple arrangements of this gasifier allow for a certain flexibility on the kinds of feedstock used. The claim in this case is that the flash point of the feedstock is

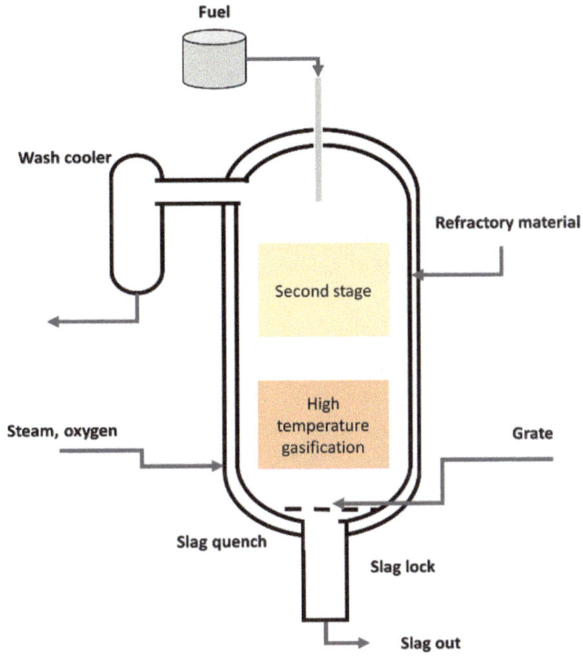


**Fig. 21.8** Description of a British Gas Lurgi Gasifier. Adapted from Krishnamoorthy and Pisupati (2015)

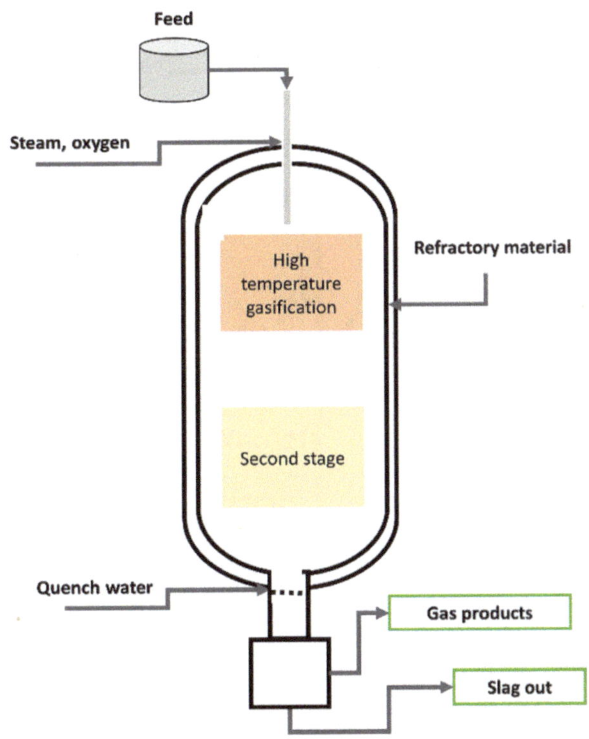


**Fig. 21.9** Lurgi MPG. Adapted from Koss and Schlichting (2005)

not limited to a certain range, the particles should be around one millimeter, and the system allows for using diverse kinds of fuels which could be inserted at the same time into the reactor even if they are not miscible. The Lurgi MPG gives the possibility of operating within two modes, the boiler mode and the quench mode; in both cases, different kinds of feedstock can be used, but there are some limitations for the case of the boiler mode. In the boiler mode, we have high pressure for the steam and higher costs, while for the quench configuration we have medium pressure steam and lower costs in case we would like to focus on the production of hydrogen. The raw gas produced needs still to have a treatment process; for instance, a desulfurization unit is then necessary.

Mitsubishi Heavy Industries has also developed gasification systems. For instance, a particular system suitable for municipal solid waste, MSW, is shown in (Fig. 21.10).

In this case, we have a fluidized bed of sand where the combustion air is injected. The gasification is taking place at relatively low temperature and between 450 and 600 °C. The system allows for the separation of heavy metals which can be recycled.

The company Synthesis Energy Systems has developed a method denominated as U-Gas® which is a gasifier with a dry feed injection. A simple schematics is shown in (Fig. 21.11).

Regarding the positive aspects of this technology, the claim in this case is that there is good mixing and the conversion is also high since the residence time is maintained at sufficient levels. The moderate temperatures allow for a longer durability.

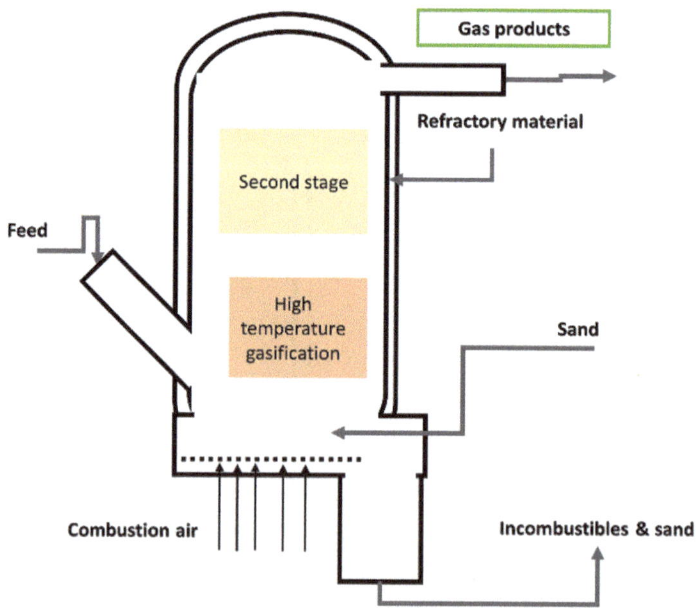


**Fig. 21.10** A system developed by Mitsubishi Heavy Industries for gasification of MSW. Adapted from Mitsubishi Heavy Industries (2019)

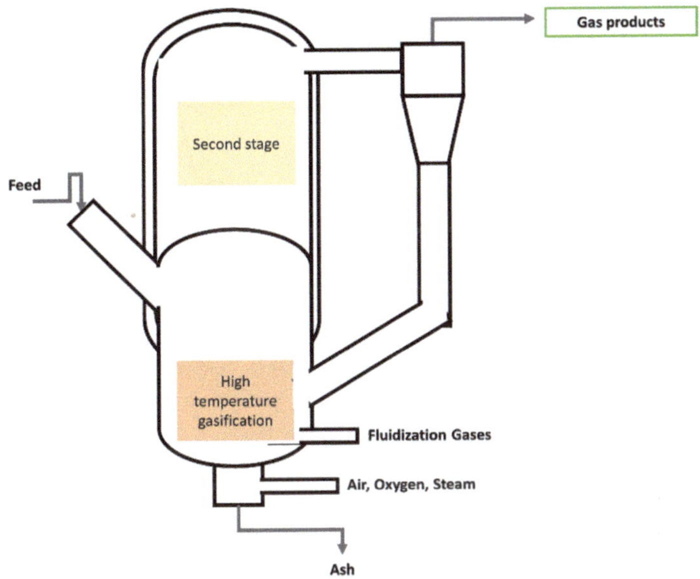


**Fig. 21.11** U-Gas method developed by synthesis energy systems. Adapted from Lau (2009)

**Table 21.1** Fuel properties range for the U-Gas gasifier

| Fuel parameter range | Tested range |
|---|---|
| M (wt%) | 1–41 |
| VM (wt%) | 3–69 |
| FC (wt%) | 6–83 |
| S (wt%) | 0.2–4.6 |
| Free swelling index | 0–8 |
| Ash (wt%) | <42% |
| Ash softening, °F | 1900–2500 |
| HHV (Btu/lb) | 5500–14,000 |

where M is the moisture, VM is the volatile matter, FC is the fixed carbon, and S is the sulfur
Adapted from Lau (2009)

The range of properties for the feedstock and the tested range values are given in Table 21.1.

The High Temperature Winkler Gasifier, HTW, is working following the principles of the Winkler Generator which was developed in the beginning of last century. The difference between the old and the new method is that the new version is pressurized. This method was developed (the new version) by ThyssenKrupp and Rheinische Braunkohlenwerke AG (Toporov and Abraham 2015) (Fig. 21.12).

The system allows for the processing of diverse kinds of feedstock, and the higher pressure (10 bar) leads to better reaction rates. On the other hand, higher temperature

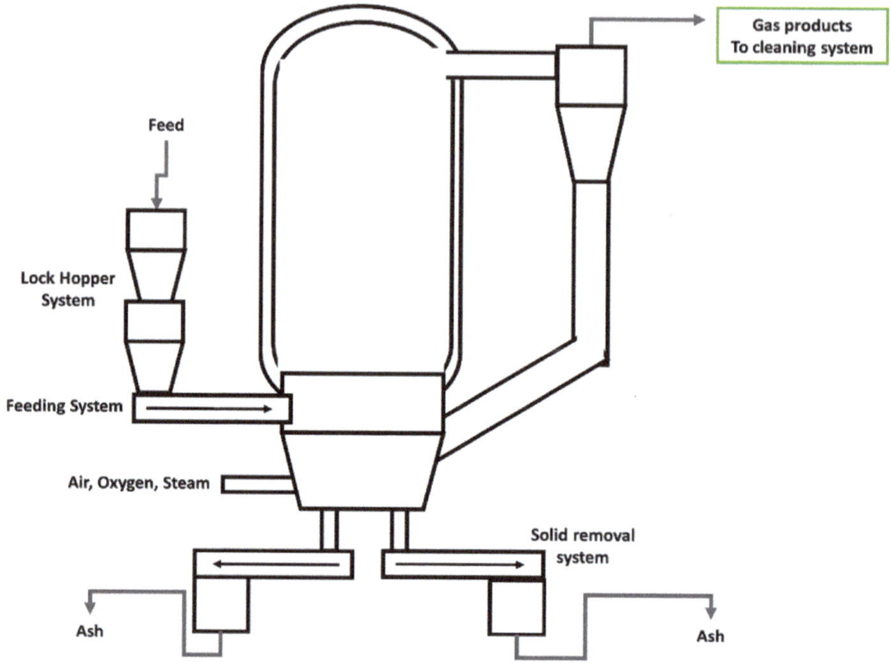


**Fig. 21.12** A scheme for the High Temperature Winkler Gasifier. Adapted from Toporov and Abraham (2015)

increases the gas yields and rates; however, in this case, the temperatures are still quite moderate with respect to other gasifiers and this gives more durability to this system. In Finland, there is one gasifier of this kind in Oulu.

The HTW and the PRENFLO® gasifiers are both described within a report published by ThyssenKrupp and available to the public (ThyssenKrupp 2019). A flow diagram of the process is reported in Fig. 21.13.

This system uses a particle size range of 0.1 mm, and the gasification media are oxygen and steam. We have here a temperature which is higher than the ash melting temperature, and this allows for the removal of the ashes as slag. In the mentioned reference, ThyssenKrupp refers that the gas produced is composed mainly of CO and $H_2$ gases. These gases are then sent to a boiler where steam is generated to produce electricity. The gas is further processed to eliminate dust and impurities.

As can be noticed, the possible solutions for producing syngas from a variety of feedstock are many. In this first edition of the manuscript, the author will limit the treatise to one type of gasification: the supercritical water gasification.

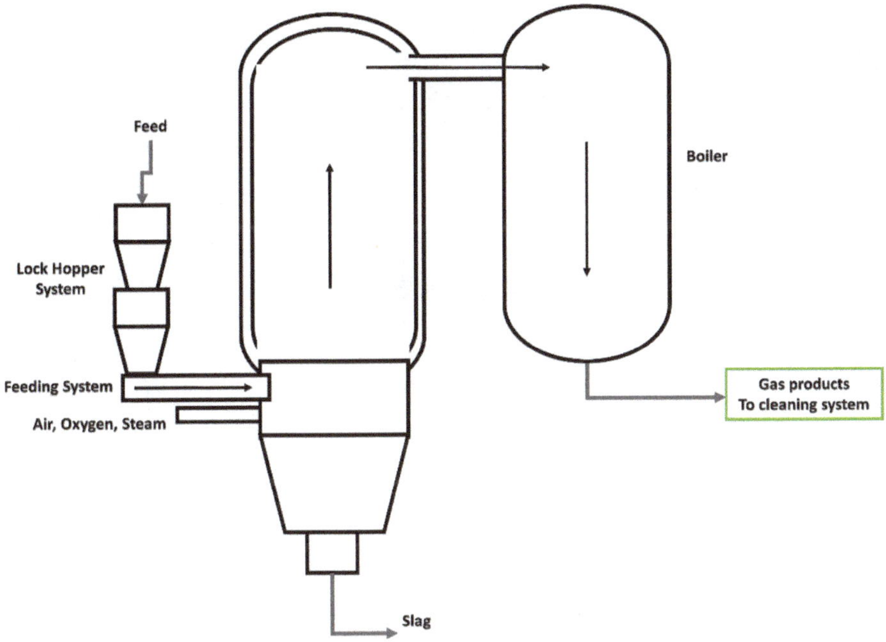


**Fig. 21.13** PRENFLO® Gasifier/Boiler. Adapted from ThyssenKrupp (2019)

## 21.2 Supercritical Water Gasification

Supercritical water, SCW, is used today already in a number of processes. SCW is used present within normal and nuclear power plants; it is normally used to drive turbines in order to generate electricity. In this case, they are nominally called supercritical water reactors (Cervi and Cammi 2018). SCW is also used for other applications; for instance, nanoparticles and new compounds are produced by these methods. When oxygen is used together with SCW, this allows for the destruction of hazardous waste by supercritical water oxydation, SCWO. SCW allows for a wide range of selection of the process conditions since parameters like temperature affect all of its properties: density, dielectric constant, specific heat capacity, diffusivity, viscosity, and more.

SCWG is quite effective at high temperatures where a very good conversion can be recorded; here, a visual example is given for the SCWG of weak black liquor at 700 °C. In the figure below, the sample before gasification is shown along with the produced liquid effluent (Fig. 21.14).

Water has indeed remarkable properties; for instance, its density can be actually reduced by decreasing its temperature, like in the case when ice is formed. Ice has a lesser density compared to the one of liquid water, and this allows for the ice to be formed on top of rivers or the sea. This is quite an important feature since it allows for the organisms in water to survive even if the weather is very cold and cold for

**Fig. 21.14** Visual example for SCWG of black liquor and liquid products. Photograph by Cataldo De Blasio

a very long time. Water has been said to be the most useful among the engineering fluids in the literature (He et al. 2015) and not only at low temperatures, but also at high temperatures and pressures, water demonstrates interesting properties. For instance, water is referred to have an unusual high temperature for its critical point (Finney 2004) and this is probably because of the strong molecular bonds of water and specifically between hydrogen and oxygen in the molecule.

At supercritical conditions, i.e., temperatures and pressures above 374 °C and 218 atm, the thermodynamic properties of water are changing drastically starting from its density which is dropping considerably at these conditions. This is shown in Fig. 21.15 where data are taken from the International Association for the Properties of Water and Steam IAPWS Formulation 1995 for the Thermodynamic Properties of Ordinary Water Substance for General and Scientific Use (Wagner and Kretzschmar 2008).

The properties of water are changing because the structure of the water itself is changing when supercritical conditions are reached. Computational molecular modeling, CMM, is in use today to explain the structure of water at these conditions. This is done mainly because it is difficult to have precise information from experimental methods. Nevertheless, considering experimental and theoretical studies, some additional information is given on what could be the structure of water at supercritical conditions. It is mentioned that the percolating network of hydrogen bonds in this system is actually broken and there is still parts of intact clusters of hydrogen atoms bonded together; this could be surrounded by water molecules which are starting to loose their bonds (Kalinichev 2017).

The variation of its density is quite an interesting feature when we consider the production of electricity by expanding it in a turbine generator. However, the conditions for this kind of application should be carefully planned. As a matter of fact, it is indeed a fact that the density is decreasing and therefore it could give an extra expansion within the turbine, but the changes in other properties have to be taken into account as well. For instance, water is referred to be quite corrosive especially

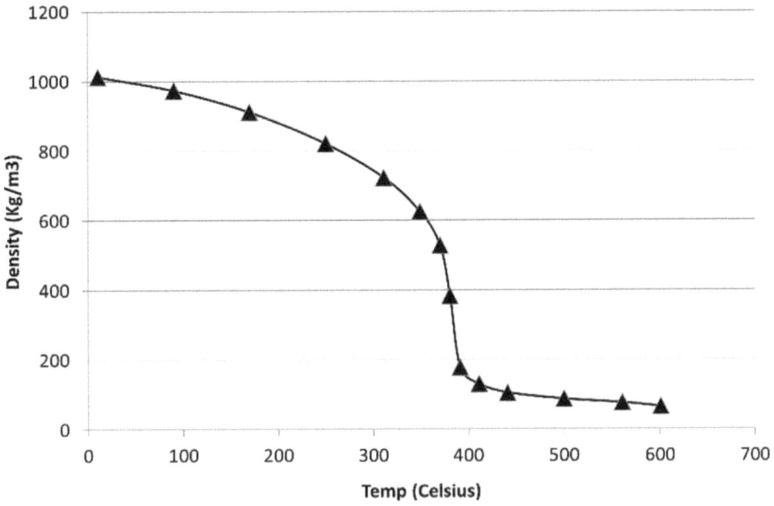


**Fig. 21.15** Density of water as a function of temperature at 250 bar. Data from Wagner and Kretzschmar (2008)

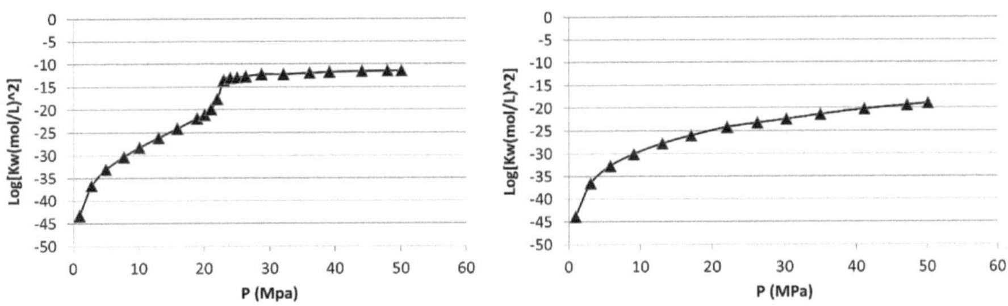


**Fig. 21.16** Isothermal variation of ionic product for water. Left, 374 °C. Right: 650 °C. Data from Wagner and Kretzschmar (2008)

in the near supercritical region and this could damage metals. The change in the heat capacity in the near supercritical region should be also considered very well.

The change in these parameters is considered more specifically in the following pages starting with the ionic product of water and its dissociation constant (Fig. 21.16).

The viscosity of supercritical water is reported usually to be low (Yang et al. 2019). In Fig. 21.17, data for the viscosity against the density of water are reported at a temperature of 386 °C.

In Fig. 21.18, the specific heat capacity against temperature is reported at two different pressures, 24 and 36 MPa. From the figure, it can be observed that the pressure does a certain "containment" effect on the heat capacity variation.

Instead, the variation of the density against the pressure is demonstrated in Fig. 21.19 where two temperatures are considered, 410 and 374 °C.

The following plot demonstrates the energy required to bring water at 700 °C at a pressure of 30 MPa (Kruse 2008). On the other hand, to bring water to the same

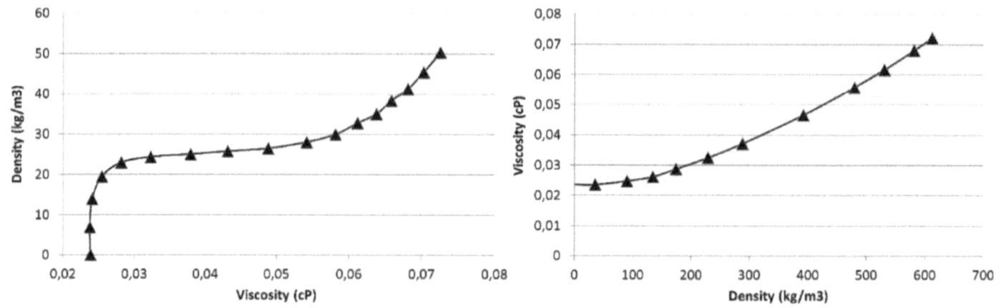


Fig. 21.17 Density versus viscosity of water at 386 °C. Data from Wagner and Kretzschmar (2008)

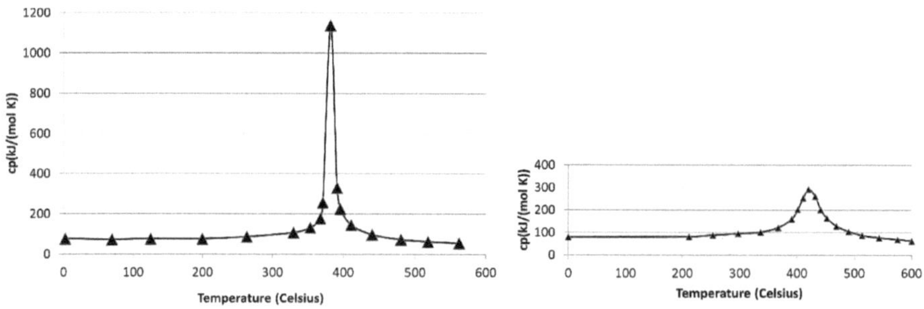


Fig. 21.18 Specific heat of SCW against temperature at 24 MPa, Left, and 36 MPa, Right

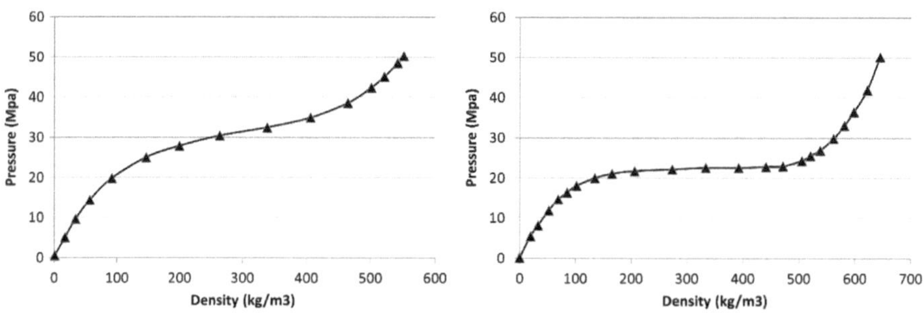


Fig. 21.19 Density of SCW plotted against pressure at 410 °C, left, and 374 °C, right

temperature, but at atmospheric pressure, it will require that we first bring the water at the boiling temperature and then wait for all the water to pass to the steam phase. This is because the boiling is taking place at constant temperature. After all the water has gone to the steam phase, then it will be possible to heat the medium further to superheated steam. This is shown in the same reference to have a higher consumption with respect of doing this at 300 bar. This is one reason why in our laboratory we choose to preheat the SCWG reactor at a sufficient pressure, for instance, 250 bar (Fig. 21.20).

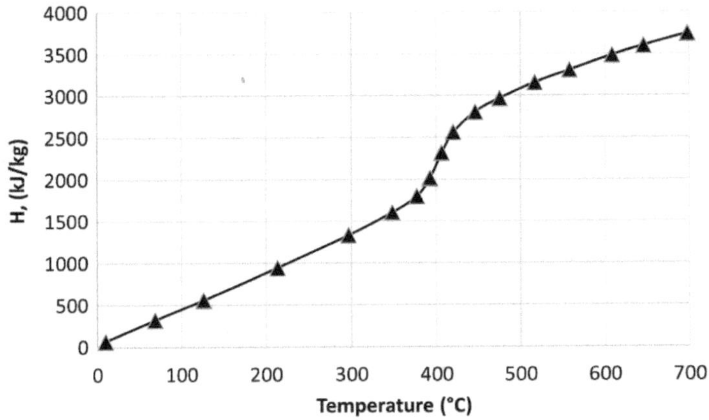


**Fig. 21.20** Enthalpy change against temperature at 30 MPa. Adapted from Kruse (2008)

### *21.2.1 SCWG Kinetics and Thermodynamics*

In the literature, it is reported that organic reactions with water can follow mainly two mechanisms depending if the water is at its normal conditions or at supercritical conditions. In fact, it has also been mentioned that the reactions involved in SCWG follow a radical mechanism of reaction.

In the case of ionic species, we can list the following phenomena, and some of them are reported to happen also in SCW:

- autoprotolysis (Farajtabar and Gharib 2009);
- protonation (Ryan et al. 1996);
- deprotonation;
- dehydration (Antal et al. 1987);
- tautomerization (Martskainen 2009);
- acetal formation or acetalization (Hoffmann and Conradi 1998);
- aldol condensation (Sasaki et al. 2002).

When there are radicals, we can list the following series of reactions:

- ionization (Portela et al. 2001);
- C–C bond beta scission (Ratkiewicz and Truong 2012);
- isomerization (for instance cis-trans) (Randolph and Carlier 1992);
- addition of radical molecules;
- dehydration (Yamamoto et al. 2018);
- substitution;
- termination.

And the following reactions related to the water molecule could also be taken into consideration:

$$H_2O \leftrightarrows OH + H \tag{21.15}$$

$$H_2O + H \leftrightarrows OH + H_2 \tag{21.16}$$

$$H_2O + O \leftrightarrows OH + OH \tag{21.17}$$

$$OH + HO_2 \leftrightarrows H_2O + R \tag{21.18}$$

$$H_2O + HO_2 \leftrightarrows OH + H_2O_2 \tag{21.19}$$

When simulating reaction conditions for SCWG of biomass, it is beneficial to have suitable relation for the related equations of state EoS. One example is the Peng–Robinson EoS which in this case is applied to a pure component and not to a mixture at supercritical conditions:

$$P = \frac{RT}{\overline{V} - b} - \frac{a}{\overline{V}^2 + 2b\overline{V} - b^2}; \tag{21.20}$$

where $\overline{V}$ is the molar volume of the medium, and it is related to the compressibility factor, which is defined as:

$$Z = \frac{P\overline{V}}{RT} \tag{21.21}$$

where $R$ is the universal gas constant, $P$ is the pressure, and $T$ is the temperature of the medium. In the Peng–Robinson formula, the coefficients $a$ and $b$ are derived from the critical temperature and pressure and also from the acentric factor. Expressions for these terms can be found in the literature (Castello and Fiori 2011), and some examples of $a$, $b$, and acentric factor are given as follows:

$$a = \frac{0.45724R^2T_c^2}{P_c}\left[1 + \left(0.37464 + 1.54226\omega - 0.26992\omega^2\right)\left(1 - \sqrt{\frac{T}{T_c}}\right)\right]^2; \tag{21.22}$$

$$b = \frac{0.07780RT_c}{P_c}; \tag{21.23}$$

where the subscript $c$ is related to the critical state and $\omega$ is the Pitzer acentric factor defined as:

$$\omega = -1 - \log_{10}\left[\frac{P^{vap}(T_r = 0.7)}{P_c}\right] \tag{21.24}$$

In the previous formula, the term $T_r$ is referring to the reduced temperature $T_r = T/T_c$.

The van der Waals' binary mixing rules are also used when there is a mixture of components in the supercritical medium. With these relations, it is possible to evaluate average coefficients for the mixture.

$$a_m = \sum_{i=1}^{N} \sum_{J=1}^{N} x_i x_j a_{ij} \tag{21.25}$$

$$b_m = \sum_{i=1}^{N} \sum_{J=1}^{N} x_i x_j b_{ij} \tag{21.26}$$

In the equations written previously, the term $x$ refers to the molar fraction of the $i$ and $j$ components. The terms $a_{ij}$ and $b_{ij}$ can be estimated from:

$$a_{ij} = \sqrt{a_{ii} a_{jj}} \tag{21.27}$$

$$b_{ij} = \frac{(b_{ii} + b_{jj})}{2} \tag{21.28}$$

where $a_{ii}$, $a_{jj}$, $b_{ii}$, and $b_{jj}$, are the coefficients described in Eqs. (21.22) and (21.23) since $a_{ii}$, for instance, is the coefficient $a$ related to one element (interaction between the element $i$ and itself) and the same reasoning can be done for the element $j$.

In real conditions, the partial pressure $P_i$ of one component is expressed in terms of the fraction of the component, the overall pressure of the system, and a proportionality coefficient called fugacity.

$$f_i = P x_i \varphi_i \tag{21.29}$$

The fugacity term for one component is related to its compressibility factor, and here, it will be shown one example of correlation from the literature (Zhou and Zhou 2001).

$$\ln f_i = Z - 1 - \ln(Z - B^*) - \frac{A^*}{B^*} \ln\left(1 + \frac{B^*}{Z}\right) \tag{21.30}$$

This is also referred as the Redlich–Kwong (SRK) equation where:

$$A^* = 0.4278 \cdot \alpha \cdot \frac{P/P_c}{\left(\frac{T}{T_c}\right)^2}; \tag{21.31}$$

$$B^* = 0.0867 \cdot \frac{P/P_c}{T/T_c} \tag{21.32}$$

$$\alpha^{0.5} = 1 + m\left(1 - T_r^{0.5}\right); \tag{21.33}$$

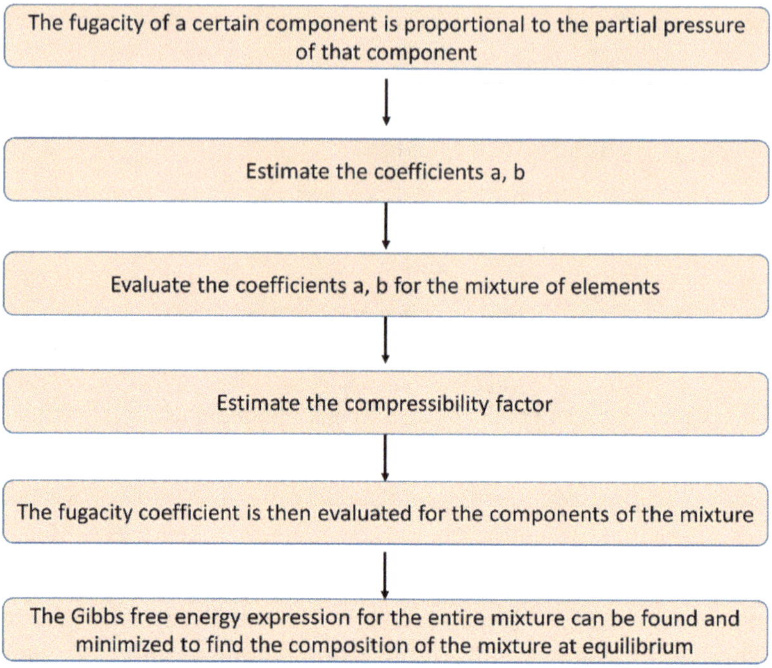


**Fig. 21.21** General procedure for evaluating the composition of a mixture at equilibrium

$$m = 0.480 + 1.574\omega - 0.176\omega^2; \tag{21.34}$$

As a summary, in order to obtain the composition of a mixture at equilibrium, one procedure could be by following the steps shown in Fig. 21.21.

## *21.2.2 Catalysis in SCWG*

In SCWG of biomass, different kinds of catalysts are utilized to promote reactions and improve the gasification rate and yield. Also in the case of SCWG, we can have homogeneous and heterogeneous catalysis. Among the heterogeneous catalyst, we have:

- nickel (Wang et al. 2018)—this catalyst can be supported also on zirconia ($Ni/ZrO_2$). This kind of catalyst is referred in the literature to have beneficial effects for the methanation of CO and $CO_2$ (da Silva et al. 2012, p. 2);
- ruthenium ($Ru/Al_2O_3$), ($Ru/ZrO_2$) (Byrd et al. 2008);
- palladium (Osada et al. 2006);
- $Ru/TiO_2$ catalyst (Osada et al. 2012);

- iron;
- copper (Gizir et al. 2003);
- zinc (Resende and Savage 2010);
- zirconium (Watanabe et al. 2001);
- Raney nickel (Ni–Al Alloy) (Nguyen et al. 2014);
- platinum (Pt/$Al_2O_3$) (Karakuş et al. 2013).

Among the homogeneous catalysts, we have:

- alkali hydroxides (Nanda et al. 2016);
- carbonates ($Na_2CO_3$) (Nanda et al. 2016);
- KOH and $KHCO_3$ (Yakaboylu et al. 2015).

Natural minerals:

- $NaHCO_3 \cdot Na_2CO_3 \cdot 2H_2O$ also known as Trona (Yanik et al. 2008);
- $K_2CO_3$ (Louw et al. 2016);
- $Ca(OH)_2$ (Wang and Takarada 2001).

It has been reported that even black liquor itself can be considered as a catalyst (Rönnlund et al. 2011), and this is because of its high content of catalyst salts.

### 21.2.3 Salts and Inorganics in SCWG of Biomass

Salts and inorganic compounds play a role as catalysts, as previously mentioned, and by SCWG, they could be even recovered, since as it will be shown in the following pages, the dielectric constant of water is fast dropping at supercritical conditions and this provoke that salts and inorganics are precipitating within the reactor.

Among the main salts to be considered in SCWG of organic fractions, the sodium and potassium salts play an important role when feedstock like black liquor is processed. With respect to the sodium salts, we can recognize:

- sodium acetate—$CH_3COONa$;
- sodium formate—HCOONa;
- sodium lactate—$C_3H_5NaO_3$;
- sodium oxalate—$Na_2C_2O_4$;
- sodium carbonate—$Na_2CO_3$;
- sodium bicarbonate—$NaHCO_3$.

Data on the sodium formate and sodium acetate reaction in SCWG can be found in the literature (Onwudili and William 2009, 2010).

$$2HCOONa + H_2O \rightarrow 2H_2 + Na_2CO_3 + CO_2 \quad \Delta H^0_{298.15K} = 50.6\ \text{kJ/mol} \tag{21.35}$$

$$2CH_3COONa + H_2O \rightarrow 2CH_4 + Na_2CO_3 + CO_2 \quad \Delta H^0_{298.15K} = -14.4\ \text{kJ/mol} \tag{21.36}$$

We can also have that there is higher production of hydrogen at high temperatures, and the following reaction can also occur

$$CH_3COONa + 3H_2O \rightarrow 4H_2(g) + NaHCO_3(s) + CO_2(g) \quad \Delta H^0_{298.15K} = 89.9 \text{ kJ/mol} \tag{21.37}$$

Salts and inorganics precipitate in SCWG, and this is an evident during the experiments. As a matter of fact, the dielectric constant of water drops to almost 1 at supercritical conditions, and this means that water is no more a polar compound. Therefore, salts and inorganics precipitate giving the possibility of recovering these compounds (Fig. 21.22).

Nevertheless, this is not very easy in practice, even if several solutions have been proposed in the literature. SCWG offers the possibility of using diverse reactor setups. For instance, in the literature, scientists have proposed the usage of diverse reactor zones and in each of these regions there could be different conditions (Susanti et al. 2010).

As an example of solubility decrease of organic salts in SCW, Fig. 21.23 shows the solubility of sodium carbonate as a function of the temperature (Fig. 21.23).

Precipitating salts and inorganics are creating also problems to the normal operations of the reactor. Plugging can occur, and because of this, we could experience different fluctuations in pressure and temperatures within the reactor. The plugging

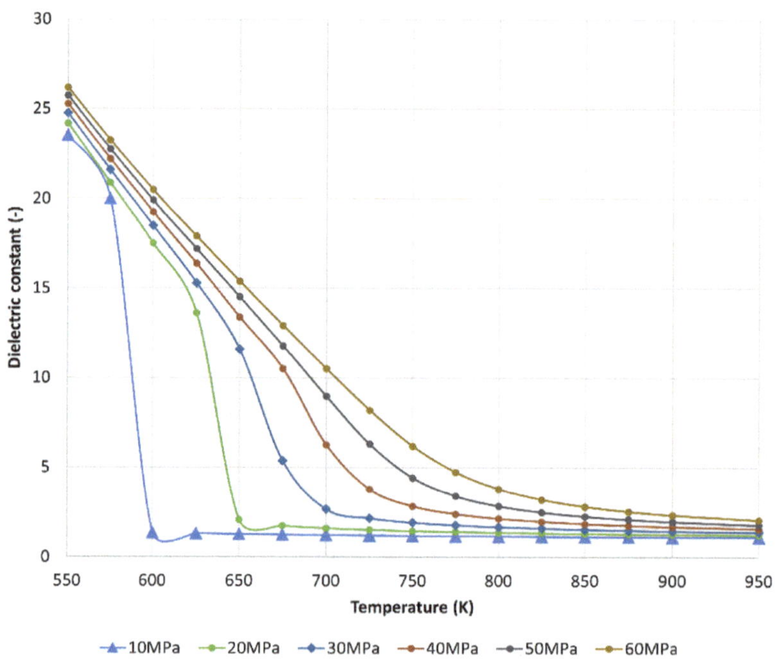


**Fig. 21.22** Dielectric constant of water against temperature at different pressures

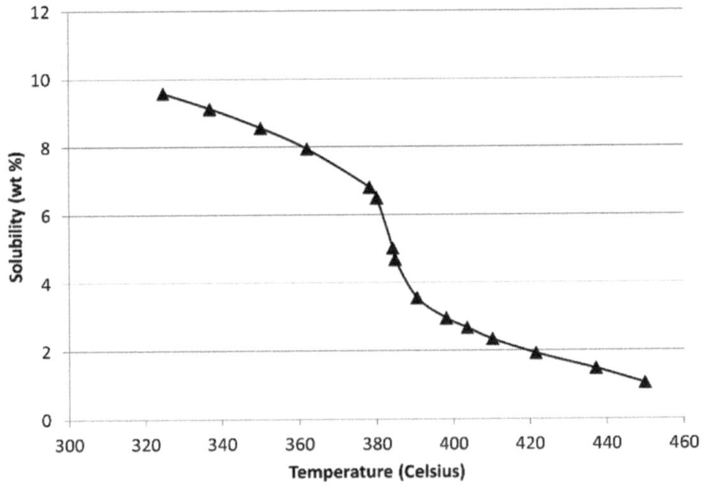


**Fig. 21.23** $Na_2CO_3$ solubility against temperature. Adapted from Jones et al. (1976)

problem is not frequently studied in the literature with regard to the SCWG of biomass even if there is some study cases (De Blasio et al. 2019).

Figure 21.24 describes the dynamics of what is taking place during the plugging of one reactor entry and the sudden release of the clogging wall due to the incoming pressure. Because of the wall release, a sudden injection of a quantity of feedstock at much lower temperature will take place. The lower temperature will bring the conditions to subcritical with a proportional increase of the volume density; however, since the reactor walls will continue to pump energy into the reactor, these conditions are rapidly changing causing stress to the reactor walls, inlet, outlet, and related valves (Fig. 21.24).

To test SCWG of biomass, several diverse solutions have been adopted. Figure 21.25 demonstrates one example of setup used in the literature (De Blasio et al. 2019).

The reactors used are listed as follows, starting from the smallest to the pilot-scale reactors configurations.

- capillary reactors, batch mode of operation (Sricharoenchaikul 2009; Potic et al. 2004);
- micro-reactors, or medium-size batch mode of operation (Castello et al. 2017);
- micro-reactors, continuous mode of operation, (Cao et al. 2011; De Blasio et al. 2015);
- fluidized bed reactors (simulation studies) (Yao and Lu 2017);
- medium-size reactors for laboratory use (Lu et al. 2012);
- Pilot-scale reactors (Yakaboylu et al. 2015).

The difference among them is the size of the reactor, but also the mode of operation. In the literature, it is not uncommon to find studies which were done in batch mode of

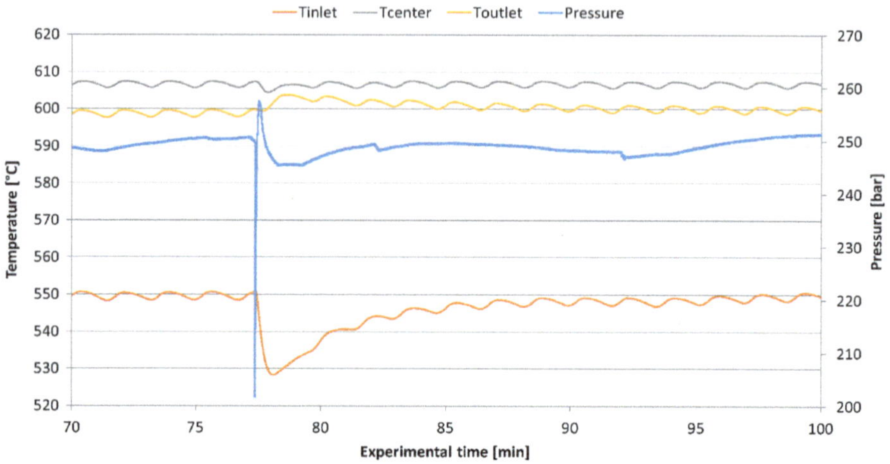


**Fig. 21.24** Pressure drop due to plugging of reactor and sudden release of the blockage wall

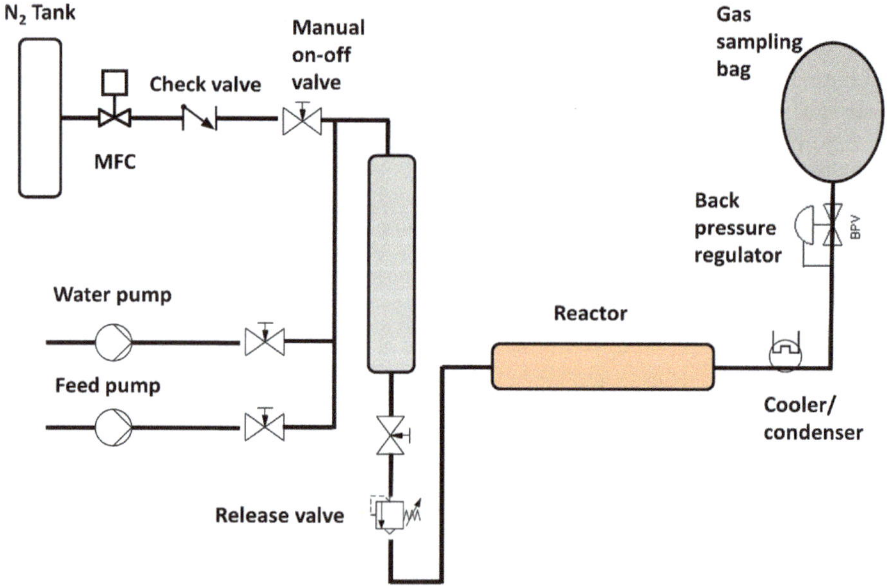


**Fig. 21.25** Schematics of a SCWG setup for laboratory-scale experiments. Adapted from De Blasio et al. (2019)

operation. However, in this case, a remark should be done on the conditions adopted. When batch conditions are adopted, the sample is normally inserted into the reactor at a certain time, and after this, the reactor should be brought to the target reactor's conditions. However, during this time, the temperature and pressure of the system will also change, and therefore, it will be difficult to really say when the experiment is started or not. In addition, during the heating of the reactor, there could have been reactions which could have changed the structure and composition of the feed sample, and therefore, when the experiment has started, we cannot talk about the same kind of sample.

## References

Antal, M. J., Brittain, A., DeAlmeida, C., Ramayya, S., & Roy, J. C. (1987). Heterolysis and homolysis in supercritical water. In *Supercritical Fluids, ACS Symposium Series. American Chemical Society* (pp. 77–86). https://doi.org/10.1021/bk-1987-0329.ch007.

Ariyapadi, S., Shires, P., Bhargava, M., & Ebbern, D. (2008). KBR's transport gasifier (TRIG)—An advanced gasification technology for SNG production from low-rank coals. In *Presented at the Twenty-fifth Annual International Pittsburgh Coal Conference*, KBR, Pittsburgh, USA.

Breault, R. W. (2010). Gasification processes old and new: A basic review of the major technologies. *Energies, 3,* 216–240. https://doi.org/10.3390/en3020216.

Byrd, A. J., Pant, K. K., & Gupta, R. B. (2008). Hydrogen production from glycerol by reforming in supercritical water over Ru/Al2O3 catalyst. *Fuel, 87,* 2956–2960. https://doi.org/10.1016/j.fuel.2008.04.024.

Cao, C., Guo, L., Chen, Y., Guo, S., & Lu, Y. (2011). Hydrogen production from supercritical water gasification of alkaline wheat straw pulping black liquor in continuous flow system. *International Journal of Hydrogen Energy, 2010 Asian/APEC BioH2, 36,* 13528–13535. https://doi.org/10.1016/j.ijhydene.2011.07.101.

Castello, D., & Fiori, L. (2011). Supercritical water gasification of biomass: Thermodynamic constraints. *Bioresource Technology, 102,* 7574–7582. https://doi.org/10.1016/j.biortech.2011.05.017.

Castello, D., Rolli, B., Kruse, A., & Fiori, L. (2017). Supercritical water gasification of biomass in a ceramic reactor: Long-time batch experiments. *Energies, 10,* 1734. https://doi.org/10.3390/en10111734.

Cervi, E., & Cammi, A. (2018). Stability analysis of the supercritical water reactor by means of the root locus criterion. *Nuclear Engineering and Design, 338,* 137–157. https://doi.org/10.1016/j.nucengdes.2018.08.004.

Chen, Z., Li, Y., Lai, D., Geng, S., Zhou, Q., Gao, S., et al. (2018). Coupling coal pyrolysis with char gasification in a multi-stage fluidized bed to co-produce high-quality tar and syngas. *Applied Energy, 215,* 348–355. https://doi.org/10.1016/j.apenergy.2018.02.023.

da Silva, D. C. D., Letichevsky, S., Borges, L. E. P., & Appel, L. G. (2012). The Ni/$ZrO_2$ catalyst and the methanation of CO and $CO_2$. *International Journal of Hydrogen Energy, 37,* 8923–8928. https://doi.org/10.1016/j.ijhydene.2012.03.020.

De Agarwal, S., Moholkar, V. S., & Thallada, B. (2018). *Coal and biomass gasification: Recent advances and future challenges* (1st ed.). Singapore: Springer.

De Blasio, C., De Gisi, S., Molino, A., Simonetti, M., Santarelli, M., & Björklund-Sänkiaho, M. (2019). Concerning operational aspects in supercritical water gasification of kraft black liquor. *Renewable Energy, 130,* 891–901. https://doi.org/10.1016/j.renene.2018.07.004.

De Blasio, C., & Järvinen, M. (2017). Supercritical water gasification of biomass. In *Encyclopedia of sustainable technologies* (pp. 171–195). Abraham M Editor.

De Blasio, C., Lucca, G., Özdenkci, K., Mulas, M., Lundqvist, K., Koskinen, J., et al. (2015). A study on supercritical water gasification of black liquor conducted in stainless steel and nickel-chromium-molybdenum reactors. *Journal of Chemical Technology and Biotechnology, 91*(10), 2664–2678. https://doi.org/10.1002/jctb.4871.

Farajtabar, A., & Gharib, F. (2009). Autoprotolysis constants determination of water-methanol mixtures and solvent effect. *Journal of Taibah University for Science, 2,* 7–13. https://doi.org/10.1016/S1658-3655(12)60002-8.

Finney, J. L. (2004). Water? What's so special about it? Philos. *Transactions of the Royal Society B: Biological Sciences, 359,* 1145–1328. https://doi.org/10.1098/rstb.2004.1495.

Gizir, A. M., Clifford, A. A., & Bartle, K. D. (2003). The catalytic role of transition metal salts on supercritical water oxidation of phenol and chlorophenols in a titanium reactor. *Reaction Kinetics and Catalysis Letters, 78,* 175–182. https://doi.org/10.1023/A:1022586505112.

Görsch, C., Hannemann, F., Hammer, T., & Meyer, B. (2011). Impact of the water quench on the syngas composition obtained from entrained flow gasification of carbonaceous fuels. In *28th international pittsburgh coal conference. Presented at the Pittsburgh Coal Conference*. Siemens, Pittsburgh, USA.

He, M., Su, C., Liu, X., Qi, X., & Lv, N. (2015). Measurement of isobaric heat capacity of pure water up to supercritical conditions. *The Journal of Supercritical Fluids, 100,* 1–6.

Hoffmann, M. M., & Conradi, M. S. (1998). Are there hydrogen bonds in supercritical methanol and ethanol? *The Journal of Physical Chemistry B, 102,* 263–271. https://doi.org/10.1021/jp9726706.

Jones, D. G., Slater, J. E., Staehle, R. W. (1976). High temperature high pressure electrochemistry in aqueous solutions. In *International Corrosion Conference Series Volume 4 of National Association of Corrosion Engineers: NACE*. National Association of corrosion engineers, University of Surrey, Guildford.

Kalinichev, A. G. (2017). Universality of hydrogen bond distributions in liquid and supercritical water. *Journal of Molecular Liquids, 241,* 1038–1043. https://doi.org/10.1016/j.molliq.2017.06.114.

Karakuş, Y., Aynacı, F., Kıpçak, E., & Akgün, M. (2013). Hydrogen production from 2-propanol over Pt/Al2O3 and Ru/Al2O3 catalysts in supercritical water. *International Journal of Hydrogen Energy, 38,* 7298–7306. https://doi.org/10.1016/j.ijhydene.2013.04.033.

Koss, U., & Schlichting, H. (2005). *Lurgi's MPG Gasification plus Rectisol® gas purification–Advanced process combination for reliable syngas production*. San Francisco, California, US: Presented at the Gasification Technologies.

Krishnamoorthy, V., & Pisupati, S. V. (2015). A critical review of mineral matter related issues during gasification of coal in fixed, fluidized, and entrained flow gasifiers. *Energies, 8,* 10430–10463. https://doi.org/10.3390/en80910430.

Kruse, A. (2008). Supercritical water gasification. *Biofuels, Bioproducts and Biorefining, 2,* 415–437. https://doi.org/10.1002/bbb.93.

Kunii, D., & Levenspiel, O. (1991). Fluidization engineering (2nd ed.). Elsevier.

Lau, F. (2009). Commercial development of the SES U-gas gasification technology. In *Presented at the gasification technologies conference, synthesis energy systems*. US: Colorado Springs.

Lisandy, K. Y., Kim, R.-G., Hwang, C.-W., & Jeon, C.-H. (2016). Sensitivity test of low rank Indonesian coal utilization using steady state and dynamic simulations of entrained-type gasifier. *Applied Thermal Engineering, 102,* 1433–1450. https://doi.org/10.1016/j.applthermaleng.2016.04.040.

Louw, J., Schwarz, C. E., & Burger, A. J. (2016). Catalytic supercritical water gasification of primary paper sludge using a homogeneous and heterogeneous catalyst: Experimental vs thermodynamic equilibrium results. *Bioresource Technology, 201,* 111–120. https://doi.org/10.1016/j.biortech.2015.11.043.

Lu, Y., Guo, L., Zhang, X., Ji, C. (2012). Hydrogen production by supercritical water gasification of biomass: Explore the way to maximum hydrogen yield and high carbon gasification efficiency. *International Journey of Hydrogen Energy, International Conference on Renewable Energy (ICRE 2011)*, 37, 3177–3185. https://doi.org/10.1016/j.ijhydene.2011.11.064

Martskainen, O. (2009). *Tautomerism and fragmentation of biologically active hetero atom (O,N)—containing acylic and cyclic compounds under electron ionization*. Painosalama OY.

Mitsubishi Heavy Industries. (2019). *Gasification process | mitsubishi heavy industries environmental & chemical engineering* [WWW Document]. URL https://www.mhiec.co.jp/en/products/recycle/city/meltingsystem/contents/gasification.html. Accessed January 24, 19.

Nanda, S., Dalai, A. K., & Kozinski, J. A. (2016). Supercritical water gasification of timothy grass as an energy crop in the presence of alkali carbonate and hydroxide catalysts. *Biomass and Bioenergy, 95,* 378–387. https://doi.org/10.1016/j.biombioe.2016.05.023.

Nguyen, H. T., Lu, H., Kobayashi, E., Ishikawa, T., & Komiyama, M. (2014). Raney-Nickel catalyst deactivation in supercritical water gasification of ethanol fermentation stillage and its mitigation. *Topics in Catalysis, 57,* 1078–1084. https://doi.org/10.1007/s11244-014-0272-x.

Onwudili, J. A., & Williams, P. T. (2009). Role of sodium hydroxide in the production of hydrogen gas from the hydrothermal gasification of biomass. International Journal of Hydrogen Energy, 2nd International Conference on Hydrogen Safety 34, 5645–5656. https://doi.org/10.1016/j.ijhydene.2009.05.082.

Onwudili, J. A., & Williams, P. T. (2010). Hydrothermal reactions of sodium formate and sodium acetate as model intermediate products of the sodium hydroxide-promoted hydrothermal gasification of biomass. *Green Chemistry, 12,* 2214–2224. https://doi.org/10.1039/C0GC00547A.

Osada, M., Sato, O., Watanabe, M., Arai, K., & Shirai, M. (2006). Water density effect on lignin gasification over supported noble metal catalysts in supercritical water. *Energy & Fuels, 20,* 930–935. https://doi.org/10.1021/ef050398q.

Osada, M., Yamaguchi, A., Hiyoshi, N., Sato, O., & Shirai, M. (2012). Gasification of sugarcane bagasse over supported ruthenium catalysts in supercritical water. *Energy & Fuels, 26,* 3179–3186. https://doi.org/10.1021/ef300460c.

Özdenkçi, K., De Blasio, C., Muddassar, H. R., Melin, K., Oinas, P., Koskinen, J., et al. (2017). A novel biorefinery integration concept for lignocellulosic biomass. *Energy Conversion and Management, 149,* 974–987. https://doi.org/10.1016/j.enconman.2017.04.034.

Portela, J. R., Nebot, E., & Martínez de la Ossa, E. (2001). Kinetic comparison between subcritical and supercritical water oxidation of phenol. *Chemical Engineering Journal, 81,* 287–299. https://doi.org/10.1016/S1385-8947(00)00226-6.

Potic, B., Kersten, S. R. A., Prins, W., & van Swaaij, W. P. M. (2004). A high-throughput screening technique for conversion in hot compressed water. *Industrial and Engineering Chemistry Research, 43,* 4580–4584. https://doi.org/10.1021/ie030732a.

Randolph, T. W., & Carlier, C. (1992). Free-radical reactions in supercritical ethane: A probe of supercritical fluid structure. *Journal of Physical Chemistry, 96,* 5146–5151. https://doi.org/10.1021/j100191a072.

Ratkiewicz, A., & Truong, T. N. (2012). Kinetics of the C–C bond beta scission reactions in alkyl radical reaction class. *Journal of Physical Chemistry A, 116,* 6643–6654. https://doi.org/10.1021/jp3018265.

Resende, F. L. P., & Savage, P. E. (2010). Effect of metals on supercritical water gasification of cellulose and lignin. *Industrial and Engineering Chemistry Research, 49,* 2694–2700. https://doi.org/10.1021/ie901928f.

Rönnlund, I., Myréen, L., Lundqvist, K., Ahlbeck, J., & Westerlund, T. (2011). Waste to energy by industrially integrated supercritical water gasification—Effects of alkali salts in residual by-products from the pulp and paper industry. In: Energy, 5th Dubrovnik Conference on Sustainable Development of Energy, Water & Environment Systems, (Vol 36, pp. 2151–2163). https://doi.org/10.1016/j.energy.2010.03.027.

Ryan, E. T., Xiang, T., Johnston, K. P., & Fox, M. A. (1996). Excited-state proton transfer reactions in subcritical and supercritical water. *Journal of Physical Chemistry, 100,* 9395–9402. https://doi.org/10.1021/jp9604204.

Sasaki, M., Furukawa, M., Minami, K., Adschiri, T., & Arai, K. (2002). Kinetics and mechanism of cellobiose hydrolysis and retro-aldol condensation in subcritical and supercritical water. *Industrial and Engineering Chemistry Research, 41,* 6642–6649. https://doi.org/10.1021/ie020326b.

Sikarwar, V. S., Zhao, M., Clough, P., Yao, J., Zhong, X., Memon, M. Z., et al. (2016). An overview of advances in biomass gasification. *Energy & Environmental Science, 9,* 2939–2977. https://doi.org/10.1039/C6EE00935B.

Sricharoenchaikul, V. (2009). Assessment of black liquor gasification in supercritical water. *Bioresource Technology, 100,* 638–643. https://doi.org/10.1016/j.biortech.2008.07.011.

Stangeland, K., Kalai, D., Li, H., & Yu, Z. (2017). $CO_2$ Methanation: The effect of catalysts and reaction conditions. In Energy procedia, 8th international conference on applied energy, ICAE2016, 8–11 Oct 2016 (Vol. 105, pp. 2022–2027), Beijing, China. https://doi.org/10.1016/j.egypro.2017.03.577.

Susanti, R. F., Veriansyah, B., Kim, J.-D., Kim, J., & Lee, Y.-W. (2010). Continuous supercritical water gasification of isooctane: A promising reactor design. *International Journal of Hydrogen Energy, 35,* 1957–1970. https://doi.org/10.1016/j.ijhydene.2009.12.157.

ThyssenKrupp. (2019). *Gasification technologies.*

Toporov, D., & Abraham, R. (2015). Gasification of low-rank coal in the high-temperature winkler (HTW) process. *Journal of the South African Institute of Mining and Metallurgy, 115,* 589–597. https://doi.org/10.17159/2411-9717/2015/V115N7A5.

Wagner, W., & Kretzschmar, H.-J. (2008). *International steam tables—Properties of water and steam based on the industrial formulation IAPWS-IF97: Tables, algorithms, diagrams, and CD-ROM electronic steam tables—All of the equations of IAPWS-IF97 including a complete set of supplementary backward equations for fast calculations of heat cycles, boilers, and steam turbines* (2nd ed.). Berlin Heidelberg: Springer.

Wang, J., & Takarada, T. (2001). Role of calcium hydroxide in supercritical water gasification of low-rank coal. *Energy & Fuels, 15,* 356–362. https://doi.org/10.1021/ef000144z.

Wang, C., Zhu, W., Chen, C., Zhang, H., Lin, N., & Su, Y. (2018). Influence of reaction conditions on the catalytic activity of a nickel during the supercritical water gasification of dewatered sewage sludge. *Journal of Supercritical Fluids, 140,* 356–363. https://doi.org/10.1016/j.supflu.2018.07.018.

Watanabe, M., Inomata, H., Smith, R. L., & Arai, K. (2001). Catalytic decarboxylation of acetic acid with zirconia catalyst in supercritical water. *Applied Catalysis General, 219,* 149–156. https://doi.org/10.1016/S0926-860X(01)00677-9.

Yakaboylu, O., Harinck, J., Smit, K. G., & de Jong, W. (2015). Supercritical water gasification of biomass: A literature and technology overview. *Energies, 8,* 859–894. https://doi.org/10.3390/en8020859.

Yamamoto, N., Murakami, K., Kimthet, C., Wahyudiono, W., Onwona-Agyeman, S., Kanda, H., et al. (2018). Utilization of sub and supercritical water reactions in resource recovery of biomass wastes. *Engineering Journal, 22,* 31–44. https://doi.org/10.4186/ej.2018.22.5.31.

Yang, X., Cheng, K., & Jia, G. (2019). The molecular dynamics simulation of hydrogen bonding in supercritical water. *Physica A: Statistical Mechanics and its Applications, 516,* 365–375. https://doi.org/10.1016/j.physa.2018.10.022.

Yanik, J., Ebale, S., Kruse, A., Saglam, M., & Yüksel, M. (2008). Biomass gasification in supercritical water: II. Effect of catalyst. *International Journal of Hydrogen Energy, 33,* 4520–4526. https://doi.org/10.1016/j.ijhydene.2008.06.024.

Yao, L., & Lu, Y. (2017). Supercritical water gasification of glucose in fluidized bed reactor: A numerical study. *International Journal of Hydrogen Energy, 42,* 7857–7865. https://doi.org/10.1016/j.ijhydene.2017.03.009.

Zhang, J., Liu, J., & Liu, R. (2015). Effects of pyrolysis temperature and heating time on biochar obtained from the pyrolysis of straw and lignosulfonate. *Bioresource Technology, 176,* 288–291. https://doi.org/10.1016/j.biortech.2014.11.011.

Zhou, L., & Zhou, Y. (2001). Determination of compressibility factor and fugacity coefficient of hydrogen in studies of adsorptive storage. *International Journal of Hydrogen Energy, 26,* 597–601. https://doi.org/10.1016/S0360-3199(00)00123-3.

# Part III
# Review of Some Concepts of Thermodynamics and Support Material

# Chapter 22
# Preliminary Concepts

*This, to be sure, deprives the work of the character of an original contribution to science, and stamps it rather as an introductory text-book on Thermodynamics for students who have taken elementary courses in Physics and Chemistry, and are familiar with the elements of the Differential and Integral Calculus.*

A sentence written by Max Planck about his book, Treatise on Thermodynamics, translated by A. Ogg in 1903. Longmans, green and Co. Publishers.

## 22.1 Definition of Mole

Let us assume that a quantity of mass is entering a reactor at a known mass flow (kg/s). In case the substance entering the reactor is known, then it would be possible to calculate the number of moles of that substance. In other words, it will be possible to calculate the number of molecules that are entering that reactor as a function of time. This is useful, as it was stated in the manuscript, since the reactor volume can vary (meaning that the volume of the matter in the reactor can actually vary during the reaction time).

By definition, one mole represents the same number of particles present in 12 g of carbon-12. For this reason, this is referred as g-mole. This number is called Avogadro's number or Avogadro's constant ($N_A$) and is equal to $6.022 \times 10^{23}$ particles.

Sometimes, it is referred in the literature to kg-mole instead of g-mole and since 1 kg = 1000 g, then the number of molecules in a kg-mole is 1000 g-mole.

From the definition, it comes natural that one mole of a pure substance has a mass in grams equal to its molecular mass, also known as molecular weight (MW). This is often referred to as the molar mass.

The volume of a substance is directly proportional to its number of moles. To give one example, one mole of an ideal gas has a volume of: 22.4 L at standard temperature and pressure, 0 °C (273 K), and 101.3 kPa (1 atm) (Prigogine 1968; Levine 2008).

 
C. De Blasio, *Fundamentals of Biofuels Engineering and Technology*,
Green Energy and Technology, https://doi.org/10.1007/978-3-030-11599-9_22

## 22.2 Conservation of Mass, Closed Systems, Degree of Advancement of Reaction

Chemical reactions are almost everywhere in engineering processes. It is therefore of outmost importance to understand some preliminary concepts concerning chemical reactions. It has been stated in this manuscript that, in case we consider a balance on the number of moles entering a reactor, the volume of the matter inside the reactor will indeed not remain the same and it is normally an assumption to consider it as constant. In this case, we will consider here the definition of the degree of advancement of a reaction. Let us define the change in mass for a component:

$$\mathrm{d}m_{\gamma} = \nu_{\gamma} M_{\gamma} \mathrm{d}\varepsilon \tag{22.1}$$

Where

$\nu_{\gamma}$ is the stoichiometric coefficient (dimensions of mole).
$\mathrm{d}m_{\gamma}$ is the change in mass of a component $\gamma$.
$M_{\gamma}$ is the molar mass.

The mass in closed systems is conserved even if there are chemical reactions involved, with changes in mole numbers.

$$\mathrm{d}m = \left(\sum_{\gamma} \nu_{\gamma} M_{\gamma}\right) \mathrm{d}\varepsilon = 0 \tag{22.2}$$

## 22.3 About Internal Energy and Enthalpy

In this section, we will take into account two important concepts normally used in energy technology and chemical engineering. First, let us consider here the enthalpy "which is the formally incorrect combination of two different kinds of magnitudes" (Astarita 1993). The enthalpy is defined as:

$$H = U + PV \tag{22.3}$$

Differentiating Eq. (22.3) we have:

$$\mathrm{d}H = \mathrm{d}U + p\mathrm{d}V + V\mathrm{d}p \tag{22.4}$$

Now, it is important to notice that the internal energy, $U$, is not a function of the volume, and therefore, the terms stating a variation of volume such as $\mathrm{d}V$ are going to be zero if the volume is not changing. We have:

$$\mathrm{d}H = \frac{\mathrm{d}U}{\mathrm{d}T}_v \mathrm{d}T + \frac{\mathrm{d}U}{\mathrm{d}V}_T \mathrm{d}V + p\mathrm{d}V + V\mathrm{d}P \tag{22.5}$$

And therefore:

$$H = c \cdot \mathrm{d}T \tag{22.6}$$

where $c$ is a generic specific heat

$$c_v \mathrm{d}T = u \tag{22.7}$$

This is because we do not have the term $P \cdot V$ in case we consider only $U$. Therefore:

$$c\mathrm{d}T = \frac{\mathrm{d}U}{\mathrm{d}T}_v \mathrm{d}T + p\mathrm{d}V + V\mathrm{d}P \tag{22.8}$$

and

$$c_p \mathrm{d}T = \mathrm{d}u + p\mathrm{d}V + v\mathrm{d}p = \mathrm{d}h \tag{22.9}$$

This is derived by following the definition of enthalpy.

## 22.4 Relations Between Specific Heat at Constant Volume and Pressure

We know that:

$$P \cdot v = R \cdot T \tag{22.10}$$

where $R$ is the universal gas constant and equal to 8.314 J/(mol K). We also have from previous derivations:

$$c_v \mathrm{d}T = u \tag{22.11}$$

In addition, we know also that:

$$c_p \mathrm{d}T = u + p\mathrm{d}v + v\mathrm{d}p = h \tag{22.12}$$

Substract Eq. (22.11) from Eq. (22.12), we obtain:

$$c_p \mathrm{d}T - c_v \mathrm{d}T = u + p\mathrm{d}v + v\mathrm{d}p - u \tag{22.13}$$

where

$$p\mathrm{d}v + v\mathrm{d}p = R\mathrm{d}T \tag{22.14}$$

To be noticed that Eq. (22.14) comes straight from Eq. (22.10) differentiating with respect to the variables $P$ and $T$. And the following relation is derived:

$$c_p - c_v = R \tag{22.15}$$

In case we consider an isoentropic process the ratio between the specific heat parameters at constant pressure and constant volume:

$$\frac{c_p}{c_v} = k \tag{22.16}$$

Equation (22.16) also has a straightforward derivation from Eqs. (22.11) and (22.12). This is because if the entropy is not changing, then there is no variation of volume or pressure on the system and also the following relation is valid:

$$PV^k = \text{constant} \tag{22.17}$$

## 22.5 The First Law of Thermodynamics

The change in energy contained in a system during a time interval is equal to the net energy transfer in the system by heat transfer during the time interval minus the net amount of energy given out of the system by work during the time interval. This is stated as:

$$\frac{\mathrm{dKE}}{\mathrm{d}t} + \frac{\mathrm{dPE}}{\mathrm{d}t} + \frac{\mathrm{d}U}{\mathrm{d}t} = \dot{Q} - \dot{W} \tag{22.18}$$

where *KE* and *PE* are the kinetic energy and the potential energy, while $\dot{Q}$ and $\dot{W}$ are, respectively, the flows of heat and work. To be noticed that the work done on a system is positive and this comes directly from the fact that when the pressure is constant and we apply this pressure on a volume, the work will be described by $W = -P \cdot \mathrm{d}V$ and since $\mathrm{d}V$ would be approximated as $(V_{\text{final}} - V_{\text{initial}})$, if we do a compression on the volume, this work will be positive.

A general balance should always be remembered when starting doing calculations on mass and energy balance:

$$\text{Input} - \text{Output} +/- \text{Generation (example: chemical reaction)} = \text{Accumulation} \tag{22.19}$$

Notice again that the term "generation" refers, for instance, to the number of moles in case the balance is done on them.

## 22.6 Some Relations Between Pressure and Temperature in Gases. Gas Laws

For an ideal gas, the following notorious formula is applied:

$$PV = nRT \tag{22.20}$$

where the term $R$ was mentioned already, and it refers to the universal gas constant:

$$R = 8.3144621 \; \frac{\text{J}}{\text{mol K}} \tag{22.21}$$

and $n$ is the number of moles. One mole is formed by: $6.02214129 \times 10^{23}$ molecules/mol$^{-1}$.

If the temperature is constant, we derive Boyle's law from Eq. (22.20)

$$P_1 V_1 = P_2 V_2 \tag{22.22}$$

where 1 and 2 refer to the states of the system. On the other hand, in case the pressure is constant, we have the Charles law:

$$V_1 / V_2 = T_1 / T_2 \tag{22.23}$$

$$P_1 V_1 / T_1 = P_2 V_2 / T_2 \tag{22.24}$$

If $n_1 = n_2$ and $V_1 = V_2$, we derive the Gay-Lussac Law. The $n$'s and the $V$'s cancel in the above original expression and we have:

$$P_1 / T_1 = P_2 / T_2 \tag{22.25}$$

## 22.7 A Remark on Reversible Processes

Processes, described by equations which are invariant to the sign we give to the variable of time, are reversible.

For instance, the following equation:

$$\frac{1}{c^2}\frac{\vartheta^2 u}{\vartheta t^2} = \frac{\vartheta^2 u}{\vartheta x^2} + \frac{\vartheta^2 u}{\vartheta y^2} + \frac{\vartheta^2 u}{\vartheta z^2} \tag{22.26}$$

Instead, processes, described by equations where the sign given to the time makes a difference, are irreversible. Such as:

$$\frac{1}{\alpha}\frac{\vartheta T}{\vartheta t} = \frac{\vartheta^2 T}{\vartheta x^2} + \frac{\vartheta^2 T}{\vartheta y^2} + \frac{\vartheta^2 T}{\vartheta z^2} \tag{22.27}$$

## 22.8 Pressure and Internal Energy, the Kinetic Model

Consider Fig. 22.1 where a simple space is given as a box. In the box, we imagine there is an object moving back and forth along a distance $L$ with a velocity $w_i$. With the letter $i$, we refer to the $i$th particle of a system.

Remember that the velocity is a vector and it has a module (the quantity that describes how big the vector is) and a direction. We consider now one molecule going back and forth and a box (cube) of length $L$.

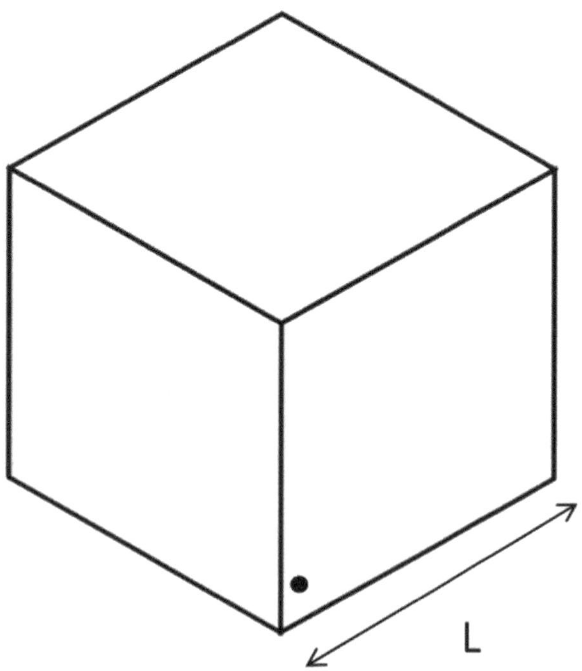


**Fig. 22.1** Simple representation of a space. The box with a moving object in it

$$\Delta\text{Momentum} = \text{m}w_i - (-mw_i) = 2mw_i \tag{22.28}$$

The period will be

$$t = \frac{2L}{w_i} \tag{22.29}$$

The frequency will be:

$$\text{Frequency} = \frac{\text{number of movement}}{\text{time}} = \frac{w_i}{2L} \tag{22.30}$$

The force associated with the particle is:

$$\text{Force} = 2m \cdot w_i \cdot \frac{w_i}{2L} = \frac{mw_i^2}{L} \tag{22.31}$$

To calculate the pressure, remember that the pressure is defined as the force/surface where the force is applied.

$$\text{Pressure of one molecule } i = \frac{mwi^2}{L} \cdot \frac{1}{3L^2} \tag{22.32}$$

In Eq. (22.32), the 1/3 is there since we are considering three pairs of faces, in other words, all the box. In case:

$$n^* = \frac{N}{L^3} \tag{22.33}$$

We have:

$$P = \frac{m}{3} \cdot n \cdot \frac{\sum_{i=1}^{N} w_i 2}{N} = \frac{1}{3} \cdot \rho \cdot (\hat{w})^2 \tag{22.34}$$

where

$n^*$ is the number of molecules/m$^3$.
$m \cdot N = m_{\text{tot}}$ in the space $L^3$.
$m \cdot n^* = \frac{\text{mass}}{\text{m}^3}$ which is the density.

## 22.9 Specific Heat and Universal Gas Constant

From previous equations, it is possible to write:

$$PV = \frac{1}{3} m_{\text{Tot}} w^2 \tag{22.35}$$

And therefore,

$$PV = \frac{2}{3}\frac{1}{2}m_{\text{Tot}}w^2 \tag{22.36}$$

We remember the ideal gas law:

$$PV = nRT \quad \text{and} \quad H = U + PV \tag{22.37}$$

This expression will be written to consider $M$ as the molecular weight:

$$\frac{1}{2}Mw^2 = \frac{3}{2} \cdot \frac{1}{n} \cdot nRT = \frac{3}{2}RT \tag{22.38}$$

In case we consider a system closed, this will give:

$$c_v\text{d}T = \text{d}U = \frac{3}{2}nR\text{d}T = \frac{1}{2}m_{\text{tot}}w^2 \tag{22.39}$$

This is the derived relation between the temperature and the kinetic energy. In case the system is able to do work at P constant:

$$\text{d}H = c_p\text{d}T = \text{d}U + p\text{d}V = \frac{3}{2}nR\text{d}T + \text{nRdT} = \frac{5}{2}nR\text{d}T \tag{22.40}$$

Now, let us consider a particle traveling a piece of space $\Delta d =$ space travelled. Its velocity is:

$$v_{av} = \frac{\Delta d}{\Delta t} \tag{22.41}$$

and the acceleration will be:

$$a_{av} = \frac{\Delta v}{\Delta t} \tag{22.42}$$

The force on average will be:

$$F_{av} = m \cdot a_{av} = m\left(\frac{\Delta v}{\Delta t}\right) \tag{22.43}$$

and its work:

$$\text{work} = F_{av}\Delta d = m\left(\frac{\Delta v}{\Delta t}\right)\Delta d \tag{22.44}$$

Or

$$\text{work} = F_{av}\Delta d = m\left(\frac{v_a - v_b}{\Delta t}\right)\Delta d \tag{22.45}$$

With the average speed over the time interval:

$$\Delta d = \frac{1}{2}(v_a + v_b)\Delta t \tag{22.46}$$

And therefore, the following expression is obtained:

$$\text{work} = \frac{1}{2}\left(v_a^2 - v_b^2\right) \tag{22.47}$$

## 22.10 Change of State for an Ideal Gas

For this section, it is good to recall the Avogadro number $N_A = 6.022 \times 10^{23}$ particles. It is also beneficial to observe that

$$K_B = \frac{R}{N_A} \approx 1.38 \cdot 10^{-23}\ \text{J/K} \tag{22.48}$$

This is a straight derivation from the ideal gas law. In case we have an isochoric process ($V = \text{const}$) (Fig. 22.2).

In this case, we necessarily have:

$$W_{1\to 2} = 0 \tag{22.49}$$

Since we do not have a variation of the volume for our system, however, a change in internal energy is still possible obviously:

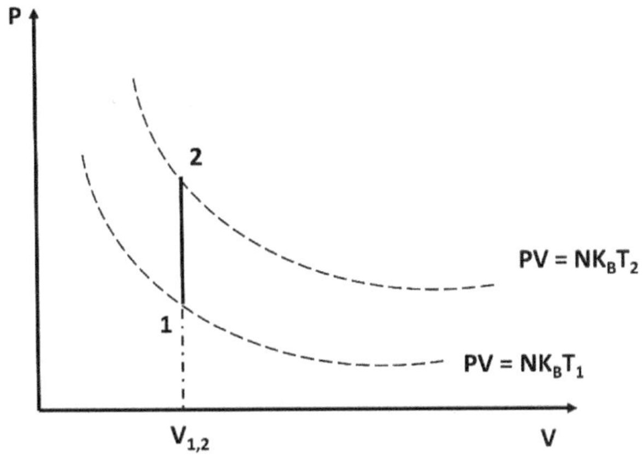


**Fig. 22.2** Constant volume system with varying pressure

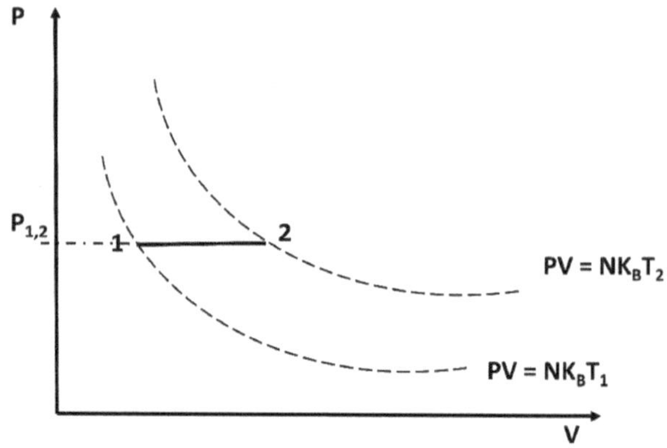


**Fig. 22.3** Simple representation of an isobaric process

$$Q_{1\to 2} = \frac{3}{2} N K_B (T_2 - T_1) > 0 \tag{22.50}$$

where

$$N(\text{molecules}) = n(\text{moles}) \cdot N_A$$

This is also more simply written as:

$$Q_{1\to 2} = C_V \Delta T \tag{22.51}$$

and since the work done on the system is zero, we have that the internal energy is just a function of the heat variation in the system considered.

$$\mathrm{d}U = Q_{1\to 2} \tag{22.52}$$

For an isobaric process, we have that the pressure is constant during the transformation and the volume is instead varying (Fig. 22.3).

In this case, the volume of the system considered is changing, and therefore, the work done on the system should be taken into account:

$$W_{1\to 2} = -\int_1^2 P(V, T)\mathrm{d}V = -P(V_2 - V_1) < 0 \tag{22.53}$$

and the change in heat content of the system will be:

$$Q_{1\to 2} = \frac{5}{2} N K_B (T_2 - T_1) > 0 \tag{22.54}$$

In this case, the quantity $Q_{1\to 2}$ is also clearly:

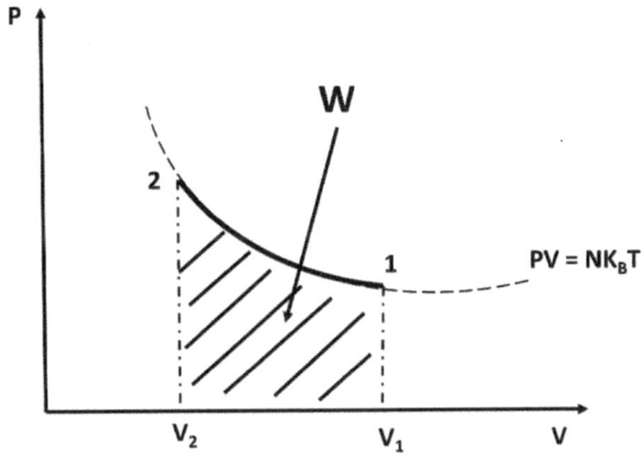


**Fig. 22.4** Isothermal process. The points 1 and 2 are on the same curve in this case: the isothermal *P*-*V* curve

$$Q_{1\to 2} = C_P \Delta T \tag{22.55}$$

and therefore, the overall internal energy of the system will be:

$$\mathrm{d}U = W_{1\to 2} + Q_{1\to 2} \tag{22.56}$$

For an isothermal process, we have a transformation where the temperature is remained constant (Fig. 22.4).

In this case, there will be no variation for the internal energy for the system considered:

$$\mathrm{d}U = 0 \tag{22.57}$$

$$W_{1\to 2} = -\int_1^2 P(V, T)\mathrm{d}V = -NK_BT \int_{V_1}^{V_2} \frac{\mathrm{d}V}{V} = -NK_BT \ln \frac{V_2}{V_1} \tag{22.58}$$

and from previously written equations, it follows that:

$$Q_{1\to 2} = -W_{1\to 2} \tag{22.59}$$

In case we generalize and describe with *i,* the initial stage and with *f* the final stage of the process:

$$W_{i-f} = -NK_BT \ln \frac{V_f}{V_i} \tag{22.60}$$

In the case, we have compression:

$$V_i > V_f;$$

and this will imply:

$$W_{i-f} > 0$$

And in case of expansion, the system does work to the environment and we have:

$$V_i < V_f$$

with

$$W_{i-f} < 0$$

## 22.11 Adiabatic Transformation of an Ideal Gas

When a system is considered to be adiabatic, this means that we are talking about a thermally isolated system.

$$Q_{1\to 2} = 0 \tag{22.61}$$

and

$$\mathrm{d}U = W_{1\to 2} \tag{22.62}$$

In order to calculate $W_{1\text{–}2}$, we need the function of the pressure which is dependent upon the volume and the temperature, $P(V, T)$, and therefore, for an adiabatic process, it is still valid the relation:

$$W_{1\to 2} = -\int_1^2 P(V, T)\mathrm{d}V \tag{22.63}$$

Let us consider the parameter $f$ which gives the degrees of freedom for a particular gas and, in other words, it gives the number of independent motions a molecule can have.

$f$ is considered to be 3 for monatomic molecules, 5 for diatomic molecules, and 6 for polyatomic. This could be simply explained by the fact that if we are considering only one atom, of course, there will be three possibilities to move along the special axes. On the other hand, for a diatomic molecule, there will be three possibilities to move (for the overall molecule) along the $x$, $y$, and $z$ axes and two possibilities for rotation. Notice that if a diatomic molecule will rotate along the axes connecting the two atoms, this will be not considered a rotation. For the case of nonlinear molecules, we have that the overall molecule can, also in this case, move along the three main axes, and furthermore, it can rotate along all the three axes. This will give a total of six configurations for the system at hand.

More generally, according to Maxwell's law of equipartition of energy the internal energy will be expressed as (Lima and Plastino 2000):

$$U = \frac{f}{2} N K_{\mathrm{B}} T \tag{22.64}$$

This is because Maxwell's law is stating that each degree of freedom has an equal amount of energy when it is in thermodynamic equilibrium, and this amount of energy will be proportional to the Boltzmann constant and the temperature of the system. A derivation for the law of equipartition for kinetic energy, thermodynamic equilibrium, and more than one dimension is given by Crawford (1986).

And this will mean:

$$\mathrm{d}U = \frac{f}{2} N K_{\mathrm{B}} \mathrm{d}T = -P \mathrm{d}V \tag{22.65}$$

From the relation:

$$PV = N K_{\mathrm{B}} T \tag{22.66}$$

we differentiate both terms with respect to the variables involved:

$$P\mathrm{d}V + V\mathrm{d}P = N K_{\mathrm{B}} \mathrm{d}T \tag{22.67}$$

and this will give:

$$P\mathrm{d}V + V\mathrm{d}P = -\frac{2}{f} P\mathrm{d}V \tag{22.68}$$

which in turn gives:

$$\frac{\mathrm{d}V}{V}\left(1 + \frac{2}{f}\right) + \frac{\mathrm{d}P}{P} = 0 \tag{22.69}$$

Now, if we consider the substitution:

$$\gamma = 1 + \frac{2}{f} \tag{22.70}$$

Integrating Eq. (22.69) we have:

$$\gamma \int_{V_1}^{V} \frac{\mathrm{d}V}{V} + \int_{P_1}^{P} \frac{\mathrm{d}P}{P} = 0 \tag{22.71}$$

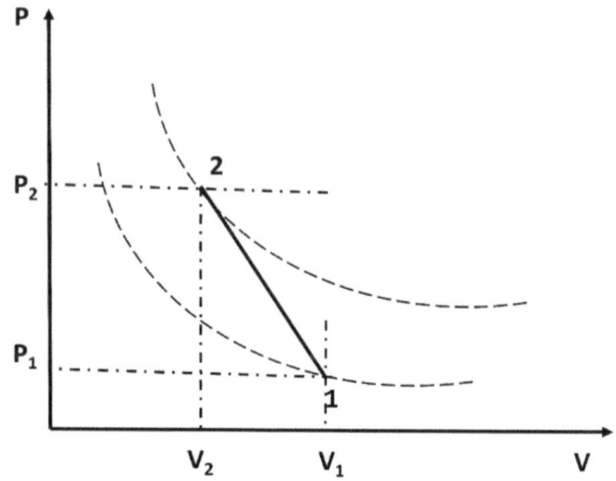


**Fig. 22.5** Adiamatic process of compression of an ideal gas

And finally:

$$\ln\left(\frac{V}{V_1}\right)^{\gamma} = \ln\left(\frac{P_1}{P}\right) \tag{22.72}$$

Which implies:

$$PV^{\gamma} = P_1 V_1^{\gamma} = \text{const} \tag{22.73}$$

In an adiabatic process (Fig. 22.5), the work done by the gas system on the environment is done at the expenses of its thermal energy, and this implies that its temperature will decrease. As a summary and in relation to the parameter $\gamma$, the following expression is obtained:

$$\begin{aligned} W_{1\to 2} &= -\int_1^2 P(V,T)\mathrm{d}V \\ &= -\int_{V_1}^{V_2} \frac{P_1 V_1^{\gamma}}{V^{\gamma}}\mathrm{d}V = -P_1 V_1^{\gamma}\frac{1}{1-\gamma}V^{1-\gamma}\Big|_{V_1}^{V_2} \\ &= P_1 V_1^{\gamma}\frac{1}{\gamma-1}\left(\frac{1}{V_2^{\gamma-1}} - \frac{1}{V_1^{\gamma-1}}\right) \end{aligned} \tag{22.74}$$

Considering Eq. (22.71), we have that $\gamma = 1 + 2/3 \approx 1.67$ for monatomic molecules, $1 + 2/5 = 1.4$ for diatomic molecules, and $1 + 2/6 \approx 1.33$ for polyatomic molecules. This is obtained not considering the vibrational degrees of freedom.

## 22.12 Energy Balance for a Given System: Macroscopic Balances

In this section (Fig. 22.6), the overall energy balance in a system will be given in accordance with what it was mentioned on a general mole/mass balance. Let us consider the following volume and always the general relation:

$$\text{In} - \text{Out} +/- \text{Gen} = \text{Accumulation}$$

The overall energy balance for the system will be:

$$\begin{aligned}\frac{\mathrm{d}(U_{\text{tot}} + K_{\text{tot}} + \Phi_{\text{tot}})}{\mathrm{d}t} &= \left(\rho_1 \widehat{U_1}\langle v_1\rangle + \frac{1}{2}\rho_1\langle v_1^3\rangle + \rho_1 \widehat{\Phi}_1\langle v_1\rangle\right) S_1 \\ &- \left(\rho_2 \widehat{U_2}\langle v_2\rangle + \frac{1}{2}\rho_2\langle v_2^3\rangle + \rho_2 \Phi_2\langle v_2\rangle\right) S_2 \\ &+ (p_1\langle v_1\rangle S_1 - p_2\langle v_2\rangle S_2 + W_m + Q \qquad (22.75)\end{aligned}$$

In Eq. (22.75), the terms $U_{\text{tot}} + K_{\text{tot}} + \Phi_{\text{tot}}$ are, respectively, the overall internal, kinetic, and potential energy change in the system. In case the balance is done within Sects. 22.1 and 22.2, the flow of energy is crossing the surfaces 1 and 2 at the inlet and outlet, respectively. $\rho_1$ is the density of the medium carrying the energy at Sect. 22.1, $\widehat{U_1}$ will be the specific internal energy of the medium at that section (notice that this overall term has to have dimensions of $U_{\text{tot}}/\text{time}$). $\langle v_1\rangle$ and $\langle v_2\rangle$ are, respectively, the velocities at Sects. 22.1 and 22.2. $S_1$ and $S_2$ are the surfaces where the flow is taking place at Sects. 22.1 and 22.2 (Biron Bird et al. 1960). $Q$ and $W$ are the rate at which heat is added/subtracted to the system and the work that is done or received by the system.

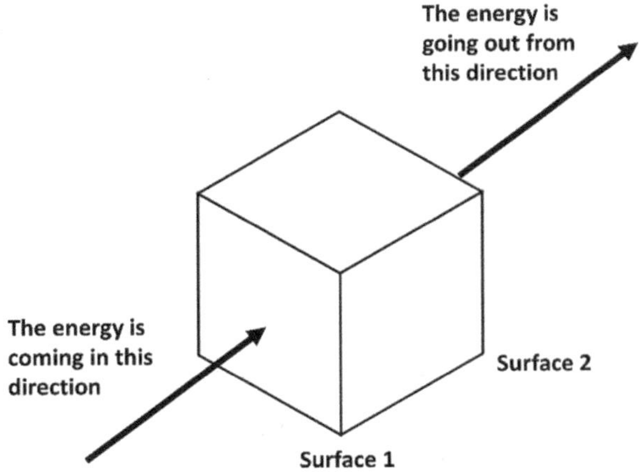


**Fig. 22.6** Volume with its boundaries

Equation (22.75) is written here with a descriptive purpose, and it is not the scope here to give a detailed treatment of the balances involved when macroscopic balances are considered. However, it is useful for the reader to notice here the analogy between heat and mass transfer.

## References

Prigogine, I. (1968). *Thermodynamics of irreversible processes* (3rd ed.). New York: Wiley.

Levine, I. N. (2008). *Physical chemistry* (6th ed.). McGraw-Hill Education.

Astarita, G. (1993). Thermodynamics: A view from outside. *Fluid Phase Equilibria, 82,* 1–14. https://doi.org/10.1016/0378-3812(93)87123-I.

Lima, J. A. S., Plastino, A. R. (2000). On the classical energy equipartition theorem. *Brazilian Journal of Physics, 30.* http://dx.doi.org/10.1590/S0103-97332000000100019.

Crawford, F. S. (1986). Elementary derivation of the law of equipartition of energy. *American Journal of Physics, 55,* 180.

Biron Bird, R., Stewart, W. E., & Lightfoot, E. N. (1960). *Transport phenomena.* Singapore: Wiley.

# Chapter 23
# Introduction to Entropy and Second Law

*Everyone knows that heat can produce motion.*
*That is, possesses vast motive-power no one can doubt, in these days when the steam-engine is everywhere so well known.*
Carnot (1897).

## 23.1 Mixing of One or Two Gases at the Same $T$ and $P$

Consider the setup given by Fig. 23.1, we have two gases in two boxes and the system formed by the two boxes is adiabatic, meaning that the system is not exchanging heat with the environment. The stage 1 refers to the gases when they are separated: The gas 1 is on the left, and the gas 2 is on the right of the figure. Additionally, let us also assume that the two gases have the same identical temperature.

In case:

$$\Delta Q = 0$$

and

$$\Delta T = 0$$

A system tends to evolve from a state at lower probability to a state at higher probability, this means that when we let the two gases to mix with each other, there will be more probability that the two gases will be completely mixed, and therefore, the uncertainty to locate a molecule in the final stage is doubled with respect to the first state.

The process described previously still involves a variation of entropy as it will be discussed later in this section.

Suppose that the heat transfer is going from a cold body to a warm body, then the first principle is still agreed; however, we know from experience that this is



C. De Blasio, *Fundamentals of Biofuels Engineering and Technology*,
Green Energy and Technology, https://doi.org/10.1007/978-3-030-11599-9_23

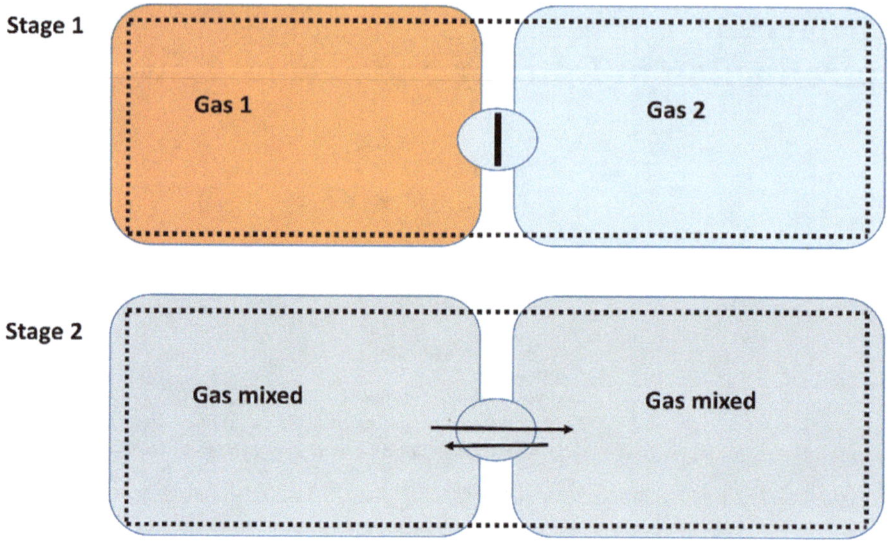


**Fig. 23.1** Adiabatic mixing of two ideal gases

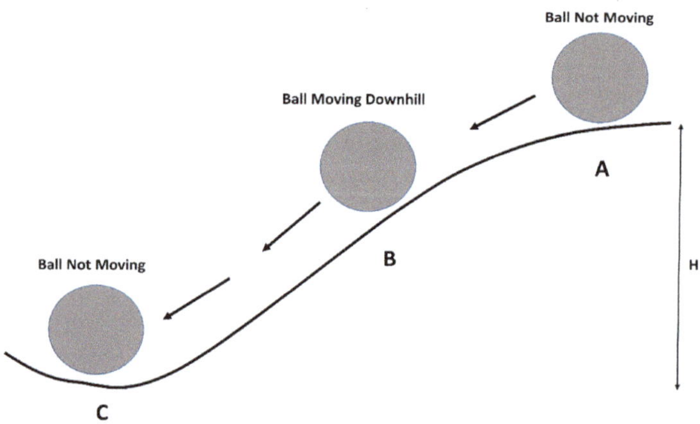


**Fig. 23.2** Object moving under the effect of a force field, gravity

impossible. The second law of thermodynamics is founded on the principle of non-conservation of a quantity called entropy.

The potential, mechanical, and thermal energies are transferred from a higher value to a lower value of the intensive variables: temperature, pressure, and coordinates. And therefore, it can be noticed the analogy between these quantities. For this purpose, let us consider the system described in Fig. 23.2.

In Fig. 23.2, we have an object (a ball) standing on a hill, the hill is at a height of $H$ with respect to the lowest available point. The ball in the initial state $A$ has potential energy, but it does not have kinetic energy since it is not moving. When the ball proceeds downhill, then it has kinetic and potential energy. At the bottom, the ball is still not moving again, and in this case, it does not have kinetic and neither potential energy.

If the system and the environment have the same value of their intensive variables, then it is impossible to have work. In addition, we want to know what is the preferable level at which the energy is.

Concerning the entropy concept, we can also think about a more simple derivation of it. Let us consider Fig. 23.3.

In Fig. 23.3, a general variable function $f(x, Y) = h$ is described as dependent upon two variables $x$ and $y$. In order to arrive to the final value, any path could be selected, and in particular, the following relations are valid:

$$\Delta h = \int_{y1}^{y2} S_y(y, x1)\mathrm{d}y + \int_{x1}^{x2} S_x(y_2, x)\mathrm{d}x \tag{23.1}$$

And

$$\Delta h = \int_{y1}^{y2} S_y(y, x2)\mathrm{d}y + \int_{x1}^{x2} S_x(y_1, x)\mathrm{d}x \tag{23.2}$$

$h$ is a function of state in case its variation does not depend on the route we choose, in addition, it has to be that the function:

$$\mathrm{d}h = H(x, y)\mathrm{d}x + F(x, y)\mathrm{d}y \tag{23.3}$$

Is an exact differential if:

$$H(x, y) = \left(\frac{\mathrm{d}g}{\mathrm{d}x}\right)_y \tag{23.4}$$

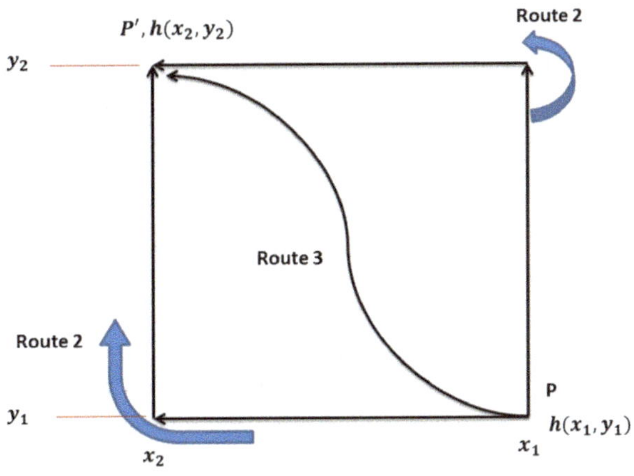


**Fig. 23.3** Initial and ending state of a variable with two possible routes to arrive to the final point

and

$$F(x, y) = \left(\frac{\mathrm{d}g}{\mathrm{d}y}\right)_x \quad (23.5)$$

Notice that this is necessary since the path described by dh in this case should be referred to the $x$ and $y$ coordinates. Of course, we could think about non-perpendicular coordinates and a different system of reference, and however, it is convenient here to keep the $x$-$y$ orthogonal axes.

**Example**

Consider the function of two variables,

$$z = x^2 y + 3y^2 \quad (23.6)$$

The differential will be then:

$$\mathrm{d}z = \left(\frac{\vartheta z}{\vartheta x}\right)_y \mathrm{d}x + \left(\frac{\vartheta z}{\vartheta y}\right)_x \mathrm{d}y = (2xy)\mathrm{d}x + \left(x^2 + 6y\right)\mathrm{d}y \quad (23.7)$$

In case we consider the points $P = (0, 0)$ and $P' = (2, 1)$. It will be possible to calculate $\Delta z$ along the Route 1 (referring to Fig. 23.3). We have:

$$(\Delta z)_1 = \int_0^1 \left(x_1^2 + 6y\right)\mathrm{d}y + \int_0^2 (2xy_2)\mathrm{d}x = 6\frac{y^2}{2}\Big|_0^1 + 2\frac{x^2}{2}\Big|_0^2 = \frac{6}{2} + 4 = 7 \quad (23.8)$$

Along the Route 2, the calculation will be as follows:

$$(\Delta z)_2 = \int_0^1 \left(x_2^2 + 6y\right)\mathrm{d}y + \int_0^2 (2xy_1)\mathrm{d}x = 0 + 4y|_0^1 + 6\frac{y^2}{2}\Big|_0^1 = \frac{6}{2} + 4 = 7 \quad (23.9)$$

In the case of the quantity heat and work, they are not variable of state, or state functions. Let us consider the following expression. In case the external pressure is not constant, we will have to consider the pressure as a variable, and it cannot be positioned outside the integral operation:

$$W = -\int_{vi}^{vf} P_{\mathrm{est}}\mathrm{d}V \quad (23.10)$$

Nevertheless, the internal energy is a state function:

$$\int_1^2 \mathrm{d}U = U_2 - U_1 = \Delta U \quad (23.11)$$

and by doing the partial differentiation for the definition of internal energy seen previously:

$$\delta q_{\text{rev}} = \mathrm{d}U - \delta w_{\text{rev}} = C_V(T)\mathrm{d}T + P\mathrm{d}V = C_V(T)\mathrm{d}T + \frac{nRT}{V}\mathrm{d}V \tag{23.12}$$

Dividing by the temperature, we have:

$$\frac{\delta q_{\text{rev}}}{T} = \frac{C_V(T)\mathrm{d}T}{T} + \frac{nR}{V}\mathrm{d}V \tag{23.13}$$

And this is a state function since the differential in this case is an exact differential. A function of this kind is also called "property."

The term:

$$\frac{\delta q_{\text{rev}}}{T} = S(V, T) \tag{23.14}$$

is called entropy and it is a function of state or state variable.

Integrating Eq. (23.14), we have:

$$S(V, T) = C_v \log T + nR \log V + \text{const} \tag{23.15}$$

## 23.2 Statistical Derivation of Entropy

In this section, the concept of entropy will be treated by its statistical meaning. Imagine two boxes filled with two different gases ($D$ and $E$) and they are put into communication by a conduct which is closed in the beginning. The two gases are at the same temperature and the system is adiabatic, meaning that there is no exchange of heat. The system boundaries are represented by the dashed line in Fig. 23.4.

In this system, there will be still variation of entropy, even if the temperature will be the same, and therefore, the definition given by Eq. (23.15) is not complete. The increase in entropy is the rise from a low-probability state to a high-probability state.

When we have the two gases mixed, there is 50% chance that one particle is in the right or left part, and for this reason, each of the particles can be considered as a Bernoulli random variable.

Each of the particles could be imagined as tosses of a coin, and in our case, the number of tosses is very high (consider the particles contained in one mole of gas). It can be written that the entropy is a function of the probability.

$$S = f(p) \tag{23.16}$$

Entropy is an extensive function of state, so referring to Fig. 23.4 it can be written:

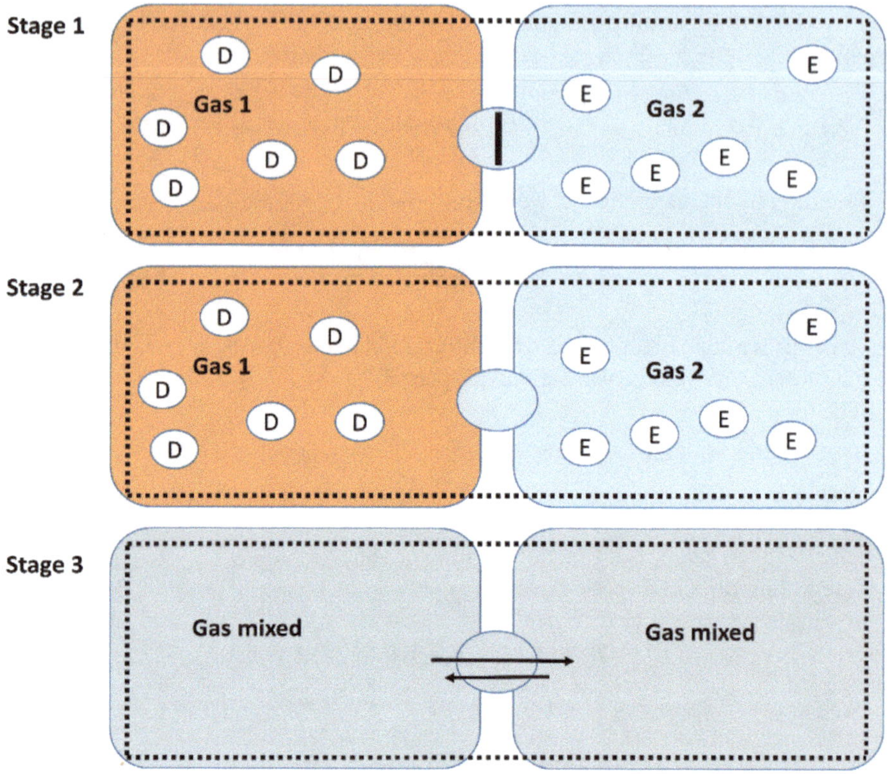


**Fig. 23.4** Mixing of two gases at the same temperature and adiabatic conditions

$$S_{1+2} = S_1 + S_2 \tag{23.17}$$

To understand this better, when we consider the overall probability for two events to happen, the overall probability is the product of the single probabilities. For example, we toss a coin and we have 0.5 chance to get one side of the coin. If we want to calculate the probability associated with two tosses, the probability to have the save side for two time is equal to $p_1 \cdot p_2 = 0.5 \cdot 0.5$. Now, let us consider the probability density function associated with the two states: The state 1 is the state when the two gases are still separated, of course, this will be impossible practically, but still the probability can be calculated. The state 2 is when the molecules are all mixed in the available space. Therefore, for the state 1 there will be a probability 1 and an associated probability density function $f(p_1)$ and the same for the state 2 with $g(p_2)$. The overall probability density function will be:

$$h(p_{1+2}) = f(p_1) + g(p_2) \tag{23.18}$$

where $p_1$ and $p_2$ are probabilities (to not be confused with pressures). According to what was said previously:

$$h(p_1 \cdot p_2) = f(p_1) + g(p_2) \tag{23.19}$$

The only way for this to happen is:

$$f(p_1) = k\ln(p_1) + a; \quad g(p_2) = k\ln(p_2) + b; \quad h(p_1 \cdot p_2) = k\ln(p_1 \cdot p_2) + c \tag{23.20}$$

for these reasons:

$$S = k \cdot \ln(p) + a \tag{23.21}$$

Consider two gases of equal volumes, we refer to 1 as the state of non-mixed gases and to 2 as the state of mixed gases, the probability of state 2 is higher and it will be considered to be the maximum possible (100%).

$$\Delta S = S_2 - S_1 = k\ln(p_2) + a - k\ln(p_1) - a = k\ln\left(\frac{p_2}{p_1}\right) \tag{23.22}$$

The probability for each of the molecules to be on one side of the system will be $0.5 = 1/2$, and the total probability that all the molecules for gas 1 and 2 will be on one side of the system will be:

$$p_1 = (1/2)^{N_d}\left(\frac{1}{2}\right)^{N_d} = \left(\frac{1}{2}\right)^{2N_d} \tag{23.23}$$

where, $N_d$, is the number of $d$ molecules. Considering the difference of these state functions, we will have:

$$S_2 - S_1 = k\ln\left(\frac{1}{\left(\frac{1}{2}\right)^{2N_d}}\right) = 2N_d k\ln(2) \tag{23.24}$$

In case we consider the Boltzmann constant, the meaning is much general

$$k = \frac{R}{N_A} = \frac{8.314\ J/(\text{mol K})}{6.022 \times 10^{23}\ \text{mol}^{-1}} \tag{23.25}$$

where $N_A$ is the Avogadro number. Notice that $N_d/N_A$ is equal to the number of moles $n_d$ and we derive the expression of the state function entropy as directly proportional to the Boltzmann constant:

$$\Delta S = 2n_d R\ln(2) \tag{23.26}$$

It is also sometimes written in the literature that for irreversible mixing:

$$\Delta S = n_a R \ln\left(\frac{V}{V_a}\right) + n_b R \ln\left(\frac{V}{V_b}\right) \tag{23.27}$$

## 23.3 Additional Remarks on Entropy with Variation in Moles, i.e., Chemical Reactions

The entropy can be considered as a function of the volume, the moles associated with each component in a reaction (the stoichiometric coefficient), and the internal energy. In this way, we have:

$$S = f(U, V, n_\gamma) \tag{23.28}$$

and for this reason, doing the total differential with respect to these variables and considering $\gamma = a, b, c, \ldots$, etc. where *a*, *b*, *c*, etc. are the different species present in the mixture, we have:

$$\mathrm{d}S = \left(\frac{\vartheta S}{\vartheta U}\right)_{Vn_\gamma} \vartheta U + \left(\frac{\vartheta S}{\vartheta V}\right)_{Un_\gamma} \mathrm{d}V + \left(\frac{\vartheta S}{\vartheta n_a}\right)_{V,U,n_{b,c\ldots\ldots}} \vartheta n_a \tag{23.29}$$

where $n_{b,c\ldots}$ indicates that all the other species *b*, *c*, *d*, etc. are constant except for the specie $n_a$ associated with the *a* component.

Let us write the more clear relation:

$$\begin{aligned} n_{\gamma'} &= n_{b,c\ldots\ldots} \\ n_\gamma &= n_a \end{aligned} \tag{23.30}$$

Following the definitions (Callen 1985; Wightman 1979):

$$\begin{aligned} S &= f(U, V) = \text{Entropy} \\ T &= \frac{\vartheta U}{\vartheta S} = \text{Temperature} \\ P &= -\frac{\vartheta U}{\vartheta V} = \text{Pressure} \end{aligned} \tag{23.31}$$

Notice that these relations were derived also previously. Equation (23.29) will be written as:

$$\mathrm{d}S = \frac{\mathrm{d}U}{T} + \frac{p}{T}\mathrm{d}V - \sum_\gamma \frac{\mu_\gamma}{T}\mathrm{d}n_\gamma \tag{23.32}$$

where the chemical potential is defined by

$$\mu_\gamma = -T\left(\frac{\vartheta S}{\vartheta n_\gamma}\right)_{U\cdot Vn_{\gamma'}} \tag{23.33}$$

Because we have from previously:

$$\mathrm{d}S = \frac{\mathrm{d}U + p\mathrm{d}V}{T} \tag{23.34}$$

we substitute this in the definition of chemical potential, Eq. (23.33) and we remember that the enthalpy is:

$$H = U + pV, \tag{23.35}$$

the Helmholtz free energy:

$$F = H - TS \tag{23.36}$$

and the Gibbs free energy

$$G = H - TS \tag{23.37}$$

In this way, we will have derived that:

$$\mu_\gamma = \left(\frac{\vartheta H}{\vartheta n_\gamma}\right)_{SPn_{\gamma'}} = \left(\frac{\vartheta F}{\vartheta n_\gamma}\right)_{TVn_{\gamma'}} = \left(\frac{\vartheta G}{\vartheta n_\gamma}\right)_{TPn_{\gamma'}} \tag{23.38}$$

## 23.4 Clausius Statement and Kelvin–Planck Statement of the Second Law

The following are two commonly referred statements of the second law of the thermodynamics which can be found in the literature. The Clausius statement and the one given by Kelvin–Planck:

Clausius: "It is impossible for any system to operate in such a way that the sole result would be an energy transfer by heat from a cooler to a hotter body."

Kelvin–Planck: "It is impossible for any system to operate in a thermodynamic cycle and deliver a net amount of energy by work to its surroundings while receiving energy by heat transfer from a single thermal reservoir." (Fig. 23.5).

$$W_{\text{cycle}} \leq 0 \text{ (for a cycle and a single reserve)} \tag{23.39}$$

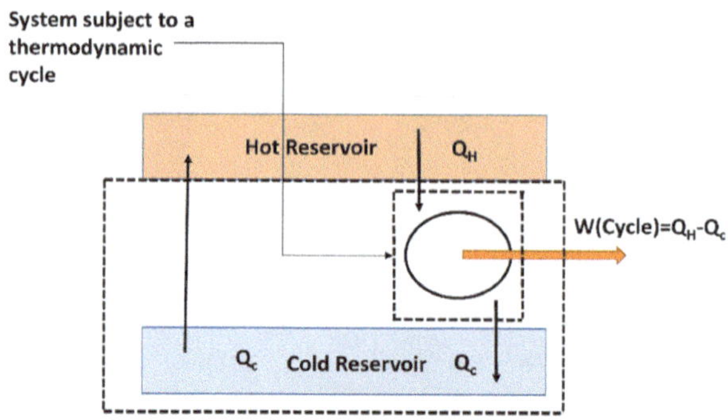


**Fig. 23.5** Simple representation of the Kelvin–Planck statement of the second law

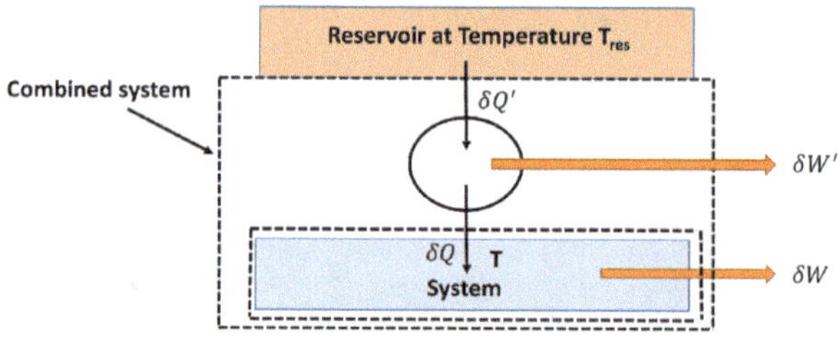


**Fig. 23.6** Example of combined thermodynamic systems

The configurations could be different, and it would be important in this case to define precisely the boundaries for the systems considered.

The Clausius inequality is expressed as:

$$\oint \left( \frac{\partial Q}{T} \right)_{\text{boundary}} \leq 0 \tag{23.40}$$

With reference to Fig. 23.6, since we have:

$$\frac{\partial Q'}{T_{\text{res}}} = \left( \frac{\partial Q}{T} \right) \tag{23.41}$$

The energy balance related to the combined cycle would be:

$$\mathrm{d}E_c = \partial Q' - \partial W_c \tag{23.42}$$

and by consequence:

$$\partial W_c = T_{\text{res}} \left( \frac{\partial Q}{T} \right) - \mathrm{d}E_c \tag{23.43}$$

Integrating over the cycle, to work produced will be:

$$W_c = T_{\text{res}} \oint \left( \frac{\partial Q}{T} \right) - \oint \mathrm{d}E_c \quad (23.44)$$

## 23.5 Entropy Balances

Entropy can be treated as any other entity, and therefore, it is possible to proceed with the related balances:

$$\dot{S}(\text{in}) + \dot{S}(\text{gen}) = \dot{S}(\text{out}) + \frac{\dot{\vartheta} S(\text{control volume})}{\vartheta \theta} \quad (23.45)$$

and more specifically, the balance on a control volume will be:

$$\sum_{\text{in}} \left| \frac{\dot{Q}}{T} \right| + \sum_{\text{in}} \dot{m}s + \dot{s}(\text{gen}) = \sum_{\text{out}} \left| \frac{\dot{Q}}{T} \right| + \sum_{\text{out}} \dot{m}s + \frac{\vartheta S(\text{control volume})}{\vartheta \theta} \quad (23.46)$$

where the first term on the left and on the right is related to the flux of thermal entropy, then we have a flux of entropy due to the convection, and finally, we have a generation of entropy.

For closed systems, the entropy balance would be as follows:

$$\sum \frac{\dot{Q}}{T} + \dot{S}(\text{gen}) = \frac{\vartheta S(\text{closed system})}{\vartheta \theta} \quad (23.47)$$

In case the heat and the entropy are specific on the unit mass:

$$\sum \frac{q}{T} + s(\text{gen}) = \Delta s(\text{closed system}) \quad (23.48)$$

In case we consider that the observation time is infinitesimal:

$$\sum \frac{\vartheta q}{T} + \vartheta s(\text{gen}) = \vartheta s(\text{closed system}) \quad (23.49)$$

and for open and stationary systems:

$$\sum \frac{\dot{Q}}{T} + \sum_{\text{in}} \dot{m}s + \dot{S}(\text{gen}) = \sum_{\text{out}} \dot{m}s \quad (23.50)$$

While for isolated systems, the difference in entropy is equal to the entropy which is generated within the system:

$$\Delta S(\text{isolated system}) = S(\text{gen} \cdot \ \text{Isolated system}) > 0$$
$$\Delta E = 0$$

## References

Callen, H. (1985). *Thermodynamics and an introduction to thermostatistics* (2nd ed.). New York: Wiley.
Carnot, N. L. S. (1897). *Reflections on the motive power of heat*. London: Wiley.
Wightman, A. (1979). Convexity and the notion of equilibrium state in thermodynamics and statistical mechanics. In *Convexity in the theory of lattice gases*. Princeton U. Press.

# Chapter 24
# Thermodynamics in Chemical Reactions Engineering

*For example, if the system contains two masses of the same substance, not in contact, nor connected by other masses consisting of or containing the same substance or its components, an infinitesimal increase of the one mass with an equal decrease of the other is not to be considered as a possible variation in the state of the system.*

On the Equilibrium of Heterogeneous Substances. J. Willard Gibbs 1878.

## 24.1 Reaction Rate and Its Dependence on Temperature

Many processes are driven by equilibrium reactions and for this reason thermodynamics is particularly useful in this kind of systems. Consider the reaction of equilibrium:

$$k_1[A][B] = k_{-1}[C][D] \tag{24.1}$$

where the terms in square brackets refer to the concentration of the different components while the constants $k_1$ and $k_{-1}$ are the forward and the backward chemical reaction constants.

The equilibrium constant for the reaction (24.1) is expressed as:

$$\frac{[C][D]}{[A][B]} = K_c = \frac{k_1}{k_{-1}} \tag{24.2}$$

The van't Hoff equation gives the relation between the equilibrium constant $K_c$ and the temperature, however, before introducing that equation, we will look here at its derivation.



C. De Blasio, *Fundamentals of Biofuels Engineering and Technology*,
Green Energy and Technology, https://doi.org/10.1007/978-3-030-11599-9_24

## 24.2 Derivation of the van't Hoff Equation

From the definition of the Gibbs free energy equation of state:

$$dG = -S\,dT + V\,dP \tag{24.3}$$

at constant pressure, this will be written as:

$$\left(\frac{\vartheta G}{\vartheta T}\right)_P = -S \tag{24.4}$$

remembering the definition of the Gibbs free energy:

$$G = H - TS \tag{24.5}$$

The following expression is derived:

$$G = H + T\left(\frac{\vartheta G}{\vartheta T}\right)_P \tag{24.6}$$

dividing by the temperature $T$:

$$\frac{G}{T} = \frac{H}{T} + \left(\frac{\vartheta G}{\vartheta T}\right)_P \tag{24.7}$$

Differentiating the term $\frac{G}{T}$ with respect to the temperature at constant pressure, we have:

$$\left(\frac{\vartheta\left(\frac{G}{T}\right)}{\vartheta T}\right)_P = -\frac{G}{T^2} + \frac{1}{T}\left(\frac{\vartheta G}{\vartheta T}\right)_P \tag{24.8}$$

Putting in evidence the term $1/T$ and considering Eq. (24.7), Eq. (24.8) will be:

$$\left(\frac{\vartheta\left(\frac{G}{T}\right)}{\vartheta T}\right)_P = -\frac{H}{T^2} \tag{24.9}$$

Considering that:

$$\left(\frac{\vartheta(G/T)}{\vartheta(1/T)}\right)_P = -T^2\left(\frac{\vartheta(G/T)}{\vartheta T}\right)_P \tag{24.10}$$

we have

$$\left(\frac{\vartheta(G/T)}{\vartheta(1/T)}\right)_P = H \tag{24.11}$$

or taking into account, the change in Gibbs free energy

$$\left(\frac{\vartheta(\Delta G/T)}{\vartheta(1/T)}\right)_P = \Delta H \tag{24.12}$$

In case, we have standard conditions:

$$\left(\frac{\vartheta(\Delta G^0/T)}{\vartheta(1/T)}\right)_P = \Delta H^0 \tag{24.13}$$

Substituting into Eq. (24.13), the expression for the standard variation of the Gibbs free energy:

$$\Delta G^0 = -RT \ln K_{\mathrm{eq}} \tag{24.14}$$

and dividing by (–$R$):

$$\left(\frac{\vartheta \ln K_{\mathrm{eq}}}{\vartheta(1/T)}\right)_P = -\Delta H^0/R \tag{24.15}$$

Now consider the following reaction

$$A + B \leftrightarrow C \tag{24.16}$$

and the case of gaseous species reacting, we have:

$$K_c = \frac{[C]}{[A][B]} = \frac{\frac{n_C}{V}}{\frac{n_A}{V}\frac{n_B}{V}} \tag{24.17}$$

and considering the ideal gas law:

$$K_c = \frac{\frac{n_C P}{nRT}}{\frac{n_A P}{nRT}\cdot\frac{n_B P}{nRT}} = \frac{\frac{x_C P}{RT}}{\frac{x_A P}{RT}\cdot\frac{x_B P}{RT}} \tag{24.18}$$

$$K_c = \frac{p_C}{p_A \cdot p_C}(RT)^{-(\upsilon_{\mathrm{Prod.}}-\upsilon_{\mathrm{reag.}})} \tag{24.19}$$

Finally:

$$K_c = K_P(RT)^{-(\upsilon_{\mathrm{Prod.}}-\upsilon_{\mathrm{reag.}})} \tag{24.20}$$

For the reaction of Eq. (24.16),

$$K_c = K_P \cdot RT \tag{24.21}$$

and for standard conditions:

$$K_P^0 = K_C^0 \cdot (RT)^{\Delta v} \tag{24.22}$$

where $\Delta v$ is the difference between the moles of products and the moles of reagents. This will be equal to:

$$\ln K_P^0 = \ln K_C^0 + \Delta v(\ln R) + \Delta v(\ln T) \tag{24.23}$$

## 24.3 Derivation of the Arrhenius Equation

Differentiating with respect to the temperature and at constant pressure, we have:

$$\left(\frac{\vartheta \ln K_P^0}{\vartheta T}\right)_P = \left(\frac{\vartheta \ln K_C^0}{\vartheta T}\right)_P + \frac{\Delta v}{T} \tag{24.24}$$

and therefore:

$$\left(\frac{\vartheta \ln K_P^0}{\vartheta T}\right)_P = \frac{\Delta H^0}{RT^2} \tag{24.25}$$

with:

$$\left(\frac{\vartheta \ln K_C^0}{\vartheta T}\right)_P = \frac{\Delta H^0}{RT^2} - \frac{\Delta v}{T} \tag{24.26}$$

this is also equal to:

$$\left(\frac{\vartheta \ln K_C^0}{\vartheta T}\right)_P = \frac{\Delta H^0}{RT^2} - \frac{\Delta v}{T} = \frac{\Delta H^0 - \Delta v RT}{RT^2} \tag{24.27}$$

Remember that the standard internal energy is related to the enthalpy by:

$$\Delta U^0 = \Delta H^0 - \Delta v \cdot RT \tag{24.28}$$

We have finally:

$$\left(\frac{\vartheta \ln K_C^0}{\vartheta T}\right)_P = \frac{\Delta U^0}{RT^2} \tag{24.29}$$

To simplify, we will use the total differential here, however, it should be noted that we are just taking into account the variation of the reaction rate constants as a function of the temperature. If we consider the equilibrium reaction as a sum of the forward and backward reactions and the reaction rate constants as $k_1$ and $k_{-1}$, then:

$$\frac{\mathrm{d}(\ln k_1)}{\mathrm{d}T} - \frac{\mathrm{d}(\ln k_{-1})}{\mathrm{d}T} = \frac{\Delta U^0}{RT^2} \tag{24.30}$$

Consider that the variation of internal energy is the difference between the energies related to the forward and the backward reaction:

$$E_1 - E_{-1} = \Delta U^0 \tag{24.31}$$

this will give:

$$\frac{\mathrm{d}(\ln k_1)}{\mathrm{d}T} = \frac{E_1}{RT^2} \tag{24.32}$$

and

$$\frac{\mathrm{d}(\ln k_{-1})}{\mathrm{d}T} = \frac{E_{-1}}{RT^2} \tag{24.33}$$

Generally, the relation between the reaction rate constant, the temperature, and the internal energy (considered for a particular direction of the reaction) is written as follows. It should be noticed that the pressure is considered constant:

$$\frac{\mathrm{d}(\ln k)}{\mathrm{d}T} = \frac{E}{RT^2} \tag{24.34}$$

which is written as:

$$\ln(k) = -\frac{E}{RT} + \mathrm{const} \tag{24.35}$$

and finally, we have derived the relation between the reaction rate constant and the temperature at a particular pressure (Arrhenius Equation):

$$k = A \cdot \mathrm{e}^{-\frac{E}{RT}} \tag{24.36}$$

## 24.4 Chemical Potential

In case, we have the van't Hoff equation with two parameters:

$$k = A'T^m \cdot e^{-\frac{E_0}{RT}} \quad (24.37)$$

Taking the logarithm on both sides:

$$\ln \mathrm{k} = \ln A' + m \ln T - \frac{E_0}{RT} \quad (24.38)$$

considering also the differential on both sides, we have the van't Hoff commonly written form:

$$\frac{\mathrm{d} \ln k}{\mathrm{d}T} = \frac{m}{T} - \frac{E_0}{RT^2} = \frac{E_0 + mRT}{RT^2} \quad (24.39)$$

and Arrhenius form:

$$\frac{\mathrm{d} \ln k}{\mathrm{d}T} = \frac{E_\mathrm{a}}{RT^2} \quad (24.40)$$

With $E_\mathrm{a} = E_0 + mRT$. And from Eq. (24.36), we remember:

$$\ln(k) = \ln(A) - \frac{E_\mathrm{a}}{RT} \quad (24.41)$$

From Eq. (24.39):

$$\ln(k) = \ln(A) - m - \frac{E_0}{RT} \quad (24.42)$$

Now, imposing and also remembering Eq. (24.38):

$$\ln(A) - m = \ln A' + m \ln T \quad (24.43)$$

we have:

$$\ln(A) = \ln A' + m \ln(T) + m \quad (24.44)$$

and finally:

$$A = A'T^m e^m \quad (24.45)$$

We have seen previously that for a three-dimensional space we have that the average kinetic energy for one molecule is:

$$\frac{1}{2}m(\text{tot}) \cdot \langle v \rangle^2 = \frac{3}{2}NK_\text{B}T \tag{24.46}$$

Recall the Boltzmann constant:

$$K_\text{B} = \frac{R}{N_A} \approx 1.38 \cdot 10^{-23}\ \text{J/K} \tag{24.47}$$

and $N$ is the total number of molecules considered (not moles). In case we consider only one direction (not three) and one molecule, we have.

$$\varepsilon = \frac{1}{2}m(\text{tot}) \cdot \langle v \rangle^2 = \frac{1}{2}K_\text{B}T \tag{24.48}$$

## 24.5 Statistical Implications of the Boltzmann Distribution Law

If we consider the Boltzmann distribution law, we have that the number of molecules having an energy state $\varepsilon_i$ is:

$$n_i = \lambda \cdot \text{e}^{-\frac{\varepsilon_i}{K_\text{B}T}} \tag{24.49}$$

lambda is the absolute activity:

$$\lambda = \text{e}^{\frac{\mu}{K_\text{B}T}} \tag{24.50}$$

and therefore, $\mu$ is the chemical potential:

$$\mu = K_\text{B}T\ln(\lambda) \tag{24.51}$$

Consider a probability density function for the values of the velocities of one molecule with respect to the center axis:

$$p(u) = \frac{1}{\left(2\pi\sigma^2\right)^{1/2}} \cdot \text{e}^{-\frac{u^2}{2\sigma^2}} \tag{24.52}$$

A representation of the probability density function for the values of the velocity of a particle is given in Fig. 24.1.

From the theory of statistics, we have that the variance is:

$$\sigma^2 = \overline{u^2} = K_\text{B} \cdot \frac{T}{m} \tag{24.53}$$

Fig. 24.1 Probability density function for the velocity values of a molecule in a box and one direction

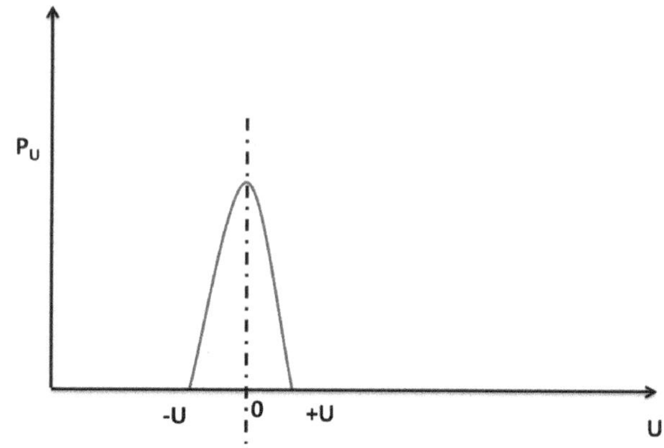


where $m$ is the mass of the particle/molecule. Equation (24.53) is valid since the mean value of the velocity is zero and then the variance becomes the mean of the squared values of the velocity, which we have in the equation written before. Therefore:

$$p(u) = \frac{m^{1/2}}{(2\pi K_B T)^{1/2}} \cdot e^{-\frac{mu^2}{2K_B T}} \tag{24.54}$$

and:

$$p(u) = \frac{m^{1/2}}{(2\pi K_B T)^{1/2}} \cdot e^{-\frac{\varepsilon_x}{K_B T}} \tag{24.55}$$

The probability that one molecule has a velocity between $u$ and $u + d(u)$ is:

$$p(u)d(u) = \frac{m^{1/2}}{(2\pi K_B T)^{1/2}} \cdot e^{-\frac{\varepsilon_x}{K_B T}}\, du \tag{24.56}$$

From the expression for the kinetic energy of a molecule:

$$\varepsilon = \frac{1}{2} m \left| \overline{u^2} \right| \tag{24.57}$$

we have:

$$u = \left( \frac{2\varepsilon}{m} \right)^{1/2} \tag{24.58}$$

and:

$$du = \frac{1}{2} \frac{1}{\left(\frac{2\varepsilon}{m}\right)^{1/2}} \frac{2}{m} d\varepsilon = \frac{1}{(2\varepsilon m)^{1/2}} d\varepsilon \tag{24.59}$$

We remember:

$$p(u)d(u) = \frac{m^{1/2}}{(2\pi K_B T)^{1/2}} \cdot e^{-\frac{\varepsilon_x}{K_B T}} du \tag{24.60}$$

and Eq. (24.59), and we obtain:

$$p(u)d(u) = \frac{m^{1/2}}{(2\pi K_B T)^{1/2}} \frac{1}{(2\varepsilon \cdot m)^{1/2}} \cdot e^{-\frac{\varepsilon_x}{K_B T}} d\varepsilon \tag{24.61}$$

which gives:

$$p(u)d(u) = \frac{1}{2(\pi \cdot \varepsilon \cdot K_B T)^{1/2}} \cdot e^{-\frac{\varepsilon_x}{K_B T}} d\varepsilon \tag{24.62}$$

and this will give:

$$\frac{dN_x}{N} = 2p(u)d(u) = \frac{1}{(\pi \cdot \varepsilon_x \cdot K_B T)^{1/2}} \cdot e^{-\frac{\varepsilon_x}{K_B T}} d\varepsilon_x \tag{24.63}$$

which is the fraction of molecules having energies from $\varepsilon_x$ and $\varepsilon_x + d\varepsilon_x$ in the direction $x$.

In two dimensions, this will give:

$$\frac{dN}{N} = \frac{1}{\left(\pi \cdot \varepsilon_x \cdot \varepsilon_y \cdot K_B T\right)^{1/2}} \cdot e^{-\frac{\varepsilon_x}{K_B T}} e^{-\frac{\varepsilon_y}{K_B T}} d\varepsilon_x \, d\varepsilon_y \tag{24.64}$$

Writing: $\varepsilon = \varepsilon_x + \varepsilon_y$ and integrating between 0 and $\varepsilon$:

$$\frac{dN}{N} = \frac{1}{(K_B T)} \cdot e^{-\frac{\varepsilon}{K_B T}} d\varepsilon \tag{24.65}$$

the fraction of molecules having energy larger than $\varepsilon^*$ is obtained by integrating between $\varepsilon^*$ and infinite giving:

$$f^* = e^{-\frac{\varepsilon^*}{K_B T}} \tag{24.66}$$

If we have two states of activation 1 and 2 and:

$$n_i = \lambda e^{-\frac{\varepsilon_i}{K_B T}} \tag{24.67}$$

This will give:

$$n_2/n_1 = e^{-\frac{(\varepsilon_2 - \varepsilon_1)}{K_B T}} \tag{24.68}$$

and finally:

$$n_2/n_1 = e^{-\frac{E_a}{RT}} \tag{24.69}$$

Considering:

$$\varepsilon^* = \varepsilon_2 - \varepsilon_1 \tag{24.70}$$

and taking into account:

$$E_a = N_a \varepsilon^* \tag{24.71}$$

where $N_a$ is the Avogadro number.

# Chapter 25
# Some Parameters and Properties of Biomass Fuels

*Life is movement.*
(Aristotle, 4th century BC), as cited by Brügemann and Gerds-Ploeger (2013).

## 25.1 Some Useful Figures on Biomass Feedstock

To have the fundamental understanding of biofuels engineering and technology, concepts like energy content, calorific values, and related properties need to be understood for a necessary assessment of the biomass feedstock first and the produced biofuel later.

The following part will give some more information on what is the composition of common biomass with particular focus on crops and plants in general; some properties will be given in terms of composition and energy content. Since this section of the manuscript is mainly consisting of support material, data on the properties of diverse kinds of biomass are given here in the form of tables, and these figures are taken from the literature.

If the intention is to utilize energy crops as one mean for energy conversion, the first step would be to assess how much of crops we are actually producing per unit time and per unit of land. To be remembered that land utilization is actually one of the biggest problems in the production of energy crops to be used in energy and production of biofuels. We have to remember also that in the world, there are large portions of population who actually do not have enough food. Fortunately, we can think about third-generation biorefinery for the future.

A common parameter used to assess the production of biomass for different species is the Net Primary Production (NPP), value and the Energy Storage Capacity (ESC), of biomass (Gholz 1982).

It is estimated that the global NPP is around 100 Gton of carbon per year (Ksenzhek and Volkov 1998). Considering that the specific energy storage capacity is around 41.3 kJ/g of fixed carbon, the energy stored in photosynthetic biomass is around

 
C. De Blasio, *Fundamentals of Biofuels Engineering and Technology*,
Green Energy and Technology, https://doi.org/10.1007/978-3-030-11599-9_25

$4 \times 10^{21}$ J/year while the total solar energy on earth is estimated around $2.75 \times 10^{24}$ J/year.

Concerning the chemical composition of wood and biomass, a suitable description of the main composition and characteristics of biomass in terms of lignin for instance can be found in the literature together with an approximate evaluation of the main constituents (Demirbaş 2005). This gives:

- 75% carbohydrates, cellulose and hemicellulose;
- 25% lignin;
- lower calorific value of about 18 GJ/ton at 100% dry matter. As comparison, conventional oils have 42–44 GJ/ton.

As an example here, the composition of common types of biomass is given in Table 25.1 in terms of cellulose, hemicellulose, lignin, and ash.

Plants can be cultivated for the purpose of energy conversion; as a matter of fact, this solution is being implemented in diverse countries in the world and Europe (Jezierska-Thöle et al. 2016). The biomass produced is assessed on the basis of its properties and also on the basis of how much of the same feedstock is produced per year and per hectare (ha) (Jankowski et al. 2016). Another estimation of how efficient is the biomass production in terms of incoming radiative energy can be done on the basis of the so-called radiation use efficiency (RUE); this gives a good estimation of how effective is a determinate kind of plant or organism in utilizing light energy. An example of RUE estimation can be found in (Chimonyo et al. 2018). It is therefore very important to assess the biomass in terms of radiation use efficiency or related quantities.

Let us assume here that the produced biomass will be suitable for combustion in power plants. Before the utilization of the feedstock, drying is strictly necessary when moisture content is higher than 65% according to industrial recommendations. However, less than 50% is suggested in practice. In fresh wood, the moisture content

**Table 25.1** Composition of common types of biomass, percentages in weight

| | Cellulose | Hemicellulose | Lignin | Ash | Other |
|---|---|---|---|---|---|
| Saccharum bagasse | 48 | 22 | 20 | 6 | 4 |
| Wheat | 54 | 26–30 | 16–18 | 7–8 | 6.8 |
| Bamboo | 34–36 | 16–16 | 26–28 | 3 | 9.6 |
| Pine | 42 | 24 | 27 | 0.2 | 7.6 |
| Poplar | 70 | 23 | 4 | 4 | 28 |
| Eucalyptus | 49 | 15 | 28 | 0.4 | 34 |
| Miscanthus | 44 | 24 | 17 | 1.5 | 4 |
| Phalaris | 28 | 22 | 14 | 8 | |
| Lucerne | 25 | 23 | 12 | 6 | |
| Betula | 41 | 33 | 21 | <1 | |

Data from Rayaprolu (2009)

is around 30–50%, and it reduces to 18–25% after one-year drying (Serup et al. 2005).

The ash content in wood is around 2.5%, and there is low amount of sulfur. The moisture content is a function of wood age and type. These information are obtained by communicating with wood-driven power plant managers in Finland. The suggestion is in this case to have the power plant quite near the origin of the feedstock and max at around 100 km.

Energy crops can be classified as "above ground" and "below ground" or for instance they can give a high-energy-specific product, for instance ethanol from sugarcane or oilseed rape and can be classified accordingly.

Common values for the radiation use efficiencies (RUE) and energy conversion efficiencies are given as follows (Pessarakli 2005):

- Sugarcane, for instance with a 18-month harvest interval, has a RUE of 2.3 g/(MJ $m^2$).
- Sorghum has a RUE of approximately 3.2 g/(MJ $m^2$).
- Maize has a RUE of approximately 2.5 g/(MJ $m^2$).
- Miscanthus has a RUE of approximately 1.4 g/(MJ $m^2$).
- *Miscanthus sacchariflorus* has a RUE of approximately 3.6 g/(MJ $m^2$).
- Willow has a RUE of approximately 1.4 g/(MJ $m^2$).
- Poplar has a RUE of approximately 1.0 g/(MJ $m^2$).
- Eucalyptus has an ECE of approximately 5.8%.
- Populus has a RUE of approximately 0.5 g/(MJ $m^2$).
- Pinus has a RUE of approximately 1.3 g/(MJ $m^2$).
- Cannabis has a RUE of approximately 2.26 g/(MJ $m^2$).
- Rice has a RUE of approximately 0.93 g/(MJ of total radiation).
- Potato has a RUE of approximately 1.6 g/(MJ of total radiation).
- Wheat has a RUE of approximately 1.3 g/(MJ of total radiation).
- Soybean has a RUE of approximately 1.20 g/(MJ of total radiation).
- Peanut has a RUE of approximately 0.98–2.24 g/(MJ of total radiation).

Table 25.2 shows some values of proximate analysis for peat, wood, paper, and coal as comparison.

**Table 25.2** Proximate values for wood, paper, peat, and coal. A comparison

| | Volatiles (wt%) | Moisture (wt%) | Carbon (wt%) | Ash (wt%) | Fuel ratio | HHV (MJ/kg) |
|---|---|---|---|---|---|---|
| Peat | 65 | 7 | 20 | 8 | 0.30 | 22 |
| Wood | 85 | 6 | 8 | 1 | 0.10 | 19 |
| Paper | 75 | 4 | 11 | 10 | 0.15 | 13 |
| Coal | 30 | 5 | 45 | 20 | 1.5 | 26 |

Data taken from Rayaprolu (2009)

When biomass is utilized for energy conversion, for instance when wood or wood chips are used for combustion in a power plant. The most important properties considered for energy recovery are as follows:

- moisture content;
- density;
- heating value;
- particle size distribution;
- ash content and properties;
- chemical composition;
- amount of volatiles.

And the main kind of analyses usually done on the original feedstock is the ultimate and proximate analyses. The lower heating value (LHV) and the higher heating value (HHV) are evaluated firstly in a calorimetric bomb as we have seen in the first part of this book. The LHV and HHV are also encountered in the literature as net calorific value (NCV) and gross calorific value (GCV).

Figure 25.1 is a simple representation of how the calorific value is determined.

When the HHV is determined in a calorimetric bomb, it should be noticed that if water is contained in the original biomass feedstock, it will be condensed also in the combustion chamber. Additionally, any water that is also formed during the combustion process will condense in the same chamber (see part 1 of the manuscript).

The net calorific value is evaluated from the GCV taking into consideration the moisture and the hydrogen content. The GCV and the NCV are expressed also by empirical formulas in the literature. For instance, an expression for the gross calorific value (GCV) is given as follows (Reed and Golden 1988):

$$\mathrm{GCV} = 34.1\,\mathrm{C} + 132.2\,\mathrm{H} + 6.8\,\mathrm{S} - 1.53\,\mathrm{A} - 12.0\,(\mathrm{O} + \mathrm{N})\ \mathrm{kJ/g} \tag{25.1}$$

$$\mathrm{GCV} = [146.6\,\mathrm{C} + 568.8\,\mathrm{H} + 29.4\,\mathrm{S} - 6.6\,\mathrm{A} - 51.5\,(\mathrm{O} + \mathrm{N})] \times 10^2\ \mathrm{Btu/lb} \tag{25.2}$$

where C, H, S, A, O, and N are expressed in wt%/100, and they refer to carbon, hydrogen, sulfur, ash, oxygen, and nitrogen present in the fuel. The latent heat of vaporization (at 25 °C) of water is reported as 2.444 kJ/g, while $m$ and $H_2$ are

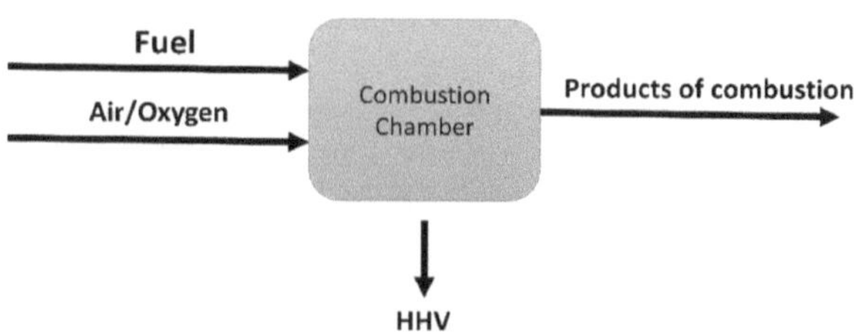


**Fig. 25.1** A simple scheme of the gross calorific value determination

the moisture and hydrogen in wt%/100. Equation (25.1) is also known as Milne's formula, and it is valid for GCV on dry basis. The HHV and the LHV can be expressed on the basis of different parameters, for instance on dry basis (d.b.), dry and ash free basis (d.a.f.), and as received (a.r.). The net calorific value (NCV) is calculated usually from the GCV as follows:

$$\mathrm{NCV} = \mathrm{GCV} - (m_{\mathrm{w}} + 9H) \cdot \text{latent heat} \tag{25.3}$$

Therefore, it is possible to write (Channiwala and Parikh 2002):

$$\mathrm{HHV}_{\mathrm{ar}} = \mathrm{HHV}_{\mathrm{dry}} \cdot (1 - x_{\mathrm{w}}) \tag{25.4}$$

$$\mathrm{HHV}_{\mathrm{dry}} = \mathrm{HHV}_{\mathrm{daf}} \cdot (1 - x_{\mathrm{ash}}) \tag{25.5}$$

$$\mathrm{LHV}_{\mathrm{dry}} = \mathrm{HHV}_{\mathrm{dry}} - 2.443 \cdot 8.936 \cdot x_{\mathrm{H}} \tag{25.6}$$

$$\mathrm{LHV}_{\mathrm{ar}} = \mathrm{LHV}_{\mathrm{dry}}(1 - x_{\mathrm{w}}) - 2.443 \cdot x_{\mathrm{w}} \tag{25.7}$$

$$\mathrm{LHV}_{\mathrm{ar}} = \mathrm{HHV}_{\mathrm{ar}} - 2.443 \cdot \{8.936 \cdot x_{\mathrm{H}}(1 - x_{\mathrm{w}}) + x_{\mathrm{w}}\} \tag{25.8}$$

In Eqs. (25.5)–(25.8), $x_{\mathrm{w}}$ and $x_{\mathrm{H}}$ are the fraction in weight of water and hydrogen, respectively.

Sometimes in the literature, it is also possible to find the correlation in the following form (Van Loo and Koppejan 2008):

$$\mathrm{NCV} = \mathrm{GCV}_{\mathrm{dry}}(1 - x_{\mathrm{w}}) - 2.443 \cdot x_{\mathrm{w}} - 2.443 \cdot x_{\mathrm{H}} \cdot 8.936(1 - x_{\mathrm{w}})\left[\frac{\mathrm{MJ}}{\mathrm{kg}}, \mathrm{w.b}\right] \tag{25.9}$$

Table 25.3 gives examples of ultimate and proximate analyses for additional kinds of biomass feedstock.

## 25.2 Overall Energy (Heat) Balances

### *25.2.1 Design Variables Affecting Thermochemical Processes*

Biomass can contain a large number of compounds; this depends on what kind of biomass is taken into consideration and what is the process where the biomass feedstock is utilized. Table 25.4 is a very general example of what are some effects of biomass compounds on thermochemical processes. These of course are just a

**Table 25.3** Proximate and ultimate analysis of some biomass feedstock with HHV values

| Material | Water (wt%) | FC (wt%, d.b.) | VM (wt%, d.b.) | Ash (wt%, d.b.) | C (wt%, d.b.) | H (wt%, d.b.) | O (wt%, d.b.) | N (wt%, d.b.) | HHV (MJ/kg, d.b.) |
|---|---|---|---|---|---|---|---|---|---|
| Coconut shell | 8.27 | 22.1 | 77.19 | 0.71 | 50.22 | 5.70 | 43.37 | 0.00 | 20.50 |
| Akhrot shell | 8.76 | 18.78 | 79.98 | 1.20 | 49.81 | 5.64 | 42.94 | 0.41 | 20.01 |
| Groundnut shell | 8.10 | 21.60 | 72.70 | 5.70 | 48.59 | 5.64 | 39.49 | 0.58 | 19.85 |
| Eucalyptus wood | 16.40 | 21.30 | 75.35 | 3.35 | 46.04 | 5.82 | 44.49 | 0.30 | 18.64 |
| Bamboo wood | 11.50 | 11.24 | 86.80 | 1.95 | 48.76 | 6.32 | 42.77 | 0.20 | 20.55 |
| Mango wood | 14.10 | 11.36 | 85.64 | 2.98 | 46.24 | 6.08 | 44.42 | 0.28 | 19.17 |
| Jujube wood | 12.50 | 14.14 | 83.63 | 2.32 | 47.63 | 6.12 | 43.78 | 0.15 | 19.77 |
| Plywood | 13.79 | 15.77 | 82.14 | 2.09 | 48.13 | 5.87 | 42.46 | 1.45 | 18.95 |
| Block wood | 12.16 | 14.59 | 83.32 | 2.09 | 46.90 | 6.07 | 43.09 | 0.95 | 18.26 |
| Millet straw | 5.06 | 16.45 | 78.28 | 5.27 | 43.71 | 5.85 | 45.16 | 0.01 | 18.05 |
| Wheat straw | 8.87 | 10.98 | 82.12 | 6.90 | 42.95 | 5.35 | 46.99 | 0.00 | 17.99 |
| Jowar straw | 7.50 | 15.15 | 75.97 | 8.88 | 42.10 | 5.60 | 43.38 | 0.04 | 17.95 |
| Rice straw | 8.10 | 13.91 | 65.70 | 20.38 | 35.68 | 4.62 | 39.14 | 0.28 | 14.85 |

(continued)

**Table 25.3** (continued)

| Material | Water (wt%) | FC (wt%, d.b.) | VM (wt%, d.b.) | Ash (wt%, d.b.) | C (wt%, d.b.) | H (wt%, d.b.) | O (wt%, d.b.) | N (wt%, d.b.) | HHV (MJ/kg, d.b.) |
|---|---|---|---|---|---|---|---|---|---|
| Rice husk bran | 8.47 | 16.95 | 61.81 | 21.24 | 38.50 | 5.20 | 34.61 | 0.45 | 14.69 |
| Coconut coir pith | 19.77 | 28.82 | 66.02 | 5.16 | 43.36 | 4.98 | 44.87 | 1.63 | 18.07 |
| Sugarcane bagasse | 51.01 | 13.15 | 83.66 | 3.20 | 45.48 | 5.96 | 45.21 | 0.15 | 18.73 |
| Millet grain waste | 5.20 | 14.50 | 77.10 | 8.40 | 40.56 | 5.24 | 45.50 | 0.40 | 15.21 |
| Cotton gin waste | 4.99 | 14.97 | 84.41 | 1.61 | 42.66 | 6.05 | 49.50 | 0.18 | 17.49 |

Adapted from Channiwala and Parikh (2002)

**Table 25.4** Properties of biomass and their effect on thermochemical processes

| Property | Effect on |
|---|---|
| Moisture | Heating value, ignition, plant design, power plant efficiency (Bahadori et al. 2014) |
| HHV, LHV | Plant design, mass flows, volumes, suitability of fuel, average burning rate, maximum solid temperature (Yang et al. 2005) |
| Volatiles | Plant design, mass flows, volumes, suitability of fuel. Fluidization (Saastamoinen 2015) |
| Ash content | Fuel suitability, disposal, safety, deposition, efficiency, safety. Technological and environmental problems (Vassilev et al. 2017) |
| Bulk density | Plant design, transport, handling. Sometimes biomass densification is necessary (Manickam et al. 2006) |
| Particles density | Plant design, transport, handling, efficiency |
| Particles size and distribution | Plant design, transport, handling, efficiency |
| Content of carbon, hydrogen, oxygen | HHV, LHV, Plant design and operations. Oxygen is decreasing the ignition time for engine fuels, emissions are reduced, however $NO_X$ and fuel consumption may be increased (Song et al. 2016) |
| Cl content | Emission control processes, corrosion, ash melting temperature. Composition of ash deposits (Liu et al. 2000) |
| N, content | $NO_X$ emissions, process design |
| S, content | $SO_2$ emissions, flue gas treatment. Composition of ash deposits (Liu et al. 2000) |
| F, content | HF emissions and corrosion |
| K, content | Corrosion, emission, ash processing and melting temperature |
| Na, content | Corrosion, ash melting temperature, ash composition |
| Mg, content | Increases ash melting temperature, ash processing |
| Ca, content | Increases ash melting temperature, ash processing |
| P, content | Ash processing |
| Heavy metals | Safety, emissions, ash processing |

minimal part of the effects that the compounds could have on a process (for instance, a biomass power plant) and therefore the effects given are not sufficient, but they should be considered as descriptive of their importance.

Let us consider here a common power plant which could utilize biomass as fuel, for instance, woodchips. The heat produced from the combustion of the fuel will be utilized to heat up water to the required conditions necessary to run a turbine, in case electricity is produced, or when district heating is in place. The main variables considered when the design of the facilities is to be performed are:

- moisture content;
- calorific value (here some empirical formulas can be seen for GCV and NCV);
- temperature, residence time;
- stoichiometry;
- mixing.

The moisture amount in the biomass varies largely on the basis of the biomass type and biomass storage. Values for biomass moisture content can be found in the literature along with the effect of moisture on biomass treatment processes (Motta et al. 2018). Commonly, it is necessary that biomass undergoes the process of drying before it can be used in thermochemical processes; this is because the moisture content can reduce the adiabatic combustion temperature, and for this reason, the residence time in the combustion chamber will be increased. One typical example is the drying of black liquor in pulp and paper industry to utilize it in recovery boilers. Nevertheless, some options have been proposed in the literature to utilize black liquor directly in gasification processes; for instance, alternative methods like supercritical water gasification (SCWG), where it would be possible to utilize the feedstock directly into the reactor with no need of drying (De Blasio et al. 2015).

Typical moisture and ash data on wood biomass are given in the literature (Van Loo and Koppejan 2008), and these are typical recommendations also given by power plants' managers. These parameters are affecting heavily the ignition front propagation speed and the bed temperature for biomass-fired power plants (Arce et al. 2013). As mentioned also previously and as a summary here:

- 30–50% is the moisture content in fresh wood; this reduces to 18–25% after one-year drying.
- Drying is strictly necessary when moisture is higher than 65%; however, less than 50% is the suggested percentage.
- The ash content in wood is normally below 2.5%.
- The nitrogen content is very low.
- Wood contains very low amount of sulfur.
- Moisture is a function of wood age and type.
- Transportation of wood is not economical for distances longer than 100 km.

To give one example, Table 25.5 gives a comparison of the moisture, ash, volatile matter, HHV, fixed carbon, and elemental analysis for diverse kind of trees.

### *25.2.2 Additional Data on Moisture and Ash Content of Fuels*

Additional data are reported here on the moisture and ash content of different types of biomass. As mentioned, these are two parameters which affect largely the HHV and LHV; some values are reported in Table 25.6 where these values are given for diverse feedstock kinds to make a comparison.

**Table 25.5** Moisture, ash, volatile matter, HHV, fixed carbon and elemental analysis for diverse kind of trees

| Specie | Clone | $H_2O$ (%) | Ash (%) | VM (%) | FC (%) | HHV (J/g) | C | H (%) | S (%) | N (%) | O (%) |
|---|---|---|---|---|---|---|---|---|---|---|---|
| Willow | Olof | 6.406 | 1.174 | 76.662 | 15.757 | 18.46 | 45.39 | 6.37 | 0.16 | 0.16 | 47.91 |
| | Tordis | 7.11 | 1.19 | 75.33 | 16.37 | 18.32 | 45.49 | 6.36 | 0.16 | 0.19 | 47.80 |
| Poplar | 1214 | 6.45 | 2.18 | 75.99 | 15.39 | 18.19 | 45.11 | 6.26 | 0.16 | 0.15 | 48.33 |
| | AF2 | 6.02 | 1.70 | 76.90 | 15.37 | 18.42 | 45.58 | 6.26 | 0.15 | 0.15 | 47.86 |
| | Beaupré | 6.19 | 1.86 | 76.63 | 15.32 | 18.39 | 45.70 | 6.32 | 0.14 | 0.11 | 47.73 |
| Chestnut | | 6.07 | 0.92 | 74.72 | 18.30 | 18.00 | 46.04 | 6.18 | 0.19 | 0.25 | 47.35 |
| Maritime pine | | 6.95 | 0.60 | 76.58 | 15.98 | 19.37 | 47.78 | 6.26 | 0.65 | 0.50 | 44.82 |

Adapted from Álvarez-Álvarez et al. (2018)

### 25.2.3 Properties of Agrofuels and Typical Air Requirements

Some of the properties related to biofuels' feedstock derived from agriculture are given in Table 25.7

Typical air requirements for agrofuel-based HHV are given in Table 25.8.

Peat represents the part of biomass which is not decomposed completely, and by some definitions, it should be composed by around 30% of dead organic material (Clarke and Rieley 2010). There has been some discussion on this particular kind of biomass and if we should consider Peat as a renewable source of energy. Peat is considered to be a form of renewable biomass by some countries; however, there is still a discussion about if considering peat as renewable source or not. In Sweden and in Finland, wood and peat are the only indigenous fuels according to some literature (Rayaprolu 2009) and in Nordic countries a large part of the ground could be utilized for peat production (Virtanen et al. 2003). The content of carbon sequestrated within

**Table 25.6** Heating values, moisture and ash content of some examples of biomass

| | HHV (MJ/kg) | LHV (MJ/kg) | Moisture (%) | Ash (fraction in mass) |
|---|---|---|---|---|
| Charcoal | 30.44 | 28.33 | 6.361 | 0.0231 |
| Firewood | 19.04 | 15.6 | 16.041 | 0.008525 |
| Croton husk briquette | 17.48 | 15.39 | 10.482 | 0.101 |
| Carbonized croton husk | 27.78 | 25.3 | 8.200 | 0.0745 |

Data from Turpeinen (2016)

**Table 25.7** Different property values for agrofuels

| Parameter | Range |
|---|---|
| Theoretical air requirement (kg/kcal) | 0.00122 |
| HHV (kcal/kg) | 4500–5500 |
| Fouling-factor | 1.005–1.015 |
| Theoretical $CO_2$ (%) | 20.2–20.75 |
| Moisture (%) | 10 |
| $H_2$ (dry and ash free) (%) | 5.6–5.8 |
| VM (d.a.f.) (%) | 72–90 |

where VM is the volatile matter and *d.a.f.* means *dry and ash free*
Adapted from Rayaprolu (2009)

**Table 25.8** Typical air requirements for agrofuel expressed as kg/kcal

| Coal | Agrofuel |
|---|---|
| 0.00131–0.00136 | 0.00122–0.00124 |

Data from Rayaprolu (2009)

peat is important considering regions like Finland. In fact, it was estimated that the total carbon content in Finnish peat was 5.5 Pg in 1950, and the total carbon sequestration into Finnish peat has increased from 2.2 to 3.6 Tg $ha^{-1}$ in 2010 (Minkkinen et al. 2002). Peat can be of course processed; one of the technologies used for this purpose is the wet carbonization process (Lau et al. 1987).

The main characteristics of peat are described as follows in Table 25.9; peat has high amount of moisture, 80–95%, and can be reduced to 35–55% when drying is applied.

The ultimate analysis of this fuel is described in Table 25.10.

Concerning the agrofuels, in this case there is a great variety of them and they can vary very much in composition. Additionally, they are also seasonal and their growth is very much affected by the climate in which they are produced. Furthermore, the degradation of this kind of feedstock is quite fast. It is also quite obvious that if a power plant should be run on agrofuels, then the power output cannot be very large but in a range that could be 5–30 MW.

**Table 25.9** Main properties of peat

| Peat property | Value (approximately) |
|---|---|
| Moisture | 90% of water (Lau et al. 1987). 10% (briquette), 50% (mulled), 25% (air dried) (Rayaprolu 2009) |
| Density (kg/m$^3$) | 240–400 (air dried), 480–960 (Briquette) |
| Heating value | 19.8 MJ/kg (dry basis) (Lau et al. 1987) |
| Ash (%) | 2–10 |
| Volatiles (%) | 55–70 |
| Carbon (%) | 50–60 |
| Hydrogen (%) | 6 |
| Oxygen (%) | 30–40 |
| Sulfur (%) | 0.3–0.5 |
| Nitrogen | 1.2–1.5 |

Data mostly taken from Rayaprolu (2009)

**Table 25.10** Ultimate analysis on dry and ash free basis of peat

| Element | % |
|---|---|
| $N_2$ | 1.2–1.5 |
| S | 0.3–0.5 |
| $O_2$ | 30.0–40.0 |
| $H_2$ | 5.5–7.0 |
| C | 53.0–63.0 |

Adapted from Rayaprolu (2009)

Agrofuels have usually a gross calorific value which is higher than 13,500 kcal/kg and for this value, the air requirement is less than 1.21 kg/1000 kcal of GCV (Gujarat Pollution Control Board, Paryavaran Bhavan 2014).

It is also quite reasonable that agrofuels are suited for stoker boilers and bubbling beds but not for bubbling fluidized bed boilers, and this has to do with their density and consistency.

They are efficient for ignition and combustion, with volatile matter (VM), of 70–90% (Rayaprolu 2009). The amount of ash is limited in this case, and their higher amount of oxygen lowers the requirement of air for combustion. The furnace exit gas temperature is higher than coal-fired boilers when grates are used; this is due to higher amount of VM, volatile matter.

With respect to the combustion properties of minor agrofuels, Table 25.11 gives a small set of data.

While, regarding the moisture content and additional parameters of wood-related parts, Table 25.12 gives some more additional data.

Table 25.13 gives further data of proximate and ultimate analysis conducted on additional kinds of samples.

While Table 25.14 gives the HHV and LHV for selected agricultural residues, crops, forest and urban residues.

As a comparison to the previous data given, a list of thermodynamic properties is reported for alkane fuels in Table 25.15.

**Table 25.11** Combustion properties of minor agrofuels

| | Coconut shell | Coffee husks | Cotton seed husks | Cotton pods | Palm nut shells | Peanut shells |
|---|---|---|---|---|---|---|
| HHV (kcal/kg) | 4220 | 3720 | 4055 | 3610 | 3720 | 4335 |
| LHV (kcal/kg) | 3900 | 3395 | 3720 | 3305 | 3840 | 4000 |
| $H_2O$ (%) | 10.0 | 12.0 | 11.0 | 12.0 | 10.0 | 10.0 |
| Ash (%) | 5.0 | 3.0 | 3.0 | 7.0 | 3.0 | 2.0 |
| Theoretical air (kg/kg) | 5.2 | 5.0 | 5.0 | 4.45 | 5.15 | 5.35 |
| Theoretical $CO_2$ by volume % | 20.4 | 20.5 | 20.5 | 20.5 | 20.5 | 20.4 |

Adapted from Rayaprolu (2009)

**Table 25.12** Moisture content, HHV, LHV, density, and calorific value for wood-related biomass

| | Moisture content (wt%) | HHV (MJ/kg, d.b.) | LHV (MJ/kg, d.b.) | Bulk density (kg, w.b./m$^3$) | Calorific value (MJ/m$^3$) |
|---|---|---|---|---|---|
| Olive residues (from three-phase production) | 53.0 | 22.6 | 8.5 | 650 | 5530 |
| Olive residues (from two-phase production) | 63.0 | 21.5 | 6.1 | 1130 | 6890 |
| Straw (winter wheat) | 15.0 | 18.7 | 14.5 | 120 | 1740 |
| High-pressure bales sawdust | 50.0 | 19.8 | 8.0 | 240 | 1920 |
| Cereals | 15.0 | 18.7 | 14.5 | 175 | 2540 |
| Bark | 50.0 | 20.2 | 8.2 | 320 | 2620 |
| Grass (high-pressure bales) | 18.0 | 18.4 | 13.7 | 200 | 2740 |
| Woodchips (softwood) | 50.0 | 19.8 | 8.0 | 350 | 2800 |
| Woodchips (softwood predried) | 30.0 | 19.8 | 12.2 | 250 | 3050 |
| Woodchips (hardwood) | 50.0 | 19.8 | 8.0 | 450 | 3600 |
| Woodchips (hardwood predried) | 30.0 | 19.8 | 12.2 | 320 | 3900 |
| Wood pellets | 10.0 | 19.8 | 16.4 | 600 | 9840 |

Adapted from Van Loo and Koppejan (2008)

An explanation of the parameters given in Table 25.15 can be found in Table 25.16. For additional information on thermodynamic data for fuels, the reader can refer also to additional literature sources (Rumble 2018; Bartok and Sarofim 1991).

**Table 25.13** Proximate and ultimate analyses for corn, straw, husks, and wood chips

| | Corn stems | Corn cobs | Corn grain | Straw | Straw pellets | Rice husk | Wood chips | Bark |
|---|---|---|---|---|---|---|---|---|
| Particle density ($kg/m^3$) | 360 | 385 | 420 | 650 | 1400 | 740 | 920 | 880 |
| Ash (%) | 2.60 | 2.40 | 1.88 | 6.00 | 6.10 | 22.70 | 0.75 | 1.81 |
| Bulk density ($kg/m^3$) | 158 | 174 | 235 | 100 | 790 | 120 | 590 | 430 |
| LHV (MJ/kg) | 15.80 | 16.30 | 15.20 | 16.60 | 16.50 | 12.90 | 9.55 | 6.52 |
| Moisture (%) | 9.20 | 8.00 | 11.90 | 8.50 | 8.50 | 4.10 | 41.50 | 58.00 |
| VM (%) | 67.30 | 69.10 | 75.00 | 61.60 | 61.40 | 59.70 | 47.94 | 31.79 |
| FC (%) | 20.90 | 20.50 | 11.20 | 24.20 | 24.00 | 13.50 | 9.82 | 8.45 |
| C (wt%) | 42.16 | 45.57 | 43.72 | 41.61 | | 37.81 | 47.62 | 48.30 |
| $H_2$ (wt%) | 6.28 | 5.93 | 7.18 | 6.70 | | 4.68 | 6.11 | 6.34 |
| $O_2$ (wt%) | 45.70 | 45.89 | 45.73 | 44.40 | | 33.52 | 44.34 | 39.76 |
| $N_2$ (wt%) | 0.71 | 0.74 | 1.24 | 0.66 | | 0.28 | 0.65 | 1.29 |
| S (wt%) | 0.01 | 0.02 | 0.01 | 0.01 | | 0.03 | 0.00 | 0.00 |
| Ash composition (wt%) | | | | | | | | |
| $SiO_2$ | | 32.74 | 43.00 | 59.52 | | 97.62 | 6.45 | 5.88 |
| $Al_2O_3$ | | 5.05 | 8.40 | 2.18 | | 0.01 | 0.66 | 0.20 |
| $TiO_2$ | | 0.02 | 0.40 | 0.02 | | 0.01 | 0.02 | 0.02 |
| $Fe_2O_3$ | | 0.70 | 5.05 | 8.76 | | 0.19 | 2.05 | 0.33 |
| CaO | | 2.63 | 5.10 | 6.52 | | 0.01 | 73.05 | 80.72 |
| MgO | | 6.83 | 8.10 | 2.26 | | 0.59 | 0.54 | 0.23 |
| $K_2O$ | | 29.24 | 15.30 | 13.16 | | 0.87 | 4.50 | 2.57 |
| $Na_2O$ | | 1.38 | 0.90 | 1.60 | | 0.11 | 0.17 | 0.10 |
| $P_2O_5$ | | 4.68 | 13.00 | 5.07 | | 0.44 | 1.90 | 1.30 |
| $SO_3$ | | 0.12 | 0.10 | 0.36 | | 0.25 | 1.93 | 0.95 |

Adapted from Rayaprolu (2009)

**Table 25.14** HHV and LHV for selected agricultural residues, crops, forest, and urban residues

| | HHV (MJ/kg) | LHV |
|---|---|---|
| Corn stalks[a, b, f] | 17.6–18.5 | 16.8–18.1 |
| Sugarcane bagasse[a, b, f] | 17.3–19.4 | 17.7–17.9 |
| Wheat straw[a, b, f] | 16.1–18.9 | 15.1–17.7 |
| Hulls, shells, prunings[b, c] | 15.8–20.5 | |
| Miscanthus[f] | 18.1–19.6 | 17.8–18.1 |
| Switchgrass[a, c, f] | 18.0–19.1 | 16.8–18.6 |
| Bamboo[f] | 19.0–19.8 | |
| Black locust[a, f] | 19.5–19.9 | 18.5 |
| Eucalyptus[a, b, f] | 19.0–19.6 | 18.0 |
| Hybrid poplar[a, c, f] | 19.0–19.7 | 17.7 |
| Willow[b, c, f] | 18.6–19.7 | 16.7–18.4 |
| Hardwood[b, f] | 18.6–20.7 | |
| Softwood[a, b, c, d, e, f] | 18.6–21.1 | 17.5–20.8 |
| Municipal solid waste[b, f] | 13.1–19.9 | 12.0–18.6 |
| Refuse-derived fuels[b, f] | 15.5–19.9 | 14.3–18.6 |
| Newspaper[b, f] | 19.7–22.2 | 18.4–20.7 |
| Corrugated paper[b, f] | 17.3–18.5 | 17.01 |
| Waxed cartons[b] | 27.3 | 25.3 |

where the apices a, b, c, d, e, and f refer to the following references:
[a]Energy Efficiency & Renewable Energy Department, USA (2018)
[b]Jenkins (1993)
[c]Jenkins et al. (1998)
[d]Tillman (1978)
[e]Bushnell et al. (1990)
[f]ECN, The Netherlands (2018)

## 25.3 Stoichiometry

When there is complete combustion of hydrocarbons (for instance), $CO_2$ is normally produced from the carbon present in the biomass fuel. Water, instead, is produced from the oxidation of the hydrogen molecules of the feedstock. However, when the combustion is not perfect, there can be formation of CO in the flue gases. In case we consider a hydrocarbon with general composition $C_nH_m$, the stoichiometric reaction of oxidation is:

$$C_nH_m + \left(n + \frac{m}{4}\right)O_2 \rightarrow nCO_2 + \frac{m}{2}H_2O \tag{25.10}$$

However, it is necessary to take into consideration also the nitrogen amount when air is utilized during combustion; in this case, the reaction is described as Eq. (25.11)

**Table 25.15** Thermodynamic properties for selected alkane compounds

| Name | Sg | $T_b$ | $p_0$ | $C_{Pg}$ | $C_{pl}$ | $T_{ig}$ | HHV | LHV | $h_{fg}$ | *a*/*f* | Res. | Mot. | $\Delta h^{f0}$ | $p_c$ | $T_c$ |
|---|---|---|---|---|---|---|---|---|---|---|---|---|---|---|---|
| Methane | 0.466 | −161 | | 2.21 | | 537 | 55.54 | 50.05 | 510 | 17.2 | 120 | 120 | −74.4 | 45.4 | 190 |
| Ethane | 0.572 | −89 | | 1.75 | | 472 | 51.90 | 47.51 | 489 | 16.1 | 115 | 99 | −83.8 | 48.2 | 305 |
| Propane | 0.585 | −42 | 12.8 | 1.62 | 2.48 | 470 | 50.32 | 46.33 | 432 | 15.7 | 112 | 97 | −104.7 | 41.9 | 370 |
| *n*-Butane | 0.579 | 0 | 3.51 | 1.64 | 2.42 | 365 | 49.51 | 45.73 | 386 | 15.5 | 94 | 90 | −146.6 | 37.5 | 425 |
| Isobutane | 0.557 | −12 | 4.94 | 1.62 | 2.39 | 460 | 49.36 | 45.58 | 366 | 15.5 | 102 | 98 | −153.5 | 36 | 408 |
| *n*-Pentane | 0.626 | 36 | 1.06 | 1.62 | 2.32 | 284 | 49.00 | 45.34 | 357 | 15.3 | 62 | 63 | −173.5 | 33.3 | 470 |
| Isopentane | 0.620 | 28 | 1.39 | 1.60 | 2.28 | 420 | 48.91 | 45.25 | 342 | 15.3 | 93 | 90 | −178.5 | 33.5 | 460 |
| *n*-Hexane | 0.659 | 69 | 0.34 | 1.62 | 2.27 | 233 | 48.67 | 45.1 | 335 | 15.2 | 25 | 26 | −198.7 | 29.7 | 507 |
| Isohexane | 0.662 | 50 | 0.503 | 1.58 | 2.20 | 421 | 48.45 | 44.89 | 305 | 15.2 | 104 | 94 | −207.4 | 30.9 | 500 |
| *n*-Heptane | 0.684 | 99 | 0.11 | 1.61 | 2.24 | 215 | 48.44 | 44.93 | 317 | 15.2 | 0 | 0 | −224.2 | 27.0 | 540 |
| Triptane | 0.690 | 81 | 0.23 | 1.60 | 2.13 | 412 | 48.27 | 44.76 | 289 | 15.2 | 112 | 101 | | 29.2 | 531 |
| *n*-Octane | 0.703 | 126 | 0.04 | 1.61 | 2.23 | 206 | 48.26 | 44.79 | 301 | 15.1 | 20 | 17 | −250.1 | 24.6 | 569 |
| Isooctane | 0.692 | 114 | 0.12 | 1.59 | 2.09 | 418 | 48.12 | 44.65 | 283 | 15.1 | 100 | 100 | −259.2 | 25.4 | 544 |
| *n*-Nonane | 0.718 | 151 | 0.01 | 1.61 | 2.21 | | 48.12 | 44.69 | 288 | 15.1 | | | 274.7 | 22.6 | 595 |
| *n*-Decane | 0.730 | 174 | 0.01 | 1.61 | 2.21 | | 48.00 | 44.6 | 272 | 15.1 | | | −300.9 | 20.9 | 618 |
| Isodecane | 0.768 | 161 | | 1.61 | 2.21 | | | | | 15.1 | 113 | 92 | | 24.8 | 623 |
| *n*-Undecane | 0.740 | 196 | | 1.60 | 2.21 | | 47.90 | 44.52 | 265 | 15.0 | | | −327.2 | 19.4 | 639 |

(continued)

**Table 25.15** (continued)

| Name | Sg | $T_b$ | $p_0$ | $C_{Pg}$ | $C_{pl}$ | $T_{ig}$ | HHV | LHV | $h_{fg}$ | *alf* | Res. | Mot. | $\Delta h^{f0}$ | $p_c$ | $T_c$ |
|---|---|---|---|---|---|---|---|---|---|---|---|---|---|---|---|
| *n*-Dodecane | 0.749 | 216 | | 1.60 | 2.21 | | 47.84 | 44.57 | 256 | 15.0 | | | −350.9 | 18.0 | 658 |
| *n*-Tridecane | 0.756 | 236 | | 1.60 | 2.21 | | | | 246 | 15.0 | | | | 17.0 | 676 |
| *n*-Tetradecane | 0.763 | 253 | | 1.60 | 2.21 | | | | 239 | 15.0 | | | | 14.2 | 693 |
| *n*-Pentadecane | 0.768 | 271 | | 1.60 | 2.21 | | | | 232 | 15.0 | | | | 15.0 | 707 |
| *n*-Hexadecane | 0.773 | 287 | | 1.60 | 2.21 | | 47.61 | 44.31 | 225 | 15.0 | | | −456.1 | 13.9 | 722 |
| *n*-Heptadecane | 0.778 | 302 | | 1.60 | 2.21 | | | | 221 | 15.0 | | | | 12.8 | 733 |
| *n*-Octadecane | 0.782 | 317 | | 1.60 | 2.21 | | 47.54 | 44.26 | 214 | 15.0 | | | | 11.8 | 748 |

Adapted from Borman and Ragland (1998)

**Table 25.16** Parameters used in Table 25.15

| Parameter | Dimensions |
|---|---|
| Sg | Specific gravity, density of the component at 20 °C specific to the density of water at 4 °C. For gases is determined at the boiling point of the liquefied gas |
| $T_b$ | Boiling point temperature at 1 atm in °C |
| $p_0$ | Vapor pressure at 38 °C and 1 atm |
| $C_{Pg}$ | Specific heat of gas at 25 °C [kJ/(kg K)] |
| $C_{pl}$ | Specific heat of liquid at 25 °C [kJ/(kg K)] |
| $T_{ig}$ | Auto-ignition temperature (°C) |
| HHV | kJ/kg |
| LHV | kJ/kg |
| $h_{fg}$ | Heat of vaporization at 1 atm and boiling point temperature (kJ/kg) |
| *a/f* | Stoichiometric air-to-fuel mass ratio |
| Res. | Research octane number |
| Mot. | Motor octane number |
| $\Delta h^{f0}$ | Enthalpy of formation at 25 °C (kJ/g mol) |
| $p_c$ | Critical point pressure (atm) |
| $T_c$ | Critical point temperature (K) |

$$C_nH_m + \left(n + \frac{m}{4}\right)(O_2 + 3.78\,N_2) \rightarrow nCO_2 + \frac{m}{2}H_2O + 3.78\left(n + \frac{m}{4}\right)N_2 \tag{25.11}$$

The mole fractions of the products in case of complete combustion of the general hydrocarbon are:

$$y_{CO_2} = \frac{n}{4.78\left(n + \frac{m}{4}\right) + m/4} \quad y_{H_2O} = \frac{m/2}{4.78\left(n + \frac{m}{4}\right) + m/4}$$

$$y_{N_2} = \frac{3.78\left(n + \frac{m}{4}\right)}{4.78\left(n + \frac{m}{4}\right) + m/4}$$

## References

Álvarez-Álvarez, P., Pizarro, C., Barrio-Anta, M., Cámara-Obregón, A., Bueno, J., Álvarez, A., et al. (2018). Evaluation of tree species for biomass energy production in northwest Spain. *Forests, 9*, 160. https://doi.org/10.3390/f9040160.

Arce, M., Saavedra, Á., Míguez, J., Granada, E., Cacabelos, A., Arce, M. E., et al. (2013). Biomass fuel and combustion conditions selection in a fixed bed combustor. *Energies, 6*, 5973–5989. https://doi.org/10.3390/en6115973.

Bahadori, A., Zahedi, G., Zendehboudi, S., & Jamili, A. (2014). Estimation of the effect of biomass moisture content on the direct combustion of sugarcane bagasse in boilers. *International Journal of Sustainable Energy, 33,* 349–356. https://doi.org/10.1080/14786451.2012.748766.

Bartok, W., & Sarofim, A. F. (1991). *Fossil fuel combustion: A source book* (1st ed.). Wiley-Interscience.

Borman, G. L., & Ragland, K. W. (1998). *Combustion engineering*. McGraw-Hill.

Brügemann, J., & Gerds-Ploeger, H. (2013). Life is movement (Aristotle, 4th century BC). *Netherlands Heart Journal, 21*(10), 427–428. https://doi.org/10.1007/s12471-013-0468-x.

Bushnell, D. J., Haluzok, C., & Dadkhah-Nikoo, A. (1990). *Biomass fuel characterization : Testing and evaluating the combustion characteristics of selected biomass fuels.* Final Report May 1, 1988–July, 1989 (No. DOE/BP-1363). Portland, OR (USA): USDOE Bonneville Power Administration; Corvallis, OR (USA): Oregon State Univ. Dept. of Mechanical Engineering. https://doi.org/10.2172/6910422.

Channiwala, S. A., & Parikh, P. P. (2002). A unified correlation for estimating HHV of solid, liquid and gaseous fuels. *Fuel, 81,* 1051–1063. https://doi.org/10.1016/S0016-2361(01)00131-4.

Chimonyo, V. G. P., Modi, A. T., & Mabhaudhi, T. (2018). Sorghum radiation use efficiency and biomass partitioning in intercrop systems. *South African Journal of Botany, 118,* 76–84. https://doi.org/10.1016/j.sajb.2018.06.009.

Clarke, D., & Rieley, J. (2010). *Strategy for responsible peatland management.* Saarijärvi: International Peat Society.

De Blasio, C., Lucca, G., Özdenkci, K., Mulas, M., Lundqvist, K., Koskinen, J., et al. (2015). A study on supercritical water gasification of black liquor conducted in stainless steel and nickel-chromium-molybdenum reactors. *Journal of Chemical Technology and Biotechnology, 91*(10), 2664–2678. https://doi.org/10.1002/jctb.4871.

Demirbaş, A. (2005). Estimating of structural composition of wood and non-wood biomass samples. *Energy Sources, 27,* 761–767. https://doi.org/10.1080/00908310490450971.

ECN, The Netherlands. (2018). *Phyllis2—Database for biomass and waste* (WWW Document). https://www.ecn.nl/phyllis2/. Accessed August 22, 2018.

Energy Efficiency & Renewable Energy Department, USA. (2018). Bioenergy Technologies Office|Department of Energy (WWW Document). https://www.energy.gov/eere/bioenergy. Accessed August 22, 2018.

Gholz, H. L. (1982). Environmental limits on aboveground net primary production, leaf area, and biomass in vegetation zones of the Pacific northwest. *Ecology, 63,* 469–481. https://doi.org/10.2307/1938964.

Gujarat Pollution Control Board, Paryavaran Bhavan. (2014). *Pollution control guidelines for conversion of boilers/utilities from natural gas to solid fuels (coal, lignite, agro fuels etc).*

Jankowski, K. J., Dubis, B., Budzyński, W. S., Bórawski, P., & Bułkowska, K. (2016). Energy efficiency of crops grown for biogas production in a large-scale farm in Poland. *Energy, 109,* 277–286. https://doi.org/10.1016/j.energy.2016.04.087.

Jenkins, B. (1993). Properties of biomass. In *Biomass energy fundamentals* (EPRI Report).

Jenkins, B. M., Baxter, L. L., Miles, T. R., Jr., & Miles, T. (1998). Combustion properties of biomass. *Fuel Processing Technology, 54,* 17–46.

Jezierska-Thöle, A., Rudnicki, R., & Kluba, M. (2016). Development of energy crops cultivation for biomass production in Poland. *Renewable and Sustainable Energy Reviews, 62,* 534–545. https://doi.org/10.1016/j.rser.2016.05.024.

Ksenzhek, O. S., & Volkov, A. G. (1998). *Plant energetics.* San Diego, California: Academic Press.

Lau, F. S., Roberts, M. J., Rue, D. M., Punwani, D. V., Wen, W. -W., & Johnson, P. B. (1987). Peat beneficiation by wet carbonization. *International Journal of Coal Geology, 8,* 111–121. https://doi.org/10.1016/0166-5162(87)90026-7.

Liu, K., Xie, W., Li, D., Pan, W. -P., Riley, J. T., & Riga, A. (2000). The effect of chlorine and sulfur on the composition of ash deposits in a fluidized bed combustion system. *Energy Fuels, 14,* 963–972. https://doi.org/10.1021/ef990197k.

Manickam, I. N., Ravindran, D. D., & Subramanian, D. P. (2006). Biomass densification methods and mechanism. *Cogeneration and Distributed Generation Journal, 21,* 33–45. https://doi.org/10.1080/15453660609509098.

Minkkinen, K., Korhonen, R., Savolainen, I., & Laine, J. (2002). Carbon balance and radiative forcing of Finnish peatlands 1900–2100—The impact of forestry drainage. *Global Change Biology, 8,* 785–799. https://doi.org/10.1046/j.1365-2486.2002.00504.x.

Motta, I. L., Miranda, N. T., Maciel Filho, R., & Wolf Maciel, M. R. (2018). Biomass gasification in fluidized beds: A review of biomass moisture content and operating pressure effects. *Renewable and Sustainable Energy Reviews, 94,* 998–1023. https://doi.org/10.1016/j.rser.2018.06.042.

Pessarakli, M. (2005). *Handbook of photosynthesis*. Boca Raton: CRC Press, Taylor & Francis Group.

Rayaprolu, K. (2009). *Boilers for power and process*. Boca Raton: CRC Press, Taylor & Francis Group.

Reed, T. B., & Golden, A. D. (1988). *Handbook of biomass downdraft gasifier engine systems*.

Rumble, J. (2018). *CRC handbook of chemistry and physics*. CRC Press, Taylor & Francis Group.

Saastamoinen, J. (2015). Release profile of volatiles in fluidised bed combustion of biomass. *Journal of Fundamentals of Renewable Energy and Applications, 5,* 1–12. https://doi.org/10.4172/2090-4541.1000148.

Serup, H., Kofman, P. D., Falster, H., Gamborg, C., Gundersen, P., Hansen, L., et al. (2005). *Wood for energy production*. Dublin: Coford.

Song, H., Quinton, K., Peng, Z., Zhao, H., Ladommatos, N., Song, H., et al. (2016). Effects of oxygen content of fuels on combustion and emissions of diesel engines. *Energies, 9,* 28. https://doi.org/10.3390/en9010028.

Tillman, D. (1978). *Wood as an energy resource*. New York, USA: Academic Press. https://doi.org/10.1016/B978-0-12-691260-9.X5001-0.

Turpeinen, T. (2016). *Developing biofuels production from food industry wastes in the rural area of Kenya*. Savonia: University of Applied Sciences Technology and Transport.

Van Loo, S., & Koppejan, J. (2008). *The handbook of biomass gasification and co-firing*. Earthscan.

Vassilev, S. V., Vassileva, C. G., Song, Y. C., Li, W. Y., & Feng, J. (2017). Ash contents and ash-forming elements of biomass and their significance for solid biofuel combustion. *Fuel, 208,* 377–409. https://doi.org/10.1016/j.fuel.2017.07.036.

Virtanen, K., Hänninen, P., Kallinen, R. L., Virtiainen, S., Herranen, T., & Jokisaari, R. (2003). *The peat reserves of Finland in 2000*. Geological Survey of Finland. Vammalan Kirjapaino Oy.

Yang, Y. B., Ryu, C., Khor, A., & Yates, N. (2005). Effect of fuel properties on biomass combustion. Part II. Modelling approach—Identification of the controlling factors. *Fuel, 84,* 2116–2130. https://doi.org/10.1016/j.fuel.2005.04.023.

# Chapter 26
# A Simple Estimation of the Efficiency for a Biomass Power Plant

*Nobody really knows about dark matter. Dark matter is a myth.*
Kary Mullis, Nobel Prize in Chemistry 1993.

## 26.1 Mass Balances

We start here with the determination of simple mass balances and some considerations on the stoichiometric coefficients when combustion reactions are involved. After this, the energetic efficiencies will be determined for the main steps in the process.

Considering the combustion process and the flue gases generated, their composition can be calculated in a easier way when there is equilibrium. However, in reality, the equilibrium is not a state commonly reached and the processes have to be studied considering gas-phase kinetics. The main calculations are related to:

- flows and composition;
- overall fuel conversion reaction;
- thermal performance and efficiency.

The mass flow of the flue gas is evaluated in case the elemental composition of the fuel and the quantity of air used for the combustion are known. In case we assume complete combustion, the oxygen required can be calculated as follows (Van Loo and Koppejan 2008):

$$\bar{m}_{\mathrm{O_2,air}}\left[\frac{\mathrm{kg\ (O_2)}}{\mathrm{kg\ (fuel,\ Waf)}}\right] = \left(X_{\mathrm{c}}\frac{M_{\mathrm{O_2}}}{M_{\mathrm{c}}} + \frac{X_{\mathrm{H}}}{4}\frac{M_{\mathrm{O_2}}}{M_{\mathrm{H}}} + X_{\mathrm{s}}\frac{M_{\mathrm{O_2}}}{M_{\mathrm{s}}} - X_{\mathrm{O}}\right)\left(1 - X_{\mathrm{H_2O}}\right)\cdot\lambda \tag{26.1}$$

where in this case Waf is referred to be *wet and ash-free fuel*, $M_i\left[\frac{\mathrm{kg}}{\mathrm{kmol}}\right]$ is the molecular mass of the element $i$, $X_i$ is the fraction of the $i$ element in dry and ash-



C. De Blasio, *Fundamentals of Biofuels Engineering and Technology*,
Green Energy and Technology, https://doi.org/10.1007/978-3-030-11599-9_26

free fuel (daf), while $X_{H_2O}$ is the water mass fraction in the fuel, and $\lambda$ is the excess air ratio.

The $N_2$ from air is calculated in the same way as:

$$\bar{m}_{N_2,\text{air}}\left[\frac{\text{kg }(N_2)}{\text{kg (fuel, waf)}}\right] = \bar{m}_{O_2,\text{air}}\frac{Y_{N_2,\text{Air}}}{Y_{O_2,\text{Air}} \cdot M_{O_2}} \tag{26.2}$$

where ($Y_{O_2,\text{Air}}$) is the volume fraction of $O_2$ in air (0.21), while waf means in this case *water and ash-free*

$$Y_{N_2,\text{Air}} = 1 - Y_{O_2,\text{Air}} \tag{26.3}$$

$$\bar{m}_{\text{Air}}\left[\frac{\text{kg (Air)}}{\text{kg (fuel, waf)}}\right] = \bar{m}_{O_2,\text{air}} + \bar{m}_{N_2,\text{air}} \tag{26.4}$$

## 26.2 Thermal Performance and Efficiency

The energy which is stored in the fuel is released during the combustion process. In this sense, a simple energy balance is set up as follows:

$$\text{Energy in fuel} + \text{Preheat energy} = \text{Energy in flue gas} + \text{Energy losses} \tag{26.5}$$

The fraction of energy which is lost is due to the heat given to the environment, and in addition, there is a fraction of energy lost because of incomplete combustion. The energy stored in the fuel is evaluated from the HHV. On the other hand, to evaluate the energy contained in the flue gas, it is necessary to evaluate the enthalpy for each of the species: for instance $CO_2$, water which could be present in the flue gases in the form of steam, $SO_2$, $N_2$, $O_2$.

The enthalpy of the substance leaving our system (in this case the power plant) is a function of the temperature at which the substance is leaving (Levine 2008):

$$h_i(T)\left[\frac{\text{J}}{\text{kg}_i}\right] = T\left(a_1 + \frac{a_2}{2}T + \frac{a_3}{3}T^2 + \frac{a_4}{4}T^3 + \frac{a_5}{5}T^4 + \frac{a_6}{T}\right)\frac{R_u}{M_i} \tag{26.6}$$

where $R = 8.3145$ J/(mol K) is the universal gas constant, $T$ = gas temperature, $K$ and the coefficients $a_i$ are depending on the temperature.

The total efficiency is a function of the calorific value for the fuel, the enthalpies of air inlet and fuel inlet, and the sum of heat losses by radiation, convection, and conduction.

Regarding Eq. (26.6), the coefficients for enthalpy calculation of the flue gas are given in Tables 26.1 and 26.2.

The standard heat of reaction is the sum of the heats of formation of the reaction products minus the heats of formation of the reactants. For instance, as an example

**Table 26.1** Enthalpy calculation and related coefficients for Eq. (26.6) for selected components. Data from Van Loo and Koppejan (2008)

| | $CO_2$ | | $H_2O$ | |
|---|---|---|---|---|
| | ≤1000 K | >1000 K | ≤1000 K | >1000 K |
| $a_1$ | 2.35677352 | 3.85746029 | 4.19864056 | 3.03399249 |
| $a_2$ | 0.00898459677 | 0.00441437026 | −0.0020364341 | 0.00217691804 |
| $a_3$ | −0.00000712356269 | −0.00000221481404 | 0.00000652040211 | −0.000000164072518 |
| $a_4$ | 2.45919022E−9 | 5.23490188E−10 | −5.48797062E−9 | −9.7041987E−11 |
| $a_5$ | −1.43699548E−13 | −4.72084164E−14 | 1.77197817E−12 | 1.68200992E−14 |
| $a_6$ | −48371.9697 | −48759.166 | −30293.7266999999 | −30004.2971 |

**Table 26.2** Enthalpy calculation and related coefficients for Eq. (26.6) for selected components

| | $SO_2$ | | $N_2$ | | $O_2$ | |
|---|---|---|---|---|---|---|
| | ≤1000 K | >1000 K | ≤1000 K | >1000 K | ≤1000 K | >1000 K |
| $a_1$ | 2.911438 | 5.254498 | 3.298677 | 2.92664 | 3.78245636 | 3.28253784 |
| $a_2$ | 0.008103022 | 0.001978545 | 0.0014082404 | 0.0014879768 | −0.00299673416 | 0.00148308754 |
| $a_3$ | −0.00000690671 | −0.0000008204226 | −0.000003963222 | −0.000000568476 | 0.00000984730201 | −0.000000757966669 |
| $a_4$ | 0.000000003329015 | 1.576383E−10 | −0.000000005641515 | 1.0097038E−10 | −9.68129509E−09 | 2.09470555E−10 |
| $a_5$ | −8.777121E−13 | −1.1204512E−14 | −2.444854E−12 | −6.753351E−15 | 3.24372837E−12 | −2.16717795E−14 |
| $a_6$ | −36878.81 | −37568.85 | −1020.8999 | −922.7977 | −1063.94356 | −1088.45772 |

Data from Van Loo and Koppejan (2008)

**Table 26.3** Coefficients for specific heat calculations related to methanol, CO, and hydrogen

| | $a$ | $b(\times 10^2)$ | $c(\times 10^5)$ | $d(\times 10^9)$ |
|---|---|---|---|---|
| $CH_3OH$ | 4.55 | 2.186 | −0.291 | −1.92 |
| CO | 6.726 | 0.04001 | 0.1283 | −0.5307 |
| $H_2$ | 6.952 | −0.04576 | 0.09563 | −0.2079 |

Adapted from Hicks and Chopey (2012)

we could consider here the following reaction taking place at 600 K and 10.13 MPa (Hicks and Chopey 2012).

$$CO(g) + 2H_2(g) \rightarrow CH_3OH(g) \tag{26.7}$$

For this reaction:

$$\Delta H^0_{298} = -48.08 - [-26.416 + 2(0)] = -21.664 \text{ kcal/(g mol)} \tag{26.8}$$

The values for Eq. (26.8) are given in the literature since these are standard heat of formation for the components taking part in the reaction. The variation for the specific heat related to the overall reaction is given by:

$$\Delta C_p = \Delta \mathrm{a} + \Delta b \cdot T + \Delta c \cdot T^2 + \Delta \mathrm{d} \cdot T^3 \tag{26.9}$$

where in reference to Table 26.3 the variation for the coefficients used in Eq. (26.9) are given by:

$$\Delta a = 4.55 - 6.726 - 2(6.952) = -16.08$$

$$\Delta b = [2.186 - 0.04001 + 2(0.04576)]10^{-2}$$

$$\Delta c = -0.61056 \times 10^{-5}$$

$$\Delta d = -0.9735 \times 10^{-9}$$

The calculation for the enthalpy of the gas at a determinate temperature follows the Kirchhoff's law:

$$\Delta H(T) = \Delta H^0(T = 298) + \int_{298}^{T} \left[\Delta \mathrm{a} + \Delta b \cdot T + \Delta c \cdot T^2 + \Delta \mathrm{d} \cdot T^3\right] \mathrm{d}T \tag{26.10}$$

And therefore, for this example, the total variation for the enthalpy of reaction is given as follows:

$$\Delta H(T = 600) = -2.39 \times 10^4\ \mathrm{cal/(g\ mol)}$$

The integral in Eq. (26.10) gives results in cal/(g mol K). Additional examples can be found in textbooks of thermodynamics or physical chemistry (Levine 2008).

## 26.3 Incomplete Combustion

The energy which is lost because of incomplete combustion is evaluated by considering the calorific value of the species taken into consideration and lost at the exit of the power plant. For the carbon monoxide, we have:

$$\bar{E}_{\mathrm{CO}}\left[\frac{\mathrm{kJ}}{\mathrm{kg(fuel,\ waf)}}\right] = \mathrm{CV}_{\mathrm{CO}}\left[\frac{\mathrm{kJ}}{\mathrm{kg(CO)}}\right]\bar{m}_{\mathrm{CO}}\left[\frac{\mathrm{kg(CO)}}{\mathrm{kg(fuel,\ waf)}}\right] \tag{26.11}$$

where $\mathrm{CV_{CO}}$ is the calorific value of CO which is 10,102 kJ/(kg CO) at 25 °C. The energy lost due to uncombusted carbon is:

$$\bar{E}_{C}\left[\frac{\mathrm{kJ}}{\mathrm{kg\ fuel(waf)}}\right] = \mathrm{CV}_{\mathrm{C}}\left[\frac{\mathrm{kJ}}{\mathrm{KgC}}\right]\bar{m}_{c}\left[\frac{\mathrm{kgC}}{\mathrm{kg\ fuel\ (waf)}}\right] \tag{26.12}$$

where the carbon, *C*, has a calorific value of 34,910 kJ/kg

The energy balance between the inlet fuel and the flue gases is estimated as:

$$\begin{aligned}&\mathrm{NCV}\cdot\dot{m}_{\mathrm{F}} + [h_{\mathrm{F}}(T_{\mathrm{F}}) - h_{\mathrm{F}}(T_{\mathrm{Amb}})]\dot{m}_{\mathrm{F}} + [h_{\mathrm{air}}(T_{\mathrm{air}}) - h_{\mathrm{air}}(T_{\mathrm{amb}})]\dot{m}_{\mathrm{air}}\\ &= [h_{\mathrm{FG}}(T_{\mathrm{FG}}) - h_{\mathrm{FG}}(T_{\mathrm{Amb}})]\dot{m}_{\mathrm{FG}} + \sum_i Q_i + \sum_j \bar{E}_j\cdot\dot{m}_{\mathrm{F}}\end{aligned} \tag{26.13}$$

where the energy necessary to heat up the air and the fuel is:

$$[h_{\mathrm{F}}(T_{\mathrm{F}}) - h_{\mathrm{F}}(T_{\mathrm{Amb}})]\dot{m}_{\mathrm{F}} + [h_{\mathrm{air}}(T_{\mathrm{air}}) - h_{\mathrm{air}}(T_{\mathrm{amb}})]\dot{m}_{\mathrm{air}} \tag{26.14}$$

where NCV is the net calorific value of the fuel, $h_{\mathrm{F}}$, $h_{\mathrm{Air}}$, and $h_{\mathrm{FG}}$ are the enthalpies of fuel, air and flue gas, respectively, and $T_{\mathrm{F}}$ is the fuel temperature. $\sum_i Q_i$ is the sum of heat loss by radiation, convection, and conduction, and $\sum_j \bar{E}_j\cdot\dot{m}_{\mathrm{F}}$ is the sum of heat losses in unburned components.

The thermal efficiency is evaluated from:

$$\eta_{\text{th}} = 1 - \frac{Q_{\text{ch}} + \sum_i Q_i}{\text{NCV} \cdot \dot{m}_{\text{F}} + [h_{\text{F}}(T_{\text{F}}) - h_{\text{F}}(T_{\text{Amb}})]\dot{m}_{\text{F}} + [h_{\text{air}}(T_{\text{air}}) - h_{\text{air}}(T_{\text{amb}})]\dot{m}_{\text{air}}} \tag{26.15}$$

where $Q_{\text{ch}}$ is defined as follows with $T_{\text{ch}}$ indicating the chimney inlet temperature:

$$Q_{\text{ch}} = [h_{\text{FG}}(T_{\text{ch}}) - h_{\text{FG}}(T_{\text{Amb}})]\dot{m}_{\text{FG}} \tag{26.16}$$

The combustion efficiency is calculated from:

$$\eta_{\text{comb}} = 1 - \frac{\sum_j \bar{E}_j \cdot \dot{m}_{\text{F}}}{\text{NCV} \cdot \dot{m}_{\text{F}}} \tag{26.17}$$

Considering the thermal and the combustion efficiency, the total efficiency is finally evaluated from:

$$\eta_{\text{tot}} = 1 - \frac{Q_{\text{ch}} + \sum_i Q_i + \sum_j \bar{E}_j \cdot \dot{m}_{\text{F}}}{\text{NCV} \cdot \dot{m}_{\text{F}} + [h_{\text{F}}(T_{\text{F}}) - h_{\text{F}}(T_{\text{Amb}})]\dot{m}_{\text{F}} + [h_{\text{air}}(T_{\text{air}}) - h_{\text{air}}(T_{\text{amb}})]\dot{m}_{\text{air}}} \tag{26.18}$$

## References

Hicks, T. G., Chopey, N. P. (2012). *Handbook of chemical engineering calculations*. McGraw-Hill.
Levine, I. N. (2008). *Physical chemistry* (6th ed.). McGraw-Hill Education.
Van Loo, S., Koppejan, J. (2008). *The handbook of biomass gasification and co-firing*. Earthscan.

# Chapter 27
# Some Data on Oxidation and Reduction States and Half-Cell Reactions

*Why the chicken crossed the road? To get on the other side.*
Peter Diamond, Nobel Prize in Economics 2010.

## 27.1 Reduction and Oxidation States, Electron Affinities, and Ionization Potentials

Taking into account the topics treated during this first part of the manuscript, it will be beneficial to demonstrate some basic concepts of oxidation and reduction. It is known that when one atom is going to be oxidized, this means that it will lost some electrons. On the other hand, when an atom or molecule receives electrons, it will be reduced. Therefore oxygen, which has an oxidation number equal to zero in its molecular state, is considered as oxidizing other types of atoms like carbon, which is therefore oxidized. Carbon will displace some of its electrons and will assume a positive charge, while oxygen will have a negative charge. In organic chemistry, this is normally described by the concept of oxidized state. Table 27.1 shows some oxidation states for different kind of molecules.

**Table 27.1** Reduced and oxidized states of some elements with examples (Drapcho et al. 2007)

| | Reduced state | | Oxidized state | |
|---|---|---|---|---|
| Element | Oxidation number | Example | Oxidation number | Example |
| S | −2 | $H_2S$ | +6 | $SO_4^{2-}$ |
| N | −3 | $NH_3$ | +5 | $NO_3^-$ |
| O | −2 | $H_2O$, $C_6H_{12}O_6$ | 0 | $O_2$ |
| H | 0 | $H_2$ | +1 | $H_2O$ |
| C | −4 | $CH_4$ | +4 | $CO_2$ |



C. De Blasio, *Fundamentals of Biofuels Engineering and Technology*,
Green Energy and Technology, https://doi.org/10.1007/978-3-030-11599-9_27

**Table 27.2** Ionization potential (first value) and electron affinities (second value) in eV (Jones and Fleming 2014)

| | | | | | | | |
|---|---|---|---|---|---|---|---|
| H<br>13.60<br>0.75 | | | | | | | He<br>24.59<br>0 |
| Li<br>5.39<br>0.62 | Be<br>9.32<br>0 | B<br>8.30<br>0.24 | C<br>11.26<br>1.27 | N<br>14.53<br>0 | O<br>13.62<br>1.47 | F<br>17.42<br>3.34 | Ne<br>21.56<br>0 |
| Na<br>5.14<br>0.55 | Mg<br>7.65<br>0 | Al<br>5.99<br>0.46 | Si<br>8.15<br>1.24 | P<br>10.49<br>0.77 | S<br>10.36<br>2.08 | Cl<br>12.97<br>3.61 | Ar<br>15.75<br>0 |

After acknowledging that atoms in molecules have an oxidation state, it is natural the question of how much energy is actually required to pull one electron from a molecule. The ionization potential is the energy required to move away one electron from the atom; this quantity is measured in eV/atom or kcal/mole, joules/mole …, etc. When the first electron is removed from an atom, this will be referenced as first ionization potential and Table 27.2 gives some values for it.

While Table 27.3 gives a more detailed representation of the ionization potentials expressed in kJ/mol.

Concerning the representation of an atom and its electrons, an atom is represented by its energy levels (shell number). The shell number describes how far we are from the center of the atom. For each of these energy levels, there are particular orbitals which have a particular number and shapes. The number of orbitals present in each of the energy levels is determined by the energy level itself. Referring to Table 27.4, if we are in the first energy level, then only one type of orbital is there and there is only one single orbital, the spherical one. In case we consider the second upper energetic level, we will have two kinds of orbitals: the type *s* and the type *p*. The type *p* can have 3 orientations in the space. This is given by the *m* number. The spin number concerns the way an electron can spin, and there are only two values for this parameter; this means that within one orbital we can have only two electrons and not more.

**Table 27.3** Periodic table of the ionization potentials (Jones and Fleming 2014)

| | | | | | | | | | | | | | | | | | |
|---|---|---|---|---|---|---|---|---|---|---|---|---|---|---|---|---|---|
| H 1312 | | | | | | | | | | | | | | | | | He 2372 |
| Li 520 | Be 900 | | | | | | | | | | | B 801 | C 1086 | N 1402 | O 1314 | F 1681 | Ne 2081 |
| Na 496 | Mg 738 | | | | | | | | | | | Al 578 | Si 786 | P 1012 | S 1000 | Cl 1251 | Ar 1521 |
| K 419 | Ca 590 | Sc 631 | Ti 658 | V 650 | Cr 653 | Mn 717 | Fe 759 | Co 758 | Ni 737 | Cu 745 | Zn 906 | Ga 579 | Ge 762 | As 947 | Se 941 | Br 1140 | Kr 1351 |
| Rb 403 | Sr 549 | Y 616 | Zr 660 | Nb 664 | Mo 685 | Tc 702 | Ru 711 | Rh 720 | Pd 805 | Ag 731 | Cd 868 | In 558 | Sn 709 | Sb 834 | Te 869 | I 1008 | Xe 1170 |
| Cs 356 | Ba 503 | La 538 | Hf 680 | Ta 761 | W 770 | Re 760 | Os 840 | Ir 880 | Pt 870 | Au 890 | Hg 1007 | Tl 589 | Pb 716 | Bi 703 | Po 812 | At 890 | Rn 1037 |
| Fr 384 | Ra 509 | Ac 499 | Rf | Db | Sg | Bh | Hs | Mt | Ds | Rg | Uub | Uut | Uuq | Uup | Uuh | Uus | Uuo |
| | | Ce 527 | Pr 523 | Nd 530 | Pm 536 | Sm 543 | Eu 547 | Gd 593 | Tb 565 | Dy 572 | Ho 581 | Er 589 | Tm 597 | Yb 603 | Lu 524 | | |
| | | Th 587 | Pa 568 | U 584 | Np 597 | Pu 585 | Am 578 | Cm 581 | Bk 601 | Cf 608 | Es 619 | Fm 627 | Md 635 | No 642 | Lr 473 | | |

**Table 27.4** Quantum representation of atoms, quantum numbers (Jones and Fleming 2014)

| $n$ (Shell number) | $l$ type of orbital, can have values (0 … $n$ − 1) | $m_l$ [orientation in space, (−l … +l)] | $s$ (spin) | Orbital type |
|---|---|---|---|---|
| 1 | 0 | 0 | $\mp\frac{1}{2}$ | 1s |
| 2 | 0 | 0 | $\mp\frac{1}{2}$ | 2s |
| 2 | 1 | $-1, 0, +1$ | $\mp\frac{1}{2}$ | 2p |
| 3 | 0 | 0 | $\mp\frac{1}{2}$ | 3s |
| 3 | 1 | $-1, 0, +1$ | $\mp\frac{1}{2}$ | 3p |
| 3 | 2 | $-2, -1, 0, +1, +2$ | $\mp\frac{1}{2}$ | 3d |

## 27.2 Electronegativity

The electronegativity coefficient is a measure of the capacity to attract an electron. For example, with reference to Table 27.5 the carbon is more electronegative than hydrogen; this is why in C–H bonding we have that the oxidation state of carbon is negative and H is positive.

We need to know the capability of a particular specie to attract or to loose electrons in case we would like to use that particular element in a fuel cell or galvanic cell, or any other kind of setup where there are redox reactions. We have seen previously in the first part of this manuscript that the generation of biomass has a lot in common with galvanic cells and the coupling of specific half-cell reactions. Here some more data are given in Table 27.6. For a more complete list of half-cell reactions for non-biological systems, the readers can refer to some of the available handbooks on chemistry or physics (Haynes 2010).

**Table 27.5** Electronegativity values for the components of the periodic table (Jones and Fleming 2014)

| | | | | | | | | | | | | | | | | | |
|---|---|---|---|---|---|---|---|---|---|---|---|---|---|---|---|---|---|
| H<br>2.20 | | | | | | | | | | | | | | | | | He |
| Li<br>0.98 | Be<br>1.57 | | | | | | | | | | | B<br>2.04 | C<br>2.55 | N<br>3.04 | O<br>3.44 | F<br>3.98 | Ne |
| Na<br>0.93 | Mg<br>1.31 | | | | | | | | | | | Al<br>1.61 | Si<br>1.90 | P<br>2.19 | S<br>2.58 | Cl<br>3.16 | Ar |
| K<br>0.82 | Ca<br>1.0 | Sc<br>1.36 | Ti<br>1.54 | V<br>1.63 | Cr<br>1.66 | Mn<br>1.55 | Fe<br>1.83 | Co<br>1.88 | Ni<br>1.91 | Cu<br>1.90 | Zn<br>1.65 | Ga<br>1.81 | Ge<br>2.01 | As<br>2.18 | Se<br>2.55 | Br<br>2.96 | Kr<br>2.9 |
| Rb<br>0.82 | Sr<br>0.95 | Y<br>1.22 | Zr<br>1.33 | Nb<br>1.6 | Mo<br>2.16 | Tc<br>1.9 | Ru<br>2.2 | Rh<br>2.28 | Pd<br>2.20 | Ag<br>1.93 | Cd<br>1.69 | In<br>1.78 | Sn<br>1.80 | Sb<br>2.05 | Te<br>2.1 | I<br>2.66 | Xe<br>2.6 |
| Cs<br>0.79 | Ba<br>0.89 | La<br>1.10 | Hf<br>1.3 | Ta<br>1.5 | W<br>2.36 | Re<br>1.9 | Os<br>2.2 | Ir<br>2.20 | Pt<br>2.28 | Au<br>2.54 | Hg<br>2.00 | Tl<br>1.62 | Pb<br>1.87 | Bi<br>2.02 | Po<br>2.0 | At<br>2.2 | Rn |
| Fr<br>0.7 | Ra<br>0.9 | Ac | Rf | Db | Sg | Bh | Hs | Mt | Ds | Rg | Uub | Uut | Uuq | Uup | Uuh | Uus | Uuo |
| | | Ce<br>527 | Pr<br>523 | Nd<br>530 | Pm<br>536 | Sm<br>543 | Eu<br>547 | Gd<br>593 | Tb<br>565 | Dy<br>572 | Ho<br>581 | Er<br>589 | Tm<br>597 | Yb<br>603 | Lu<br>524 | | |
| | | Th<br>587 | Pa<br>568 | U<br>584 | Np<br>597 | Pu<br>585 | Am<br>578 | Cm<br>581 | Bk<br>601 | Cf<br>608 | Es<br>619 | Fm<br>627 | Md<br>635 | No<br>642 | Lr<br>473 | | |

**Table 27.6** Standard electrochemical potentials for biological half-cell reactions

| Half cell | $E^0(V)$ |
|---|---|
| Succunate + $CO_2$ + $2H^+$ + $2e^-$ $\leftrightarrow$ $\alpha$ − ketoglutarate + $H_2O$ | −0.670 |
| Acetate + $2H^+$ + $2e^-$ $\leftrightarrow$ Acetaldehyde | −0.580 |
| $2H^+ + 2e^- \leftrightarrow H_2$ | −0.421 |
| $\alpha$ − ketoglutarate + $CO_2$ + $2H^+$ + $2e^-$ $\leftrightarrow$ isocitrate | −0.380 |
| Cystine + $2H^+$ + $2e^-$ $\leftrightarrow$ 2cysteine | −0.340 |
| $NAD^+ + 2H^+ + 2e^- \leftrightarrow NADH + H^+$ | −0.320 |
| $NADP^+ + 2H^+ + 2e^- \leftrightarrow NADPH + H^+$ | −0.324 |
| Acetaldehyde + $2H^+$ + $2e^-$ $\leftrightarrow$ ethanol | −0.197 |
| Pyruvate + $2H^+$ + $2e^-$ $\leftrightarrow$ lactate | −0.185 |
| Oxaloacetate + $2H^+$ + $2e^-$ $\leftrightarrow$ malate | −0.166 |
| $FAD + 2H^+ + 2e^- \leftrightarrow FADH_2$ | +0.031 |
| Fumarate + $2H^+$ + $2e^-$ $\leftrightarrow$ succinate | +0.031 |
| Ubiquinone + $2H^+$ + $2e^-$ $\leftrightarrow$ ubiquinol | +0.045 |
| 2cytochrome $b_{ox}$ + $2e^-$ $\leftrightarrow$ 2cytochrome $b_{red}$ | +0.070 |
| 2cytochrome $c_{ox}$ + $2e^-$ $\leftrightarrow$ 2cytochrome $c_{red}$ | +0.254 |
| 2cytochrome $a_{3(ox)}$ + $2e^-$ $\leftrightarrow$ 2cytochrome $a_{3(red)}$ | +0.385 |
| $\frac{1}{2}O_2 + 2H^+ + 2e^- \leftrightarrow H_2O$ | +0.816 |

Adapted from (Karp, 2009)

## References

Drapcho, C. M., Nhuan, N. P., Walker, T. H. (2007). *Biofuels engineering process technology*. McGraw-Hill Professional.

Haynes, W. M. (2010). *CRC handbook of chemistry and physics* (91st ed.) CRC Press, Taylor & Francis Group.

Jones, M., Fleming, S. A. (2014). *Organic chemistry* (5th edn). W. W. Norton & Company.

Karp, G. (2009). *Cell and molecular biology: Concepts and experiments* (6th ed.). New York: Wiley.